DEVELOPMENTAL AND CELL BIOLOGY SERIES
EDITORS
P. W. BARLOW D. BRAY P. B. GREEN J. M. W. SLACK

MEIOSIS

MEIOSIS

BERNARD JOHN
Research School of Biological Sciences
Australian National University
Canberra, Australia

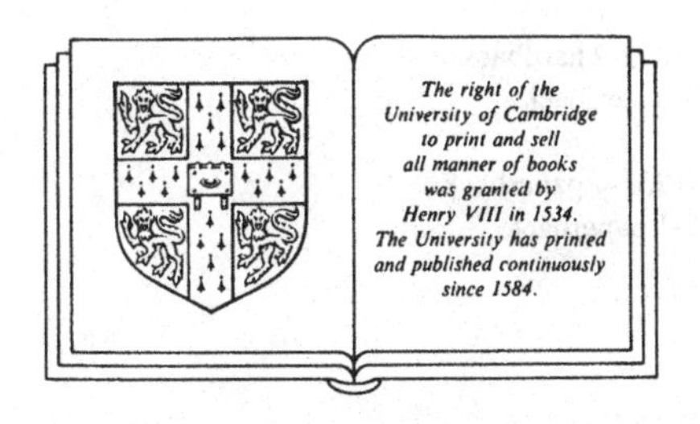

CAMBRIDGE UNIVERSITY PRESS
Cambridge
New York Port Chester
Melbourne Sydney

CAMBRIDGE UNIVERSITY PRESS
Cambridge, New York, Melbourne, Madrid, Cape Town, Singapore, São Paulo

Cambridge University Press
The Edinburgh Building, Cambridge CB2 2RU, UK

Published in the United States of America by Cambridge University Press, New York

www.cambridge.org
Information on this title: www.cambridge.org/9780521350532

First published 1990
This digitally printed first paperback version 2005

A catalogue record for this publication is available from the British Library

Library of Congress Cataloguing in Publication data

John, Bernard.
Meiosis Bernard John.
p. cm. -- (Development and cell biology series)
Bibliography : p.
Includes index.
ISBN 0-521-35053-0
1. Meiosis. I. Title. II. Series.
QH605.3.J64 1990
574.3′2 -- ac20 89-36087 CIP

ISBN-13 978-0-521-35053-2 hardback
ISBN-10 0-521-35053-0 hardback

ISBN-13 978-0-521-01752-7 paperback
ISBN-10 0-521-01752-1 paperback

Epigraph

Meiosis is still a potential battleground where dead hypotheses litter the field or rest uneasily in shallow graves, ready to emerge and haunt any conscientious scientist who tries to consolidate victory for any particular thesis.

J. Herbert Taylor

Contents

Acknowledgements

It is a pleasure to acknowledge the secretarial skills of Elizabeth Robertson, who prepared the manuscript for publication, and the artistic skills of Gary Brown, who provided all the illustrations. Additional thanks go to Graham von Schill who assisted in the production of the text in a variety of ways.

Prologue

Meiosis is a method of nuclear division leading to an orderly reduction of the chromosome number. It is coextensive with sexual reproduction and in the sexual cycle it compensates for fertilization. These statements provide us with a minimum definition of the process.
Cyril Darlington

Every living eukaryote either is, or at some time has been, a single cell. New cells arise only by the division of existing cells and this involves the division of both nucleus and cytoplasm. In eukaryotes nuclei divide in only two ways, by mitosis or by meiosis. Meiosis represents a unique form of cellular differentiation and, unlike mitosis, is most usually initiated only once in the life cycle of a eukaryote. Moreover, while mitosis is associated with uniparental, asexual, systems, it is meiosis that has made sex and biparental inheritance possible. It is then a unique and distinctive event in the life of an organism.

It is also a meticulously exact event normally involving both an accurate quantitative reduction in chromosome number, on the one hand, and a precise partitioning of genetic material, on the other hand. Meiosis thus fulfills two interrelated functions both of which are connected with the process of sexual reproduction. It ensure the production of a haploid phase in the life cycle of an organism (reduction) and, in one form or another, provides for the production of genetically distinct offspring (recombination). Deviations from this strict schedule of behaviour are, with few exceptions, eventually either lethal or sublethal since a precisely structured chromosome complement is an essential prerequisite for basic cell function during development.

Because meiosis is such a highly ordered process, the essential genes and proteins that control it can be reasonably anticipated to show considerable conservation throughout eukaryotes. While the meiotic mechanism does indeed display conservatism in its basic features it is not simply a unitary process but rather a complex of processes geared into a unitary sequence and its variations, as we shall see, involve considerably more than mere matters of detail. In animals the meiotic machinery is much more varied than in plants. On the other hand, the diversity of reproductive mechanisms is far greater in plants, including as it does systems of self-fertilization, which are rarely found in animals, and hermaphroditism on a much more extensive scale. An understanding of the nature and mechanisms of meiosis is basic to the analysis of many biological problems for while its causes are rooted to cell biology its consequences lie ultimately in evolution.

The essential characteristics of meiosis have been understood in gross descriptive terms for almost a century but it is only within the last 20 years that information has become available on the ultrastructural and molecular aspects of the process. One of the aims of the present text is to present an overview of what we now know, and do not know, about meiosis. While we still lack a complete theory of meiosis the facts now available at least provide a basis for reformulating the unresolved problems in more specific terms and this is the second aim of this monograph.

1

Introduction – multiplication and division

Reproduction of a cell is making two from one. One set of chromosomes makes two sets of chromosomes. Two sets of chromosomes make two nuclei.
Daniel Mazia

A LIFE CYCLES AND SEXUAL CYCLES

All eukaryotes are constructed out of cells or their products. The simplest forms of life are single cells which are capable of performing all life activities. More complex organisms are composed of collections of cells which individually perform only specific functions. Each eukaryotic cell has a defined and compartmentalized structure in which a specialized, internal and double membrane creates a distinctive environment, the nucleus. Within this organelle the major DNA component of the cell is associated with basic histone proteins to form a molecular complex referred to as chromatin. This chromatin is organized into a series of subunits, termed nucleosomes, each of which contains a combination of 200 base pairs (bp) of DNA together with nine histone molecules consisting of two each of H2A, 2B, 3 and 4 plus one H1 molecule. The four pairs of histone molecules are bound together as a core which is closely associated with *c.* 140 bp of DNA, with the remaining 60 bp forming a linker unit to which H1 is bound (reviewed in Matthews, 1981). Collectively the nucleosomes are organized around a protein scaffold to form a system of two or more individual threads, the chromosomes, which are highly diffused within the nuclear area.

The nuclear membrane also serves to separate the machinery of RNA transcription, carried out at the chromosome level, from the machinery of RNA translation, carried out at the ribosome level within the cell cytoplasm. Consequently, while the synthesis of proteins is a function of cell cytoplasm, the instructions for this synthesis emanate from the nucleus. Cells maintain their organization by carrying out a continual re-synthesis of the enzymes and structural proteins that determine this organization. DNA plays a key role in this enterprise since it ultimately specifies both the nature and the quantity of proteins, and hence also all the other molecules, which a cell synthesizes.

Conventionally, the life cycle of a eukaryote incorporates two principal phases of activity – development and growth, on the one hand, and maturation and reproduction, on the other, followed in turn by senescence and death. In multicellular eukaryotes both phases of activity involve cell division, for cells, like individuals, have a finite existence. Such division provides for the transmission of the biological components necessary for further division. Reproduction of a cell is not possible without the multiplication of the molecules out of which that cell is composed. In the simplest and most common category of cell division, the major molecules of the cell are first duplicated and then separated, some precisely, others less precisely, into two products formed by the partitioning of the original cell into two. Cells thus reproduce their structure at each division. This process, known as mitosis, takes its name from the greek *mitos*, meaning thread. At the outset of mitosis condensed and individually identifiable chromosome threads appear within the nucleus. The appearance of these threads, which signals the transition from interphase to mitosis, depends on the imposition of a higher order of coiling on the primary chain of nucleosomes which constitutes the basic chromosome thread.

In the simplest eukaryotes, which are unicells, mitosis is, of necessity, a form of organism reproduction, referred to as asexual. Mitosis is also the basis of all forms of asexual reproduction in multicellular animals and of vegetative propagation in multicellular plants because this class of cell division underlies the development of all multicells. Division is thus necessary to multiply the number of cells – a mathematical contradiction but a biological necessity.

From the point of view of the chromosomes, the mitotic cycle involves two phases. First a phase of replication (the synthetic or S phase) during which each chromosome is precisely copied within the intact nuclear membrane so that it comes to consist of two identical sister threads or chromatids. Different regions of the chromosome replicate at distinct times and/or at different rates during the S phase, though each region normally replicates only once in each mitotic cycle. Particular chromosome regions may be more condensed than others (heteropycnotic) during interphase giving rise to deeply staining chromocentres. Some of these heteropycnotic regions are composed of a distinctive category of DNA referred to as repetitive since it is built out of tandem arrays of unit runs of the same sequence. These regions define the constitutive heterochromatin of the genome (reviewed in John, 1988). Histone synthesis also occurs during the S phase and is loosely coupled with DNA synthesis. Like DNA synthesis itself, the products of histone replication also segregate semi-conservatively so that the new histone molecules are associated with the new DNA strands during chromosome replication (Matthews, 1981).

The second phase of the mitotic cycle is the distributive, or mitotic (M),

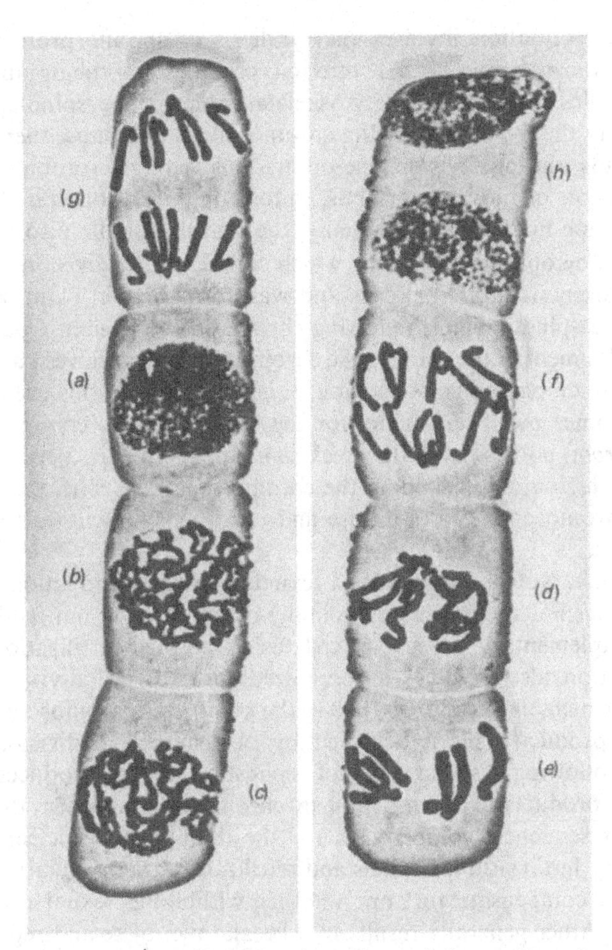

Fig. 1.1 Mitosis in root tip cells of the plant *Crocus balansae* ($2n = 2x = 6$). The sequence includes; (*a*) interphase; (*b*) early, (*c*) mid and (*d*) late prophase; (*e*) metaphase; (*f*) early and (*g*) late anaphase; and (*h*) the return to interphase. The spindle on which the chromosomes orient at metaphase and subsequently move at anaphase is not visible in squash preparations of this kind but in all the relevant cells (*e*–*g*) runs in a north–south direction.

phase when the sister chromatids of each condensed chromosome are separated, accurately and equally, into two nuclei (Fig. 1.1). This, most commonly, occurs in conjunction with the division of the cell itself and both events depend on the assembly of a bipolarized system of proteinaceous microtubules which collectively form the mitotic apparatus or spindle. The sister chromatids of a chromosome orient to opposite spindle poles, by each forming a microtubule attachment to a different pole, and then subsequently migrate to the pole toward which they are oriented.

Spindle formation involves the synthesis of specific proteins termed tubulins, the assembly of these into microtubules and the organization of microtubules into a bipolar cytoskeletal system. The spindle serves to coordinate the movement of the chromosomes, and subsequently of the chromatids, during division, the ordered array of microtubules defining the direction of movement. Thus, mitosis requires both bipolarity and chromosome mobility. Additionally, the mitotic spindle also determines the site of cytoplasmic cleavage, which completes the division of the cell, since cytokinesis in animal cells and wall formation in plant cells occur at the mid-spindle region following chromatid separation. Coupled with the development of the spindle, the chromosomes themselves pass through a cycle of coiling and condensation. It is, in reality, to this event that the chromosomes owe their name, for they are visible as colour (from the greek *chrom*) bodies (from the greek *soma*) only when in a compact state. In cells which are not dividing the chromosomes are, with the exception of the chromocentres, in a diffuse and uncoiled state within the nuclear membrane.

Most eukaryotes also practise a sexual form of reproduction. Here the new organism is the product not simply of cell division but, additionally, of a complementary process of cell fusion, termed fertilization. Sexual reproduction also involves a specialized form of cell division, termed meiosis, which, unlike mitosis, halves the number of chromosomes in each division product. This is achieved by two successive divisions of the nucleus following a single phase of replication and so produces not two but four products. The reduction in chromosome number, that results from this sequence, compensates for the doubling of nuclear contents effected by fertilization. Meiosis and fertilization thus regularly alternate with, and compensate for, one another within the sexual cycle and a failure in either generally results in a breakdown of sexual reproduction (see however Chapter 7 A.3). Asexual reproduction gives rise to offspring which share genomes identical to those of the single parent organism. By virtue of fertilization, sexual reproduction, most commonly involves a mixing of genomes from two different parents.

The cells carrying the nuclei which fuse at fertilization are termed gametes and they are specified as containing a haploid number of unreplicated chromosomes and a 1C content of DNA. It is important to recognize that the term haploid (from the greek *haplo* for simple and *id* for idant) is used in two quite different ways in Biology. First, to describe the reduced chromosome number produced by meiosis, that is the chromosome number of the haplo-phase of the life cycle. Second, to define the base chromosome number of an organism, that is the number of chromosomes within a single set of homologues. In practice, as Levan & Müntzing pointed out as long ago as 1963, confusion can be avoided by the use of two different symbols to denote the two forms of usage. The

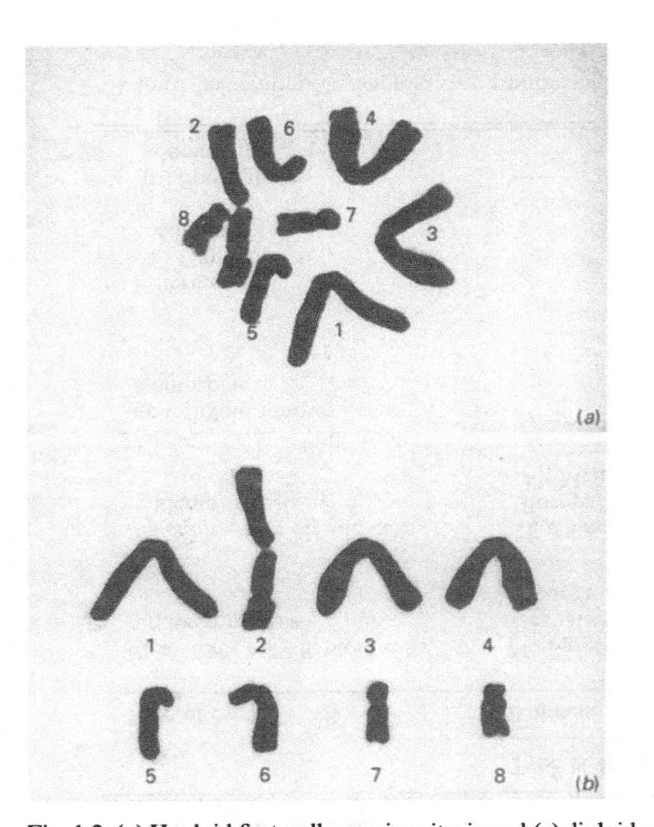

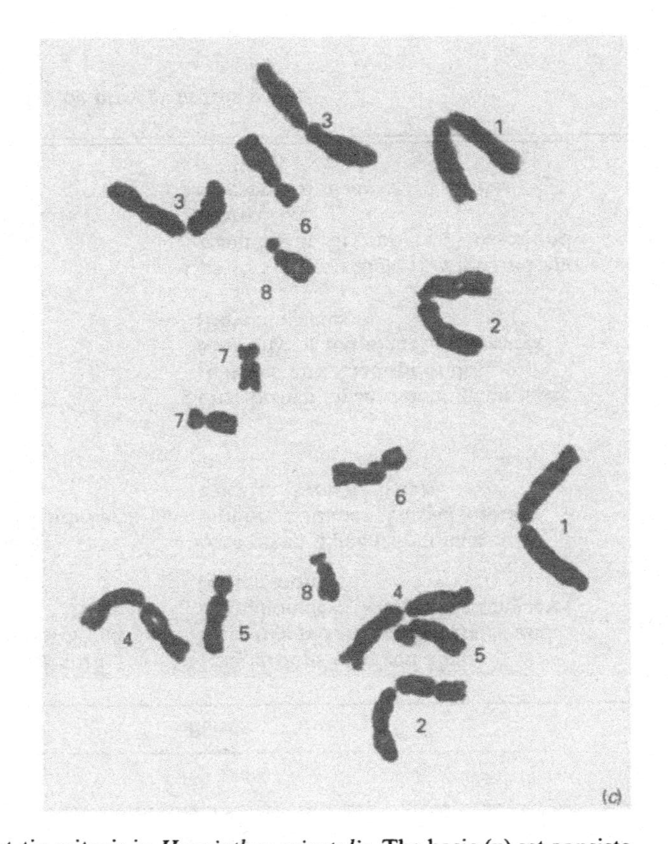

Fig. 1.2 (*a*) Haploid first pollen grain mitosis and (*c*) diploid root-tip mitosis in *Hyacinthus orientalis*. The basic (x) set consists of eight chromosomes which can be individually distinguished in terms of their relative lengths and their centromere positions (*b*) so that homologous pairs are readily identified in the diploid division ($2n = 2x = 16$).

Table 1.1 *Patterns of sexual cycles in eukaryotes*

Type of cycle	Type of eukaryote	
	Animals	Plants
Haplontic – only haploid cells undergo mitosis	Non-green flagellates with sexual reproduction. Some Sporozoa	Many groups of green algae. Primitive red algae (Bangiales and Nemalionales). Many Phycomycetes (water moulds).
Diplontic – only diploid cells undergo mitosis	All ciliates. Probably Formanifera and Radiolaria. All Metazoa	Some green algae (Siphonales, Siphonocladiales, Dasycladales). Fucales (brown algae).
Haplo-diplontic – both haploid and diploid mitoses occur		
Isomorphic	—	Some brown algae. Some green algae (Ulvales and Cladophorales). Majority of red algae[a]. *Allomyces* (Phycomycetes).
Heteromorphic		
(a) Gametophyte dominant	—	*Cutleria* (brown algae). A few red and green algae. Bryophyta (Mosses and liverworts).
(b) Sporophyte dominant	—	Some brown algae. All vascular plants.

[a] A third 'generation', a haploid carposporic stage, may also be present in this group.

symbol n is then used to designate the reduced chromosome number of the haplo-phase so that 2n identifies the unreduced diplo-phase of the life cycle. The symbol x, on the other hand, signifies the base chromosome number so that in a diploid organism 2n = 2x. Triploids, however, have 2n = 3x, tetraploids 2n = 4x and so on. In haploid organisms 2n = x so that whereas x is an algebraic term, n is not. This is the terminology that will be employed throughout the monograph.

In those diploid, 2n = 2x, species where members of the haploid (x) chromosome set can be individually distinguished on cytomorphological grounds it is possible to confirm that each chromosome in the diploid set has a homologue which is identical to it in terms not only of gross morphology (i.e. length and shape, see Fig. 1.2) but also in genetic content (i.e. in DNA sequence composition and arrangement).

Because sexual reproduction involves both meiosis and fertilization it necessarily also involves an alteration of haploid and diploid phases. These phases, however, are very variably represented in the life cycle of different eukaryotes. Darlington (1958, 1978) and Stebbins (1960) have both argued that ancestral eukaryotes, like present day prokaryotes, were haploid. This implies that, in those ancestral eukaryotic forms which reproduced sexually, the only diploid nucleus in the entire life cycle would have been that of the zygote. Such a condition is currently found among extant eukaryotes in the flagellates, the sporozoans, many groups of filamentous green algae, the simpler red algae and in the water moulds (Phycomycetes) among the fungi. That is, in certain of the unicellular Protozoa among animals and in some of the structurally simpler plants. In all of these organisms, meiosis immediately follows fertilization and only haploid mitosis occurs during the remainder of the life cycle. By contrast, in most animals, whether unicellular or multicellular, it is the diploid phase that dominates and haploidy is restricted to the gametes. Here all mitosis is diploid in character since fertilization immediately restores the diploid state. (Table 1.1).

In plants, predominantly diploid cycles are known in some green and brown algae but, in the majority of algae, there is a cycle with morphologically similar or even identical haploid and diploid phases, sometimes referred to as generations. Two types of heteromorphic haplo-diploid life cycles have been secondarily derived from the isomorphic category (Fig. 1.3). In the one, a large and long-lived diploid phase alternates with a small, evanescent haploid phase, as in the case of the more complex brown algae and all of the vascular plants. In the other, the haploid phase is dominant and the diploid phase correspondingly reduced. Such cycles occur in some algae but are found especially in mosses and liverworts (Bryophyta).

All organisms with a dominant haplo-phase are relatively undifferentiated morphologically, have a short and simple development and

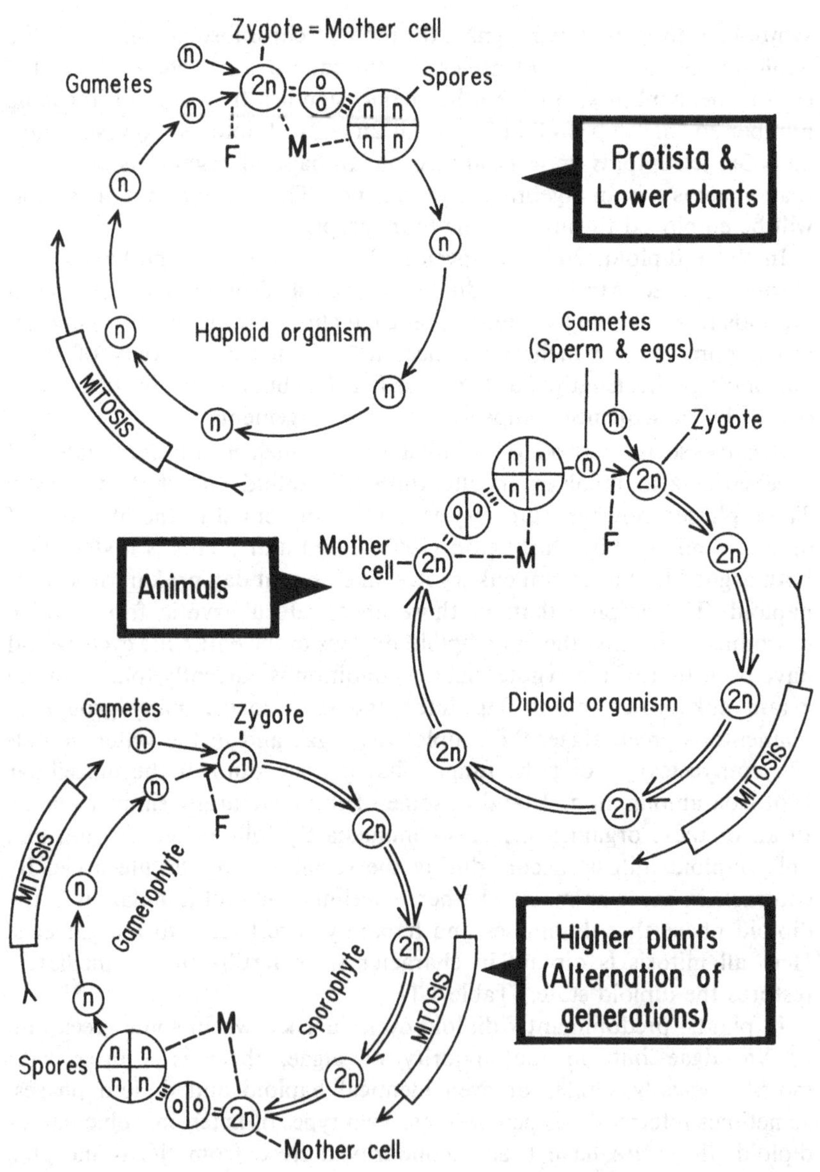

Fig. 1.3 The three basic categories of sexual cycles in eukaryotes. F = fertilization (or its equivalent in forms with cell conjugation), M = meiosis (after Darlington, 1978 and with the permission of *Nature*).

multiply rapidly. Only following the adoption of diploidy were complex developmental sequences possible, though it is worth noting that both classes of predominantly diploid cycles can still be found in some organisms with rapid reproduction and a relatively simple development.

Precisely why and how the diploid state should have facilitated the development of complex organisms is not clear.

B GAMETES AND SPORES

Gametes are haploid so that in diplontic life cycles, where there is no haploid mitosis, the gametes must be produced by the meiotic division of cells of the diplo-phase. In diplontic animals of this kind, sexual dimorphism is a common feature of sexually-reproducing forms. In its simplest form such a dimorphism may be restricted to differences in the physiology, or may also extend to the morphology, of the gametes, one of which is larger and immobile while the other is smaller and mobile. The smaller, mobile gamete is referred to as the male gamete in contrast to the larger, immobile female gamete. More commonly, however, sexual dimorphism extends to morphological and biochemical differences between the gamete-producing parents, as well as to the gametes themselves. Sexuality is thus a product of the gametes and the gamete-producing generation and, whereas reproduction involves a change in the number of cells or individuals, sex represents a change in the genetic status of cells or individuals.

In multicellular animals sexuality is also regularly, though not invariably, associated with the occurrence of differentiated sex chromosomes. Here one sex is heteromorphic for one, or sometimes more than one, pair of chromosomes so that two, chromosomally different, kinds of gamete are produced by that sex, which is thus also heterogametic. Where the male is the heterogametic sex and the difference is restricted to one pair of chromosomes these are referred to as X and Y while the female is then XX in constitution (Fig. 1.4). The functioning of this XY mechanism depends on the two sex chromosomes segregating regularly at male meiosis so that equal numbers of X and Y containing gametes are produced. Where the female is heterogametic, as in birds, butterflies and moths, males are ZZ and females ZW in constitution. In some organisms, and especially in many insects, males are X0 (i.e. *nullo* X) and females XX (Fig. 1.5). Equivalent Z0/ZZ states are also known. In some XY and ZW systems the heteromorphic sex homologues may have obvious regions of genetic homology in addition to regions which lack such homology. In other cases there may be no homology between the heteromorphic sex homologues which are completely differential in terms of their genetic composition.

While the diploid phase is also dominant in ferns, conifers and flowering plants, the haploid products of meiosis in these cases are not gametes but spores. On germination such spores give rise to a reduced haplo-phase with a relatively small number of nuclei and it is this structure that subsequently produces gametes by mitosis. For this reason it is termed the

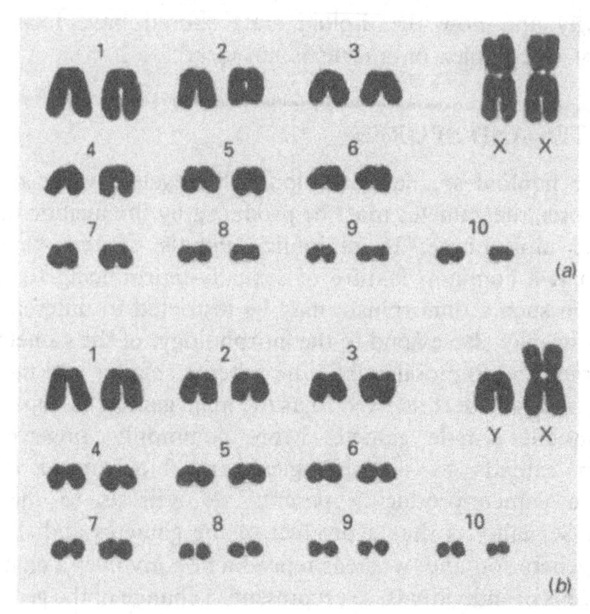

Fig. 1.4. (*a*) Female and (*b*) male mitotic karyotypes of the grasshopper *Tolgadia infirma*. The autosomal pairs of homologues are numbered 1–10 in order of decreasing size. The sex chromosomes are homomorphic (XX) in the female but heteromorphic (XY) in the male. Both have $2n = 2x = 22$.

gametophyte in contrast to the diploid, spore-producing, sporophyte. In such cases the gametophyte develops within sporophytic structures and the gametophyte itself is very much reduced.

Ferns produce only one kind of spore which, however, germinates to produce a haploid gametophyte. This, in turn, produces two kinds of sex organs – antheridia, responsible for the formation of male gametes, and archegonia, responsible for the production of female gametes. Conifers and flowering plants, on the other hand, produce two kinds of spores. Microspores give rise only to male gametes whereas megaspores form only female gametes. Here, then, the sex difference extends to the haploid spores (Lewis & John, 1967). Microsporogenesis in flowering plants occurs within anthers, organs differentiated into a central mass of sporogenous cells surrounded by several peripheral cell layers, the innermost of which, the tapetum, functions as a nutritive tissue. The tetrad of cells produced by the meiosis of the pollen mother cells (PMCs), which develop in the sporogenous area, separate into four haploid microspores. In each of these an asymmetrical mitosis produces a large vegetative and a small generative cell. The vegetative cell is the precursor of the pollen tube that emerges when the microspore germinates. Further division of the generative nucleus produces two sperm nuclei. Megaspores

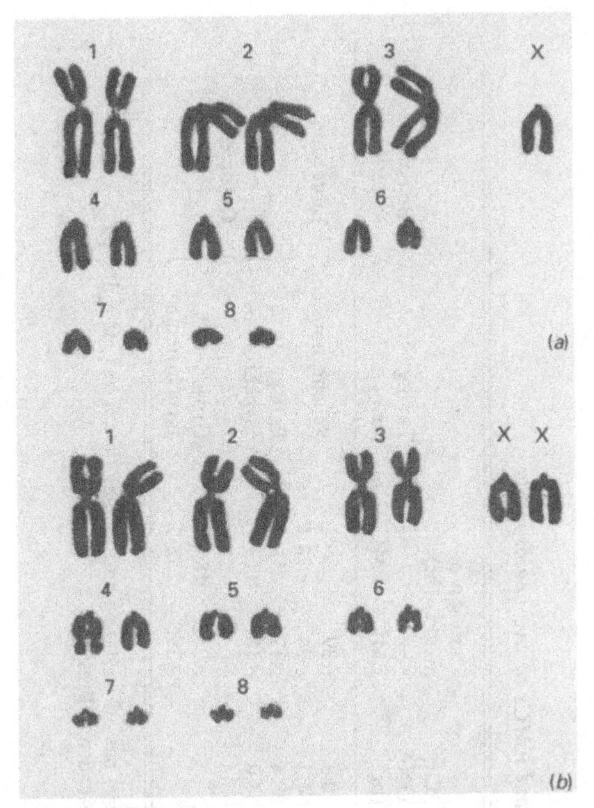

Fig. 1.5.(*a*) Male and (*b*) female mitotic karyotypes of the grasshopper *Chorthippus parallelus*. The autosomal pairs of homologues are numbered 1–8 in order of decreasing size. The male is monosomic for an unpaired sex chromosome (2n = 2x = 17, X0) whereas the female is homomorphic for a pair of X chromosomes (2n = 2x = 18, XX).

are formed in ovules, which again incorporate a central sporogenous tissue and a surrounding nutritive region. Of the four haploid nuclei produced by the meiosis of each embryo sac mother cell (EMC), three degenerate. The sole survivor then undergoes three mitoses to produce the female gametophyte or embryo sac. One of the eight nuclei of the female gametophyte eventually functions as the female gametic nucleus. In flowering plants, one of the male gametic nuclei fertilizes the ovule nucleus while the other fuses with a different female nucleus to produce a nutritive tissue referred to as the *endosperm*.

By analogy with diplontic animals, the terms male and female are also applied to diploid staminate and ovulate plants. Strictly speaking this is incorrect since their difference depends on heterospory not heterogamety. This practice can, however, be justified on genetical grounds since both heterospory and heterogamety are controlled by similar mechanisms.

Table 1.2. *The pattern of meiosis in PMCc and EMCs of five angiosperms*

Taxon	2n	3C DNA (pg)	Duration of meiosis (hr) EMC	Duration of meiosis (hr) PMC	Timing of meiosis	Breeding system
Hordeum vulgare	2x = 14	20.3	*c*.39	*c*.39	Synchronous	Inbreeding
Triticum aestivum	6x = 42	54.3	*c*.24	*c*.24		
Secale cereale	2x = 14	28.4	*c*.124	51	Partly asynchronous	Outbreeding
Tradescantia paludosa	2x = 12	54.0	*c*.188	113		
Lilium hybrids						
c.v. Black Beauty	2x = 24	[a]	*c*.252	180	Grossly asynchronous	
c.v. Sonata	2x = 24	[a]	*c*.396	252		

[a] Not determined; in other *Lilium* species, *L. henryi* has 3C = 100 pg and *L. longiflorum* has 3C = 106 pg.
After Bennett, 1976 and with the permission of Academic Press.

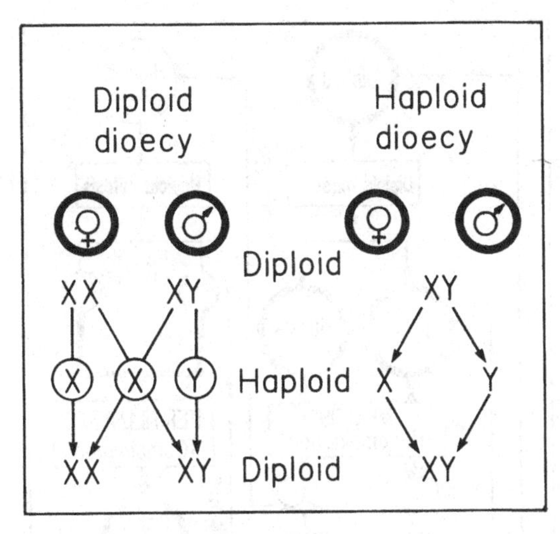

Fig. 1.6. Diploid and haploid dioecy compared. In diploid dioecy the segregation of the sex chromosomes involves the production of haploid gametes in which there is no gene expression. In haploid dioecy the segregation of the sex chromosomes involves the production of haploid spores in which there is a complete expression of genes.

Most flowering plants are *hermaphrodites*, producing both male and female gametes – a situation which is far less common in animals where it is found chiefly in simpler invertebrates (coelenterates, flatworms and annelids). However, in some flowering plants and in cycads, a given individual will produce either anthers or ovules, but not both, a situation reminiscent of the diploid dioecism seen in most animals. In a few such plants there are also differentiated sex chromosomes (*Melandrium*, *Humulus*, *Rumex*). Even in hermaphrodites, however, the two kinds of gametes may not be produced at the same time, a device which clearly precludes self-fertilization. In flowering plants, Bennett (1976) draws attention to the fact that the synchronicity of meiosis in PMCs and EMCs may be related to the breeding system of the species. In inbreeders it is synchronous and this, presumably, facilitates self-pollination. In outbreeders it is clearly advantageous for EMCs and PMCs to mature at different times and here it is at least partially asynchronous since the duration of meiosis is invariably longer in the EMC (Table 1.2).

Some mosses have a distinctive system of haploid dioecism with separate haploid sexes each carrying a differentiated sex chromosome (Lewis, 1961). Here diploid sporophytes are all of one kind, XY, but haploid gametophytes are either X or Y in constitution (Fig. 1.6). This contrasts with the diploid dioecism of flowering plants where diploids are of two kinds, XX or XY. In haploid dioecism the sexes are haploid and

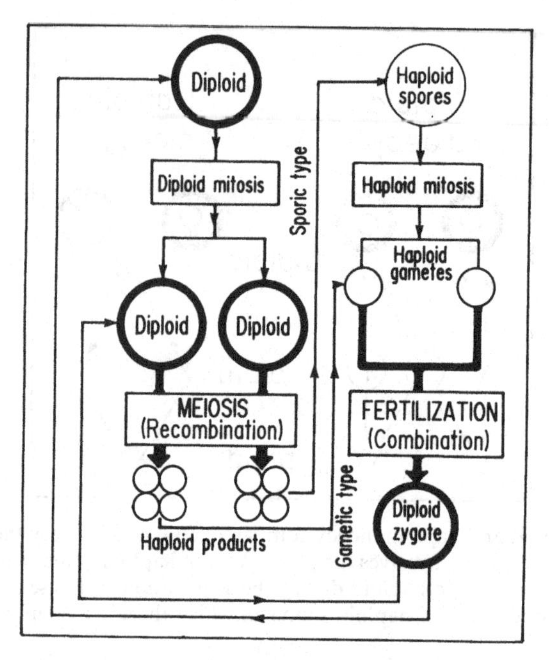

Fig. 1.7. The interrelationship of meiosis and fertilization in the sexual cycles of eukaryotes.

of two kinds. Females have an X chromosome and males a Y. The diploid zygote, XY, produces haploid spores which give rise to the sexual phases.

Sexual reproduction thus involves cycles of meiosis, haploidy, fertilization and diploidy, the precise combination determining the nature of the life cycle (Fig. 1.7).

C MITOSIS AND MEIOSIS

In multicellular eukaryotes the most active phase of mitosis occurs during the early cleavage divisions of the fertilized egg. Here cells divide without growing. Consequently, they halve their size at each division and the cell number increases geometrically while the size of the embryo remains unchanged. The cell cycle time is also much shorter. In adult tissues the only sites of active mitosis are those which are subject to attrition or loss and hence require regular replacement. Here, characteristically, *stem cell* systems are present. The term stem cell identifies cells with an extensive self-renewal capacity which persists throughout all, or most, of the life span of an organism (Lajtha, 1983). Relative to the zygote itself, all stem cells are differentiated to some extent though many are certainly capable of yielding further differentiated descendants and are then either *unipotential*, that is ancestral to one differentiated line, or *pluripotential*, yielding more than one line of descent. The first cells of the spermatogenic line are also stem cells giving rise to spermatogonia as well as maintaining their own numbers by self renewal. Likewise, in higher plants, mitotic activity occurs primarily within the confines of a complex tissue, the *meristem*, which includes a stem cell component.

Dividing populations of adult cells usually pass through three synthetic phases before entering mitosis (Prescott, 1976). The first of these, the G_1 (*first gap* phase, when no DNA synthesis occurs and no tritiated thymidine is taken up by the nucleus in autoradiographic experiments), follows on immediately from mitosis and involves the production of the ribosomal RNAs and the enzymes which are subsequently required for the replication of the chromosomes during the second phase, the *S phase*. In this phase, there is an activation of the genes which control mitosis. Coupled with this, additional copies of the DNA, and of both the histone and non-histone proteins of the chromosomes, are synthesized. New histones are then assembled into chromatin as the DNA replicates. The DNA itself is synthesized in short segments which are then ligated into longer chains until the entire duplex which constitutes each chromosome is duplicated. Thus, each DNA duplex includes multiple replication units, or *replicons*, some 10–100 μm long and arranged into early and late replicating clusters. The net result of the S phase is the production by each chromosome thread of an identical sister strand so that the replicated chromosome consists of two sister chromatids. The third, G_2 (*second gap*) phase, which follows the completion of S, involves a heightening of the synthesis of the protein tubulin which is the principal component of the mitotic spindle and is produced throughout the entire interphase period.

In embryonic cell populations of multicellular animals there may be no G_1 or G_2, the requisite syntheses having been carried out by the maternal genome in the production of the egg. Here, the cytoplasm of the unfertilized egg contains a large store of ribosomes, RNAs, tubulins and the necessary enzymes required to promote both S and M phases of the cell cycle. Likewise some unicells, like *Amoeba*, and even simple multicells like *Physarum*, lack a G_1 phase in their mitotic cell cycle. Here the G_1 events required for the onset of S must have been completed in the previous cell cycle.

The synthetic sequences which characterize interphase are followed by a distributive phase during which the two sister chromatids of each chromosome are coordinately separated, usually in conjunction with the division of the cell cytoplasm. Both these events depend on the activity of a transient, polarized and spindle-shaped mitotic apparatus composed of microtubules (MTs) which are constructed out of the tubulin molecules synthesized during interphase. The protein tubulin is a heterodimer consisting of two subunits, α and β, each approximately 55000 in molecular mass. The spindle is thus a dynamic cytoskeletal organelle

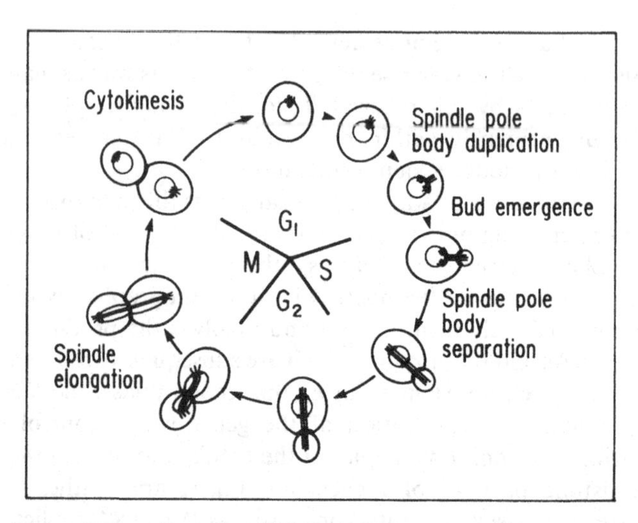

Fig. 1.8. The mitotic cell cycle of the budding yeast *Saccharomyces cerevisiae* (after Hartwell *et al.*, 1974 and with the permission of *Science*).

composed of microtubules which undergo reversible assembly and disassembly during cell division. These MTs are formed through the polymerization of tubulin monomers and are assembled from their ends, not by the insertion of subunits along their length. Moreover, since assembly is from one end only, the microtubule is itself a polarized structure. While this structural polarity is not visible using conventional microscopic methods it can be demonstrated using the flagellar enzyme dynein. This binds *in vitro* to all kinds of MTs and, under appropriate conditions, asymmetry in the morphology of the complex produced reveals the polar orientation of the dynein-decorated MT (McIntosh, 1985). Spindle formation may, or may not, involve disruption of the nuclear membrane. Enclosed spindles, in which the nuclear envelope is retained intact, so that the entire division process is intranuclear, are common in the green algae, in many Protozoa and in the lower fungi (Kubai, 1975; Heath, 1980). Here, pronounced associations may develop between the chromosomes and the nuclear envelope, on the one hand, and between the MTs and the nuclear envelope, on the other hand. Many closed spindles undergo a limited amount of nuclear envelope breakdown in the polar region while in the budding yeast, *Saccharomyces cerevisiae*, the spindle arises from a specialized spindle pole body, or *spindle plaque*, which is inserted into the nuclear membrane. This body grows steadily until it divides and the spindle then develops between the diverging products of plaque replication as these products separate on the persistent nuclear membrane (Fig. 1.8).

In most eukaryotes, however, the breakdown of the nuclear membrane appears to be essential for spindle formation and open spindles are present in basidiomycete fungi, all higher plants and a majority of animal cells. Microtubule assembly in these forms depends on the presence of seeding structures or *microtubule organizing centres*, commonly referred to as MTOCs. These centres regulate the nucleation, orientation and anchorage of individual microtubules into a cohesive apparatus (Borisy & Gould, 1977). The most obvious MTOC, and the prototype for all other centres, is the centrosome of animal cells. This consists of a pair of centrioles and their satellites, which occupy an orthogonal arrangement and are enclosed within an amorphous pericentriolar region referred to as the *centrosphere*. During interphase the centrosome lies outside, but generally close to, the nuclear membrane and maintains a minimal MT system associated with it. These centrosomal MTs tend to radiate out in all directions and their disposition is limited only by the constraints imposed by cell and nuclear shape. Starting at the S phase, short procentrioles form from each of the two centrioles. These, however, do not mature until well after mitosis has begun. Coincident with this centriolar replication during interphase, the centrosome also duplicates. This event is accompanied by a rapid increase in MT number which leads to the daughter centrosomes moving apart and taking up positions on opposite sides of the nucleus. Consequently, by the time the nuclear membrane breaks down there is already an incipient bipolarity to the cell. This is enhanced as the number of centrosome-associated MTs rapidly increases at the disruption of the nuclear envelope, giving rise to two interdigitating sets of MTs with the centrosomes occupying the polar regions of the mitotic spindle.

It was initially assumed that the centrioles played a role in spindle formation since they are known to provide precursors for the basal bodies of both cilia and flagella and so are certainly capable of functioning as MTOCs, independently of the centrosome, during cell differentiation (reviewed in Brinkley, 1985). Using centrosomes lysed from Chinese hamster ovary (CHO) cells, Gould & Borisy, as long ago as 1977, had provided substantive evidence that it is the pericentriolar material, and not the centrioles, which is associated with the production of spindle MTs. Centrioles alone can certainly nucleate a few MTs into characteristic flagellar axoneme-like bundles in lysed CHO cells but do not give rise to the larger, divergent array of MTs which characterize the intact centrosome. Additionally, a number of situations have now been described where centrioles have become dissociated either naturally or by experiment, from the poles but where normally functioning spindles still develop. In euglenoids and dinoflagellates, for example, centrioles remain dissociated from the spindle. Centrioles have also been shown to be absent in the 1182-4 cell line of *Drosophila melanogaster* which has been maintained in culture for several years (Szöllösi *et al.*, 1986). Similarly, the mouse egg has no centrioles at the second meiotic division but does exhibit a broad band of material at each pole which can be identified as pericentriolar by immunostaining using a human antibody from a scleroderma patient (Szöllösi, Calarco & Conahue, 1972). This band persists through the first four mitotic divisions of the embryo though it is not observable at interphase (Calarco-Gillam *et al.*, 1983). Thus the pericentriolar material can evidently disperse, reorganize and function in spindle development quite independently of the centriole itself. Finally, by flattening living prophase-I meiocytes of the crane fly *Pales ferruginea*, Dietz (1959, 1966) was able to prevent the centrioles from occupying their normal position at the spindle poles. Despite this a normal spindle formed.

Precisely how nucleation occurs at the centrosome, and how the MTs are linked to it, is still unknown. Electron microscope studies have consistently failed to identify any regular structure to the pericentriolar material. Yet, despite its amorphous nature, this material is able to nucleate MTs with a uniform polarity in which the growing (+) end of the tubule lies distal to the pole itself and is largely confined to one area of the spherical centre. In this way the MT lattice that eventuates is organized into a highly specific geometry (Evans, Mitchison & Kirschner, 1985). Higher plant cells lack both centrioles and centrosomes but are characterized by large and more dispersed vesicles with MTOC capabilities.

The interpolar system formed by the centrosome in multicellular animals, and its counterpart in higher plants, thus consists of two interdigitating sets of MTs which originate at the poles and overlap at the centre of the spindle. It is within this system that chromosome orientation occurs in conjunction with the development of specialized structures within the chromosome themselves. These structures, termed *kinetochores*, are only visible in ultra-thin electron microscope sections and are most commonly restricted to a localized site which appears in the light microscope as a visible constriction in the condensed chromosome thread. This constriction is known as the *centromere*. The terms kinetochore and centromere were initially coined as synonyms and are still used as such by some authors. There are, however, now sound reasons for restricting the term kinetochore specifically to that region of the chromosome to which, at the ultrastructural level, spindle MTs become attached, while retaining the term centromere for the region of the chromosome with which the kinetochore is associated (Rieder, 1982). According to the location of the centromere, chromosomes can then be defined in broad terms as *metacentric*, *acrocentric* or *telocentric* (Fig. 1.9).

With the switch from interphase to division the products of chromosome replication condense (*prophase*) so that each chromosome can be seen to consist of two sister chromatids, the kinetochore faces of which lie back-to-back. Consequently, each pair of sister chromatids orients towards

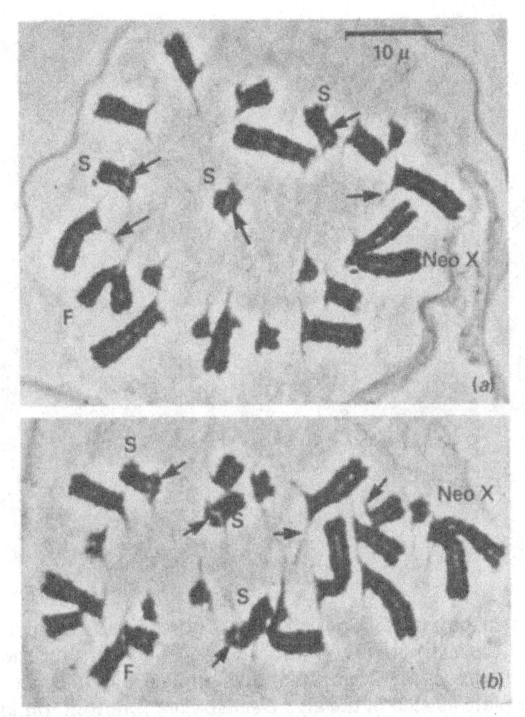

Fig. 1.9. Variation in centromere position at mitotic metaphase in male individuals of the grasshopper *Oedaleonotus enigma* heterozygous for a centric fusion (2n = 21 = 19+XY and compare with Fig. 2.35). The single fusion chromosome (F) and the neoX chromosome, which is also a fusion product, are both metacentric with the centromere located towards the mid region of the chromosome which is, therefore, biarmed. In both cells, three of the autosomes are heterozygous for centric shifts (S, arrowed) and these are also biarmed chromosomes. In five of these the arms are markedly unequal in length so that the centromeres lie close to one end of the chromosome (acrocentric). All the remaining chromosomes are uniarmed telocentrics, with strictly terminal centromeres. With the phase contrast microscope used to photograph these cells the terminal centromeres are commonly extended on the spindle (arrows).

(*pro-metaphase*, *metaphase*), and subsequently separate to (*anaphase*), opposite spindle poles. The poles then define the sites at which two new nuclei reform (*telophase*) by the return of the chromosomes to a diffuse state and the redevelopment of the nuclear membrane. Coincident with this there is a division of the cytoplasm of the parent cell at the mid-spindle region (animals) or else the formation of a new cell wall at that region (plants).

At the ultrastructural level the kinetochore takes one of two basic forms (Fig. 1.10). In many plants and some invertebrate animals (*Tetrahymena*, Orthoptera) it has a ball and cup arrangement with a less dense

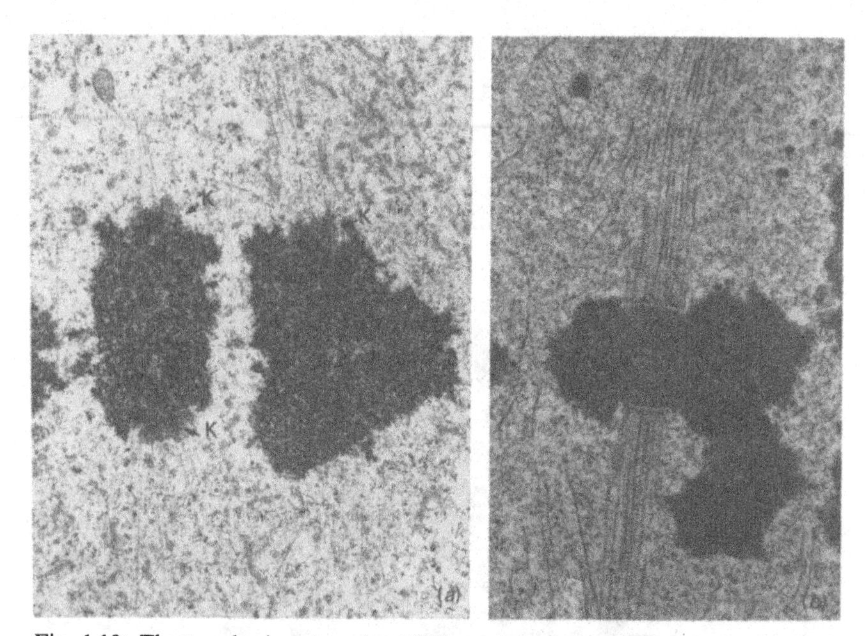

Fig. 1.10. The two basic categories of kinetochore organization to be found in eukaryotes: (*a*) the ball and cup arrangement as seen in the grasshopper *Melanoplus differentialis* (photograph kindly supplied by Dr Bruce Nicklas); and (*b*) the tripartite plate as seen in the alga *Oedogonium* (photograph kindly supplied by Dr M. J. Schibler). In (*a*) the kinetochore microtubules simply insert into a less dense region of chromatin whereas in (b) the kinetochore consists of dense outer and inner zones separated by a clear mid-zone and here the microtubules insert into the outer zone.

kinetochore sphere, to which MTs are attached, inserted into a recess in the more dense chromatin of the centromere region. In most mammals, and some invertebrates (*Drosophila*) and plants (the moss *Mnium* and some algae), it appears as a trilaminar plate consisting of two electron dense layers, each of 40–60 nm, separated by a lightly stained mid-zone of 25–50 nm. The inner layer is firmly attached to the centromeric chromatin while the MTs insert into the outer layer which is also characterized by a diffuse corona. The number of MTs which terminate in the outer layer appears to be related to the surface area of the kinetochore rather than to chromosome size *per se* (Pepper 1988).

In a few protozoans (*Amoeba*, *Blepharisma*) no morphologically differentiated kinetochore region exists and here MTs simply insert directly into the chromatin. Likewise neither centromeres nor kinetochores are evident in yeast where the centromeric DNA is organized into a small 220–250 bp chromatin segment (Clarke & Carbon, 1985) with which a single MT is associated (Petersen & Ris, 1976). Yeast chromosomes are some 100-times smaller than those of higher eukaryotes in which the

Table 1.3. *Kinetochore (KT) microtubule (MT) number variation in eukaryotes*

Species	Average number of MTs: per mitotic KT	Average number of MTs: per half bivalent KT	Reference
Saccharomyces cerevisiae	1		Petersen & Ris, 1976
Thraustotheca clavata	1		Heath, 1974
Dictyostelium discoideum	1–2		McIntosh, 1985
Homo sapiens	18		Brinkley, 1985
Locusta migratoria	21	25	Moens, 1979
Rat kangaroo PtK_1 cell line	21–45		McIntosh, 1985
Oedogonium cardicum	31–49		Schibler & Pickett-Heaps, 1987
Dissosteira		*c.*30	Nicklas *et al.*, 1979
Nephrotoma ferruginea		20–40	Forer & Koch, 1973
Melanoplus differentialis		37	Nicklas & Gordon, 1985
Chloealtis conspersa			
telocentric chromosomes		42	Moens, 1979
metacentric chromosomes		67	Moens, 1979
Neopodismopsis abdominalis			
metacentric chromosomes		67	Moens, 1979
Haemanthus katherinae	90–145		Jensen & Bajer, 1973

kinetochores are correspondingly larger and bind increased, though varying, numbers of MTs (Table 1.3). Whether this means that the simple DNA sequence which is essential for correct chromosome disjunction in yeast is simply repeated, so that the number of copies is proportional to the number of kinetochore MTs, is not known.

In some simple eukaryotes the chromosomes attach to a preformed spindle but in most eukaryotes it has long been assumed that kinetochores served as MTOCs and actually generated the MTs which link the kinetochores to the poles. McGill & Brinkley (1975) and Telzer, Moses & Rosenbaum (1975) succeeded in initiating the *in vitro* assembly of exogenous brain tubulin on the kinetochores of mammalian chromosomes. Initially, this was taken as good evidence that the kinetochore contained the nucleating intermediates required for initiating the assembly of tubulin into MTs. In the presence of MT subunits, the kinetochores of isolated chromosomes give rise to small and randomly oriented MTs (Mitchison & Kirschner, 1985a). Such a localized, but disorderly, assembly makes it difficult to see how MTs nucleated by kinetochores could become incorporated into an ordered monopolar kinetochore fiber. The idea, that the kinetochores actually generate MTs connecting chromatids to the poles is also not supported by studies on a variety of different monopolar mitoses (Mazia, 1984). In these cases a single pole is present and this organizes a half spindle on which the chromosomes orient. Under these circumstances only kinetochores which actually face the pole become engaged to it by MTs. Sister kinetochores, which face away from the pole, do not develop MTs. The implication of this behaviour is that kinetochores are not MTOCs but function by capturing and attaching to pre-existing MTs generated by the polar centrosomes or their equivalents (Tippit, Pickett-Heaps & Leslie, 1980; Pickett-Heaps, Tippit & Porter, 1982). Centrosome-associated MTs are known to be individually polarized with the fast growing or (+) ends extending away from the MTOC. The polarity of the MTs associated with the kinetochore reverses this arrangement. Here, (+) ends insert into the kinetochore (Euteneuer & McIntosh, 1981). This provides supporting evidence that kinetochores capture the ends of MTs which extend from the poles.

Mitchison & Kirschner (1985b) propose, therefore, that all MTs initially grow out of centrosomes and probe the cell area by their active (+) ends during spindle development. Some of these ends interact laterally with equivalent ends of MTs generated by the opposite pole. This leads to the stabilization of such ends through lateral association and so blocks any further MT assembly in the region of overlap. In the presence of ATP, kinetochores, too, may capture and stabilize (+) ends (Mitchison, 1986). The kinetochore MTs and the interdigitated MTs of the two half-spindles, both of which have (+) and (−) ends capped, then constitute the functional units of the anaphase spindle (Pickett-Heaps *et al.*, 1986).

When living meiocytes of crane flies are flattened just prior to the breakdown of the nuclear membrane, the centrosomes are displaced so that one or both of them are prevented from playing a role in spindle formation. Nevertheless, a functional bipolar spindle forms at the breakdown of the nuclear membrane except in cases where the displaced centrosomes remain sufficiently close to one another and to the membrane itself to give rise to a tripolar spindle. Displaced centrosomes may thus fail to be distributed to daughter cells so that secondary spermatocytes can sometimes completely lack them. Despite this a bipolar spindle is again formed and it too functions normally. The orientation of MTs within half spindles which lack centrosomes has been shown to be identical to that of opposing half spindles with a centrosome as well as with that of normal spindles. Moreover, in tripolar spindles, no kinetochore MTs orient to the dislocated centrosomes which form the third pole though such centrosomes do influence the orientation of non-kinetochore MTs (Steffen *et al.*, 1986).

Using an immunostaining combination of anti-tubulin IgG and scleroderma 5051 serum against pericentriolar material (PCM), Bastmeyer, Steffen & Fuge (1986) confirmed that there was no trace of auto-PCM at the centrosome-free poles. There was, however, strong anti-PCM fluorescence at the kinetochores though this, too, terminated once the chromosomes had reached the poles. Thus, centrosome-free spindle poles in crane flies show neither PCM nor aster MTs whereas displaced centrosomes show both. These results confirm the opinion of Dietz (1959, 1966) that in male cranefly meiocytes the kinetochores alone can make a functional spindle and centrosomes are not necessary. Additional supporting evidence has been provided by the experimental work of Czaban & Forer (1985). Colcemid depolymerizes spindle MTs. Consequently, when pro-metaphase spermatocytes of the crane fly were treated with 10^{-6}M colcemid, the asters disappeared and the spindle became barrel-shaped with individual chromosomal spindle fibers lying in parallel rather than converging towards a pole. These fibers then shortened and subsequently they too disappeared. Following this the chromosomes clumped together and often moved to the cell periphery. When whole, colcemid-treated, spermatocytes were then irradiated with a microbeam of near-ultraviolet light, short birefringent spindle fibers re-formed, though asters did not. These fibers appeared near the kinetochores and elongated away from them, often in different directions rather than being focussed to a common pole. Even so, first anaphase separation proved to be normal. Identical kinetochore fibers also re-formed in association with all three autosomes following partial cell treatment in which only one or two of the autosomes actually received irradiation.

The mitotic behaviour of chromosomes in the green alga *Oedogonium cardicum* is somewhat reminiscent of the experimental situation described by Czaban & Forer. In the closed spindle system of this organism, Schibler & Pickett-Heaps (1987) found that, as the cell progresses from prometaphase to metaphase, the overall distribution of MTs shifts from the polar region to the equator. As a result of this shift, metaphase spindles are virtually devoid of MTs in the polar regions and kinetochore fibers do not converge to a single focus but lie parallel to one another. Here, while the MTs which attach to the kinetochores originate from polar regions, they subsequently lose contact with the poles.

In grasshopper spermatocytes, the number of bivalents, and hence of kinetochores, present in the meiotic spindle can be altered by detaching bivalents from it with a microneedle and then removing them permanently from the cell by rupturing the cell membrane which subsequently reseals leaving the cell still able to function and to divide (Marek, 1978). Using this approach, Nicklas & Gordon (1985) showed that the total length of non-kinetochore MTs in a half spindle was dependent on the number of chromosomes present on the spindle (Fig. 1.11). This suggests that kinetochores may also in some way influence the number of non-kinetochore MTs within a spindle.

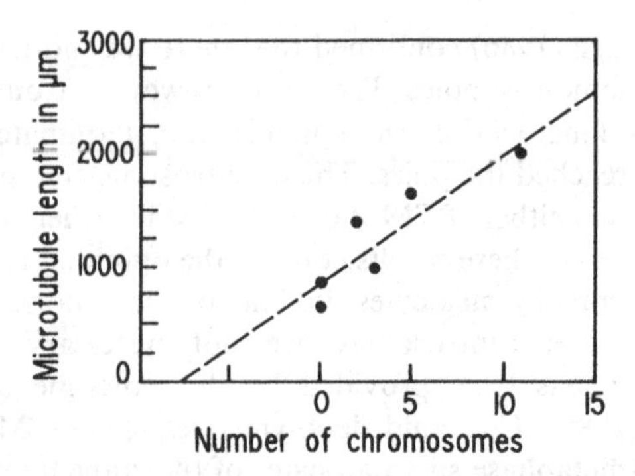

Fig. 1.11. The linear regression relationship between spindle microtubule length (in μm) and the number of chromosomes present in five different classes of male meiocytes of the grasshopper *Melanoplus differentialis*. Normal spermatocytes have 11 autosomal bivalents plus an unpaired X-chromosome (2n = 23, X0) but the number of autosomes can be experimentally reduced to 5, 3, 2 or in the extreme to 0, leaving only the X univalent (after Nicklas & Gordon, 1985 and with the permission of the *Journal of Cell Biology*).

Though an electron microscope examination of chromosomes detached from a spindle by micromanipulation indicates that they lack kinetochore MTs, such chromosome will normally reattach to the spindle if they are allowed to re-enter it at a later time (see Chapter 5A). In *Drosophila melanogaster*, however, the male meiotic spindle is isolated from the cytoplasm by several layers of membranes (see Chapter 2A.5) and, in this case, chromosomes detached from the spindle and placed outside these membranes by micromanipulation are unable to rejoin it. Instead, a miniature spindle forms around the displaced chromosomes and they proceed to divide normally on this mini-spindle (Church, Nicklas & Lin, 1986). Thus, the presence of chromosomes in the cytoplasm of male *Drosophila* meiocytes is able to trigger the formation of a functionally normal, though miniature, spindle at a site where normally no spindle forms. This, too, supports the assumption that, in some cell types of some organisms, kinetochores alone seem to be able to manufacture a spindle in the absence of centrosomes. On the other hand, chromosomes alone are not capable of producing a spindle in fertilized sea-urchin eggs in the absence of a centrosome (Sluder, Rieder & Miller, 1985).

Kinetochores also induce MT assembly in their immediate vicinity, and independent of centrosome proximity, in living cells released from either a colcemid or a nocodazole block, (De Brabander, 1982). In these cases, however, the newly assembled MTs are not initially anchored onto the kinetochore plate but form a disorganized mesh in the vicinity of the kinetochores. In the case of the more easily reversible inhibitor, nocodazole, endwise anchoring was found to occur later when the kinetochore-associated MTs started to interact with the centrosomal

MTs. These observations indicate that MT nucleation and MT anchoring are distinct events and that kinetochores may serve not only as MT attachment sites but also as independent initiators of MT assembly. Although the initial nucleation of MTs at kinetochores released from a nocodazole block is independent of their spatial disposition, these newly formed kinetochore MTs have their (−) ends free. Consequently, they are intrinsically unstable and can only elongate under the stabilizing influence arising from the proximity of either a centrosome or of centrosomal MTs.

There seems little doubt that, in most eukaryotes, the kinetochore is essential for the linking of spindle MTs to the chromosome and, hence, both for the mechanical attachment of each chromosome to the spindle and for its directed movement on that spindle. Thus, chromosomes which lack a kinetochore do not attach to the spindle and so are unable to move in a directed manner on it. It was formerly assumed that chromosomes were pulled passively to the poles via their kinetochore attachment sites. More recently this has been replaced by the view that kinetochores participate actively in their own movement and that chromosomes move poleward along stationary kinetochore MTs that disassemble from their kinetochore ends (Gorbsky, Sammak & Borisy, 1987; Nicklas 1987). Pickett-Heaps (1986) draws attention to a deviant system in pennate diatoms. Here the mitotic apparatus includes two separate sets of MTs that function differentially. There is a central component composed of two interdigitating half spindles with parallel MTs. These slide apart at anaphase and then disassemble in a unidirectional fashion from their (+) ends and backwards to the poles. There is also a separate peripheral component with which the chromosomes are associated. Whereas in conventional spindles some of the MTs terminate at the kinetochores this is not the case in diatoms. Here kinetochores are attached to a pole by means of an elastic collar which while structurally separated from the MTs is associated with them since it is stretched over one set of polar MTs. Consequently, when the collar contracts at anaphase it pulls the chromatids poleward by sliding along the associated MTs. No structural analogue of this collar has ever been found in conventional spindles though here too anaphase behaviour involves both mid-spindle elongation and chromosome movement.

Mammalian kinetochores are not evident as trilaminar entities before pro-metaphase. Sera of human patients suffering from the autoimmune disease progressive systematic sclerosis contain antibodies against the macromolecular components of cells. In the case of the CREST variant of scleroderma, which involves calcinosis, this includes an antibody to the kinetochore. Using this kinetochore-specific antiserum in an immunofluorescent probe it has been possible to identify pre-kinetochores in interphase nuclei. These are single at G_1 but double at G_2 (Brenner & Brinkley, 1982). This suggests that kinetochores in fact exist throughout

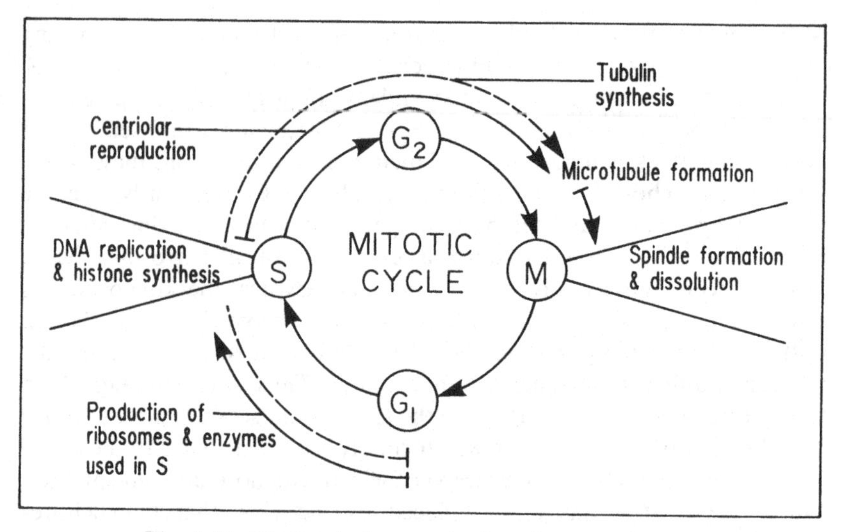

Fig. 1.12. Biochemical events in the mitotic cell cycle.

the entire cell cycle but in an unorganized, or at least differently organized, state at interphase and prophase.

The DNA of both the interphase nucleus and the mitotic chromosome is now believed to be dispersed as a series of chromatin loops organized around a substructure referred to as the chromosome scaffold. An antigenic component of the kinetochore is known to be present in this scaffold which consists of numerous discrete, non-histone, protein-anchoring complexes located at the base of the loops (Earnshaw, 1988). This structure provides a ready explanation for mitotic chromosome condensation through the clustering of neighbouring anchoring complexes and their associated loop domains. The anchoring proteins include DNA topoisomerase II (Earnshaw & Heck, 1985), an enzyme with the ability to unravel tangled DNA loops. The activity of this enzyme is blocked by mutations which result in incomplete chromosome condensation. Anaphase separation also fails in yeast cells where topoisomerase II is inactivated *in vivo* by mutation (Uemura *et al.*, 1987). Two other scaffold components, the INCEP proteins, have been localized between sister chromatids of mitotic chromosomes and especially at their centromere regions (Chaly *et al.*, 1984). These proteins dissociate from the chromosomes at the onset of anaphase and this may facilitate the separation of sister chromatids at this time.

In summary (Fig. 1.12), the two main events of the mitotic cycle are:

(1) The induction of chromosome replication. That this is a distinctive event is confirmed by the fact that fusion of a cell in the G_1 phase with one in S, immediately results in the onset of replication in the G_1

nucleus (Rao & Johnson, 1970). Similarly in F_1 hybrids between species with different cell cycle schedules the two chromosome sets start replication synchronously, though chromosomes terminate replication according to their own schedule (Mazia, 1978).

(2) The temporary bipolarization of the dividing cell. This requires that the mitotic apparatus is both a self-reproducing entity and a special kind of MTOC which organizes the microtubule lattice that subsequently regulates chromosome orientation and movement.

These two mitotic events can be dissociated. When unfertilized sea urchin eggs are treated with ammonia, DNA synthesis proceeds normally and the chromosomes condense and move to the periphery of the nucleus. However, the nuclear membrane does not breakdown. Consequently no spindle forms and no poles are present. Despite this, sister chromatids still fall apart so that their initial separation does not require any traction of the kinetochores to the poles. Autonomous centromere and chromatid separation has also been described in endosperm cells of the plant *Haemanthus* perfused with colchicine from early prophase and analysed *in vivo* by time lapse cinematography (Lambert, 1980). Thus, intrinsic centromere separation is spindle-independent and presumably serves as a trigger for anaphase movement. Indeed there are natural endomitotic cell cycles with an essentially similar behaviour (Tschermak-Woess, 1971) so that the two main events of mitosis appear to be triggered by distinct signals or switches.

The processes involved in meiosis are, in principle, similar to those of mitosis but their outcome is quite different. In meiosis, while chromosome replication, chromosome condensation, spindle organization, chromosome movement and cytoplasmic division are all involved, their interaction is much more complex. While the bulk of the nuclear DNA is replicated at pre-meiosis, in a small, though highly significant portion, replication is delayed and occurs during meiosis itself (see section 4D.2.2). Moreover, DNA replication precedes not one, but two, successive sequences of chromosome movement and nuclear division and is also disengaged from histone synthesis. The first of these sequences (*meiosis-1*) involves a prolonged prophase with a series of substages whose names reflect the morphological appearance of the chromosomes within the nucleus. Here, the individual chromosomes which are present at the onset of meiosis (*leptotene*, from the greek *leptos* for fine and *tene* for thread) are not the units of orientation and movement since, in a diploid (2x) species, homologous chromosomes undergo parallel pairing or synapsis (*zygotene*, from the greek *zygon* for coupled) to form a haploid (x) number of bivalents. Thus, chromosomes at diploid meiosis, like sexes in mating and gametes at fertilization, behave in pairs. These homologous chromosome pairs shorten and thicken (*pachytene*, from the greek *pachy* for thick) and, from this point on, each bivalent can often be seen to be composed of four chromatids arranged in two sister pairs (*diplotene*, from the greek *diplos* for double). Some authors use the suffix 'nema' in preference to 'tene' (i.e. leptonema, zygonema, pachynema and diplonema) in defining these stages. While this is etymologically more correct, since it does not mix Greek and Roman roots, most biologists continue to use the 'tene' suffix. Accepting that usage compensates for inaccuracy of derivation, these are the terms that will be employed in the present text.

While chromosome separation involves a similar spindle mechanism in both mitotic and meiotic cells the two consecutive sequences of meiotic spindle activity involve quite distinct systems of orientation behaviour which subsequently separate each of the quadripartite bivalents into their four component chromatids. This takes place in conjunction with the production of four haploid (x) nuclei and, usually, four cells. Thus, whereas mitosis maintains a constant chromosome number, meiosis is a reducing division and it is from that it derives its name (from the greek verb *meioun* – to reduce). From the point of view of DNA content, the nucleus is changed from a 4C to a 1C state as a consequence of this reduction division. Finally, in female meiosis not all the products always survive as gametes. In animal oocytes, for example, three of the four meiotic products regularly abort as polar nuclei so that only one egg is formed per meiocyte. This can be related to the need in many oogamous organisms to produce a large egg equipped with a store of molecules manufactured during oogenesis. The amount and kind of stored maternal gene products varies in different organisms but invariably includes key components of the synthetic machinery of the cell, namely ribosomes, tRNAs, mRNAs, polymerases and histones, all of which are required to maintain the synthetic activities of the fertilized egg during the early cleavage divisions and until the zygote genome is activated. The extrusion of the polar nuclei and their subsequent disintegration reflects the fact that the oocyte is characterized by a highly differentiated and heterogeneous cytoplasm in which different regions have different epigenetic fates.

2

Modes of meiosis

Repetition is the only form of permanence that nature can achieve.
George Santayana

The essential requirement of meiosis is the regular segregation of homologous chromosomes or chromosome regions. Only in this way is it possible to produce the genetically balanced gametes necessary to sustain development. There are three principal means of achieving such a segregation and these define three rather distinctive modes of meiosis: *chiasmate meiosis*, *achiasmate meiosis* and *inverted meiosis*.

A CHIASMATE MEIOSIS

This is the most common category of meiosis and it occurs, and recurs, in a reasonably conserved form, in by far the vast majority of diploid eukaryotes. It thus has high evolutionary stability. Even so, the precise details may differ in diploids and polyploids and between diploids with a structurally homozygous set of chromosomes compared to diploids which are heterozygous for structural chromosome changes.

A.1 Meiosis in diploids

As in the mitotic cycle, the replication of the greater part of the genome occurs before the onset of meiosis though, as will become apparent later (see Chapter 4D.2.2), a small amount of replication also occurs during prophase of the first meiotic division. Each chromosome entering meiosis thus consists of two sister chromatids. Unlike the situation in mitosis, however, these are not readily resolvable by light microscopy and it is for this reason that the first substage of prophase-1 is named *leptotene*. Added to this, individual chromosomes are not distinguishable since they are long and tangled (Fig. 2.1*a*). The one exception involves the sex chromosomes of many male animals, and especially the single X-chromosome of male orthopterans which is compact and heteropycnotic. Correlated with its heteropycnotic behaviour, the X-univalent, unlike the

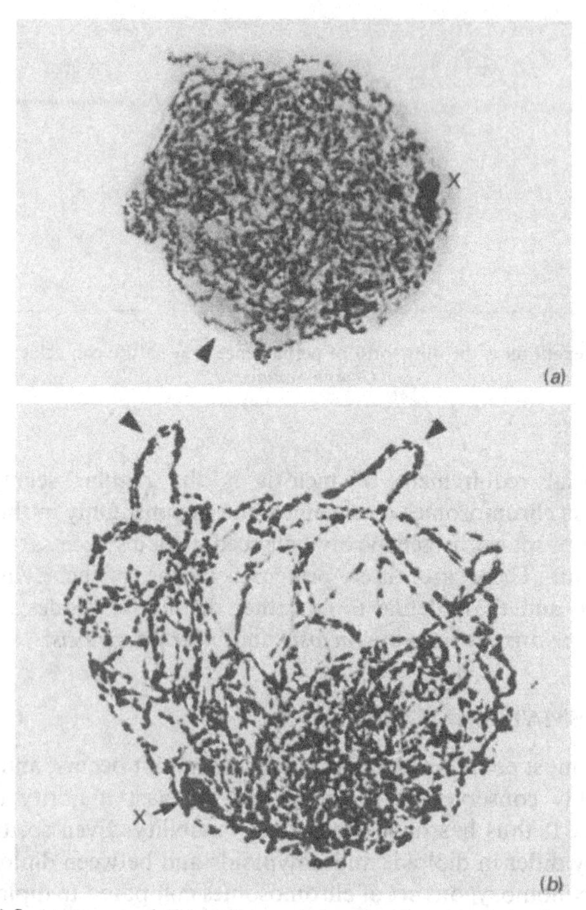

Fig. 2.1. (*a*) Leptotene and (*b*) early zygotene of male meiosis in the grasshopper *Chorthippus parallelus* (2n = 17, X0♂ and see Fig. 1.5). At leptotene the nucleus consists of a tangle of fine chromosome threads representing the dispersed autosomes, which are not individually identifiable, and a single condensed chromatin mass formed by the heteropycnotic X-chromosome. With the onset of zygotene the homologous autosomal threads pair with one another and in such paired regions the visible threads are now clearly double (arrowheads). The X univalent remains heteropycnotic.

euchromatic autosomes, is inactive in RNA synthesis throughout the first meiotic prophase as judged by its failure to incorporate tritiated uridine (Henderson, 1964). Equally, at leptotene, heterochromatic segments of chromosomes, whether autosomal or sex-linked, also appear as condensed, deeply staining structures. Electron microscope (EM) analysis of leptotene chromosomes confirms that they are each divided into two chromatids. Indeed, a dense filament is present between the two chromatids forming an axial core which is resistant to DNase but which

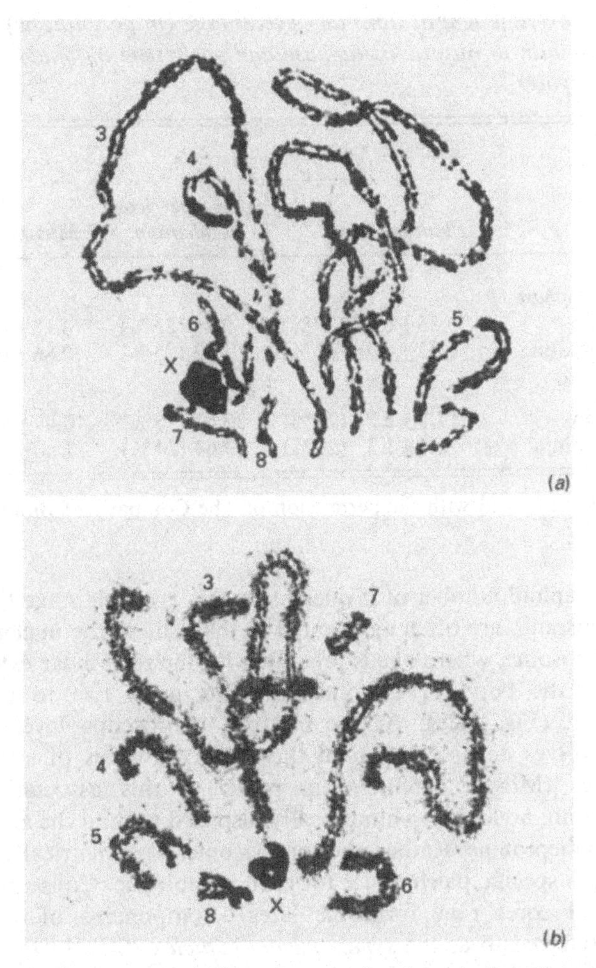

Fig. 2.2. (*a*) Full zygotene and (*b*) pachytene of male meiosis in the grasshopper. *Chorthippus parallelus*. Homologous autosomes are now fully paired resulting in eight bivalents plus the unpaired X univalent. Because centromere positions in the two largest metacentric pairs cannot be distinguished at this time only homologous pairs 3–8 can be individually identified (compare with Fig. 1.5). The X-chromosome is still heteropycnotic and is folded on itself. The bivalents are longer and thinner at zygotene than at pachytene. Additionally, the zygotene bivalents are arranged in a well defined bouquet polarity with their terminal regions clustered toward the base of the nucleus. A remnant of this arrangement is still evident at pachytene.

can be stained specifically with ammoniacal silver ions (Westergaard & von Wettstein, 1970).

The first striking event of meiosis is the association of the two homologous sets of chromosomes which synapse individually (Fig. 2.1 *b*)

Table 2.1. *Average length (μm) and percentage (in parenthesis) of eu- and heterochromatin at mitotic metaphase and pachytene of 'male' meiosis in three eukaryotes*

	Species		
Stage	*Plantago ovata*	*Lycopersicon esculentum*	*Mus musculus*
Mitotic metaphase			
euchromatin	0.85 ± 0.32 (39 %)	0.73 (25 %)	3.18 ± 0.87 (79 %)
heterochromatin	1.33 ± 0.26 (61 %)	2.23 (75 %)	0.86 ± 0.17 (21 %)
Pachytene			
euchromatin	17.0 ± 5.2 (73 %)	30.67 (76 %)	10.47 ± 2.94 (91 %)
heterochromatin	6.4 ± 1.3 (27 %)	9.64 (24 %)	1.06 ± 0.28 (9 %)

After Stack, 1984 and with the permission of The Company of Biologists Limited.

to form a haploid number of bivalents. By this *zygotene* stage the ends of the chromosomes are often aggregated to that side of the nucleus nearest to the centrosome, where one is present, with the remainder extending as loops into the body of the nucleus. This gives rise to a bouquet arrangement (Fig. 2.2*a*). At the electron microscope level, zygotene pairing involves an association of the two axial cores of each pair of homologues (Moses, 1958). As a result of this association, sister chromatids are no longer symmetrically disposed around the axial core as they were at leptotene. Rather, the core becomes asymmetrically arranged so creating a specific pairing face for each homologue. Consequently, the paired axial cores now form the lateral components of a tripartite structure referred to as the *synaptonemal complex* (SC). In this complex the two lateral elements are joined by a series of cross-bars which unite in a central element. As in mitosis (see Chapter 1.C), the chromatin of each prophase chromosome is folded into a series of radial loops during meiosis. The size of these loops is uniform for all the chromosomes of a given species but varies between species with different genome sizes. Loops are attached at their base to a distinctive meiotic structure, the chromosome core, and the cores of homologous chromosomes are aligned in parallel along the SC (Moens & Pearlman, 1988).

Once completed, pairing persists for some time, during which bivalents contract considerably in length and also thicken (Fig. 2.2*b*). In this contraction process, heterochromatic and euchromatic chromosome regions behave differentially. Stack (1984) draws attention to the fact that, on average, constitutive heterochromatin is some 2–5 times under-represented in pachytene chromosomes compared to its representation at mitotic metaphase while euchromatin is from 3–30 times longer (Table

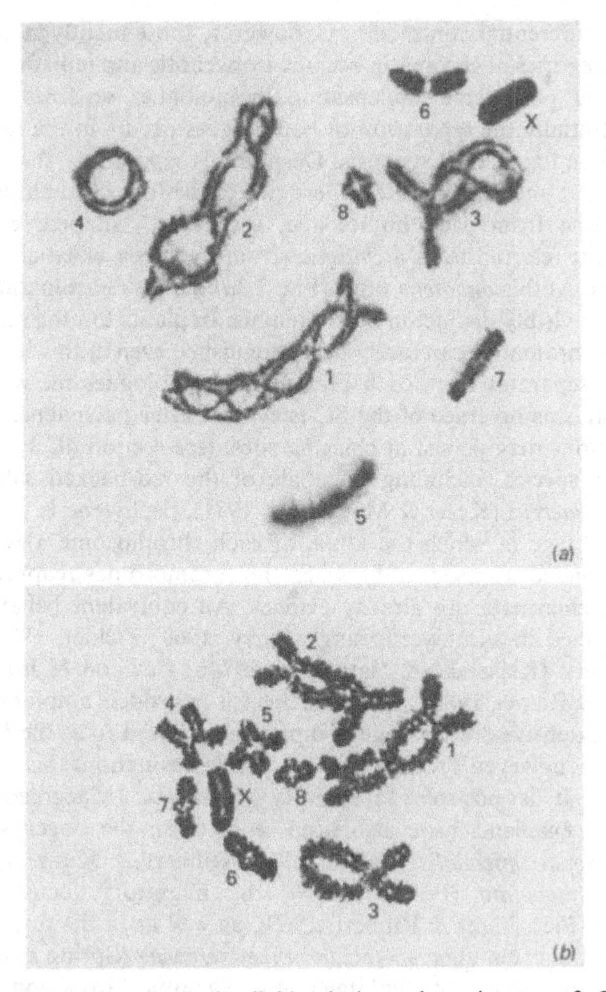

Fig. 2.3. (*a*) Diplotene and (*b*) diakinesis in male meiocytes of *Chorthippus parallelus*. Each homologue is now visibly two stranded so that each bivalent consists of four chromatids. Partner homologues within a given bivalent are separated except at one or more points along each bivalent where one strand from one homologue lies criss-crossed with the other forming chiasma sites. The number of chiasmata present per bivalent depends on the length of that bivalent. In the diplotene meiocyte illustrated the four larger bivalents have respectively 5 (bivalent 1), 4 (bivalent 2), 3 (bivalent 3) and 2 (bivalent 4) chiasmata. In three of these a rotation of the chromosome arms has led to the formation of an open cross, in place of a criss-crossed arrangement of the chromatids, at one (bivalents 2 and 3) or both (bivalent 4) of the most distal chiasmata. This has also occurred in the single interstitial chiasma of bivalent 8 and in the near terminal chiasmata of bivalents 5 and 7. Notice that in bivalent 6 the two homologous are terminally associated, with no indication of any chiasma and with a well defined gap between them. By diakinesis the bivalents have contracted and thickened considerably and in the cell illustrated there are 3 (bivalents 1 and 2), 2 (bivalent 3) and 1 (bivalents 4–8) chiasmata, those in bivalents 5–8 having been converted into the open cross form. The X univalent has now straightened out and at diakinesis can be seen to consist of two chromatids.

2.1). This differential contraction is, however, subsequently resolved and the two categories of chromatin become isopycnotic and indistinguishable.

Following pachytene condensation, homologues separate from one another. Initially the separation of homologues occurs in the same plane as they came together at synapsis. Despite this separation, they retain an association at one or more sites where two of the four chromatids in each bivalent, one from each homologue, form an X-shaped, criss-cross, arrangement referred to as a *chiasma* (from the greek word cross; plural chiasmata). At this *diplotene* stage (Fig. 2.3*a*) the four chromatids of each bivalent are visibly distinct in most animals. In plants, on the other hand, individual chromatids can rarely be distinguished even at this stage. Much of the SC separates from each bivalent as homologues move apart. In some organisms no trace of the SC is evident after pachytene. In others small portions may persist at chiasma sites. (see section 4C.1)

In some species, including the male of the red-backed salamander, *Plethodon cinereus* (Kezer & Macgregor, 1971), pachytene is followed by a 'diffuse' stage in which the DNA of each chromosome axis becomes greatly extended. Consequently, when the chromosomes reappear at late diplotene, chiasmata are already evident. An equivalent behaviour has been described in ascomycete fungi (Barry, 1969; Zickler, 1973) and in several plants (Klášerská & Natarajan, 1974b; Cawood & Jones, 1980; Klášerská & Ramel, 1980). At female meiosis in urodele amphibians there is also an extensive diffuse stage at diplotene, referred to as the *lampbrush* phase. Here, however, bivalents remain visible throughout this lampbrush phase and it is possible to identify chiasmata (Macgregor, 1980). Lampbrush bivalents have also been reported in the oogenesis of the molluscs *Sepia officinialis* (Callan, 1957; Ribbert & Kunz, 1969) and *Bythinia tentaculata* (Bottke, 1973), the migratory locust, *Locusta migratoria* (Bier, Kunz & Ribbert, 1969), as well as in the primary giant nucleus of the green alga *Acetabularia mediterranea* (Spring *et al.*, 1975). Thus, at the onset of diplotene, the synaptic attraction between homologous chromosomes suddenly ends as the SC disappears and homologues move apart except where they are held together by one or more chiasma (Fig. 2.4). The criss-cross arrangement of chromatids at each chiasma is subsequently converted into an open cross as the bivalents continue to contract. This depends on a rotation, through 90°, of all four bivalent arms on each side of the chiasma relative to one another (*diakinesis*, Fig. 2.3*b*). Bivalents with a single chiasma now adopt an open-cross shape in which all the arms are similar in size if the chiasma is interstitial (Fig. 2.4*b*) but which exist in two markedly different size classes if the chiasma occurs near one end (Fig. 2.4*c*). Where two chiasmata are present the bivalents are ring-shaped by diakinesis and include two open crosses. Bivalents with three or more chiasmata form multiple rings in which the successive loops come to lie at right angles to one another

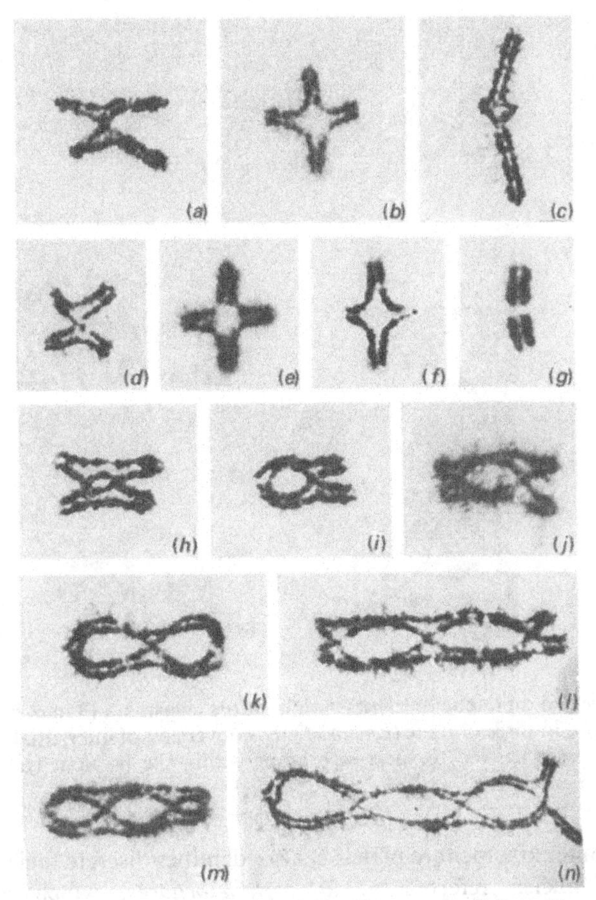

Fig. 2.4. Selected diplotene bivalents of *Chorthippus parallelus* with respectively 1 (*a–f*), 2 (*h–j*), 3 (*k* and *l*) and 4 (*m* and *n*) chiasmata. The bivalent in (*g*) again shows a clear separation between the terminally associated homologues (compare with Fig. 2.3*a*).

(Fig. 2.5). This, too, has the effect of converting criss-cross arrangements of chromatids into open-cross configurations.

By continued chromatid contraction the distinction between sister chromatids is lost, as too is the distinction between eu- and heterochromatin. At this point in the meiotic cycle the nuclear membrane breaks down giving rise to a pro-metaphase I stage (Fig. 2.6*a*) characterized by a developing spindle apparatus and the appearance in the electron microscope of discrete kinetochore regions. There is evidence to suggest that the kinetochore is either derived from, or at least organized by, the chromatin of the centromere (Pepper, 1988). Additionally, immunofluorescent analysis indicates that sera containing autoantibodies from humans with the CREST variant of scleroderma bind specifically to the

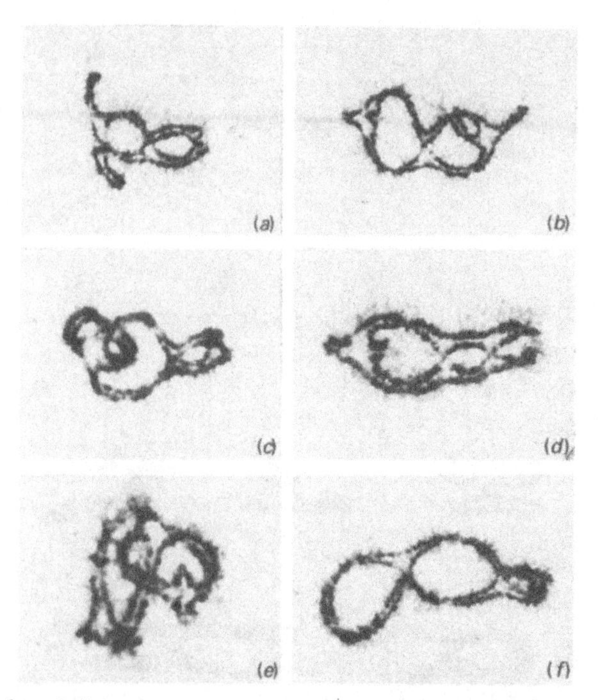

Fig. 2.5. Selected diplotene bivalents with multiple chiasmata (3 in *a*, *b* and *e* and 4 in *c*, *d* and *f*). In three of these (*c*, *e* and *f*) the conversion of interstitial chiasmata into open crosses has led to successive loops within the bivalent lying at right angles to one another.

inner and outer kinetochore plates. It also identifies discrete foci in mitotic interphase nuclei, referred to as *prekinetochores*, which not only correspond to the diploid chromosome number but which each double during G_2 of the mitotic cell cycle. The antigens recognized by these kinetochore-specific antibodies range in molecular weight from 14 to 140 which suggests that multiple proteins are involved in kinetochore structure (Kingwell & Rattner, 1988).

The use of kinetochore-specific autoantibodies, in conjunction with fluorescence (see Chapter 1C), has served not only to identify prekinetochores in mitotic interphase nuclei but also to follow their development and duplication during meiosis. With this approach, Brinkley *et al.* (1986) have shown that in the male mouse (2n = 40, XY) there are 21 fluorescent spots at pachytene. Two of these occur in the unpaired sex chromosomes which means that in each autosomal bivalent the two prekinetochores of each pair of sister chromatids are so closely apposed as to be individually indistinguishable. The number of fluorescent spots then doubles at diplotene and 20 double spots are also present in each metaphase-2 nucleus. Brinkley *et al.* (1986) conclude that, in meiosis, kinetochore duplication takes place between pachytene and diplotene. Ultrastructural

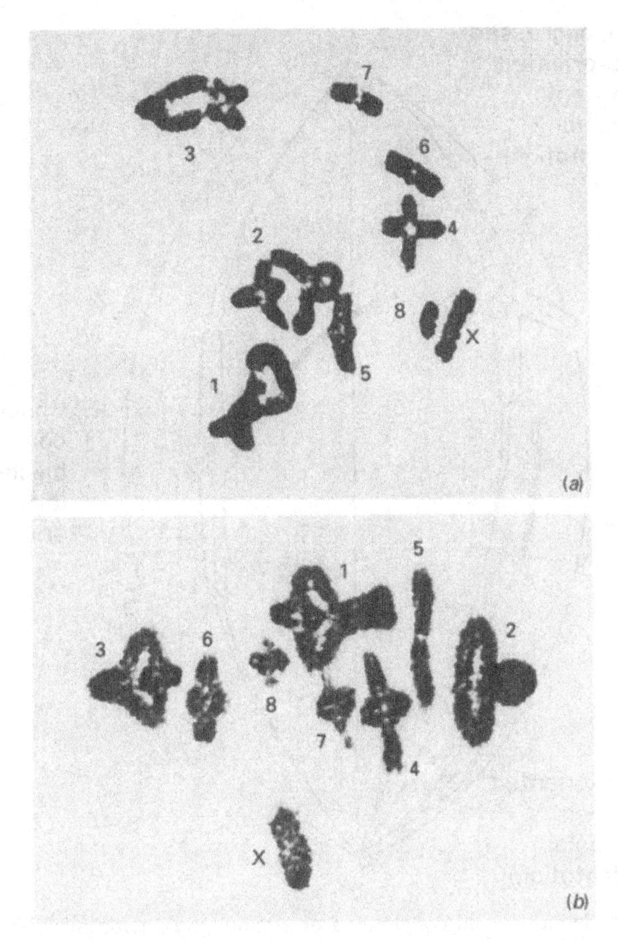

Fig. 2.6. (*a*) Pro-metaphase-1 and (*b*) metaphase-1 in male meiocytes of *Chorthippus parallelus*. The pro-metaphase bivalents are in the process of orienting on a spindle in which the poles run north–south. By metaphase all eight bivalents lie congressed with their sister centromere pairs equidistant above and below the spindle equator. The X-univalent is displaced toward the lower pole to which it is also oriented.

studies of meiotic chromosomes have, likewise, regarded the kinetochore of each half bivalent as a single structure. In males of *Drosophila melanogaster*, for example, the kinetochore of each pair of sister chromatids is a single trilaminar structure at pro-metaphase-1 which is not transformed into a double disc until anaphase-1 (Goldstein, 1981; Church & Lin, 1982).

Clarke & Carbon (1985) have proposed that a replication block at the centromere accounts for each pair of sister chromatids sharing a single kinetochore at both mitotic and first meiotic metaphase. Murray & Szostak (1985) make an even more radical proposal, claiming that both

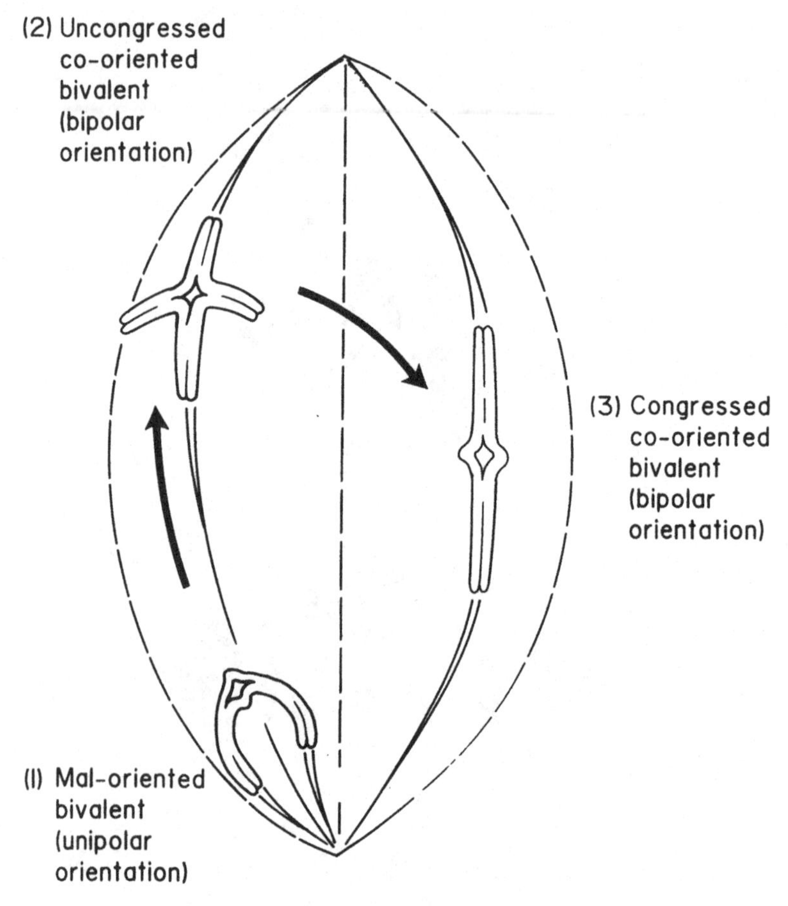

Fig. 2.7. Reorientation and congression of a maloriented telocentric bivalent at meiosis-1.

'kinetochore duplication and centromere replication is delayed until the completion of meiosis-1'. Light microscope observations leave no doubt that the centromere of each chromosome has a duplicate organization, consistent with normal replication, at both these stages (Lima-de-Faria, 1956; John & Hewitt, 1966b; Luykx, 1970), while Braselton (1981) has identified a distinct kinetochore region for each chromatid in electron micrographs of the plant *Luzula echinata*. Added to this, as we shall see shortly, the behaviour of univalent chromosomes at the first meiotic division sometimes indicates that each chromatid must have is own functional kinetochore. It is, of course, possible that kinetochore development differs in different species but, whatever the true situation, there is general agreement that half-bivalent kinetochores are double by anaphase-1.

The two kinetochore systems present in each bivalent most commonly

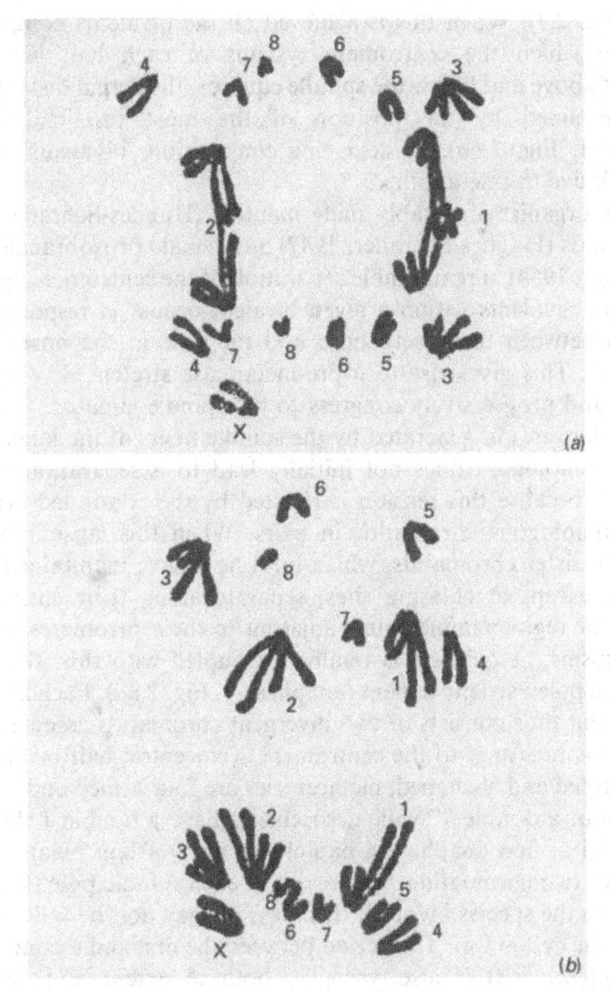

Fig. 2.8. (*a*) Early and (*b*) late anaphase-1 in male meiocytes of *Chorthippus parallelus*. Notice how the largest bivalents are delayed in their separation and how bivalent separation is correlated with a lapse of association between homologous chromatids leading to V-shaped half bivalents in the case of telocentric chromosomes and ψ shaped half bivalents in the case of metacentrics. The univalent X-chromosome passes to the pole to which it faces.

adopt a bipolar orientation on the pro-metaphase-1 spindle with the two half bivalents of each bivalent facing opposite poles. It is probable that the development of kinetochores only on the 'outer' faces of the chromatid pairs in a given bivalent (i.e. opposite to the pairing faces) predisposes such a bipolar arrangement. While this pattern of orientation may be favoured it is certainly not guaranteed. However, any initial orientation pattern other than bipolar is usually unstable and leads to a reorientation during pro-metaphase-1 until the bipolar orientation of each bivalent is secured (Fig. 2.7). When this is achieved all the bivalents congress to a position in which the centromere systems of each half bivalent lie equidistant above and below the spindle equator, the actual distance apart being determined by the position of the most proximal chiasma (metaphase-1, Fig. 2.6*b*). In achieving congression, bivalents orient at random relative to one another.

In a few organisms, notably male mantids (Hughes-Schrader, 1943), male phasmids (Hughes-Schrader, 1947) and female prosobranch gastropods (Staiger, 1954), a remarkable separation of the centromere regions of the two half bivalents within a given bivalent occurs in response to the interaction between the kinetochore and the pole at the onset of prometaphase-1. This gives rise to a pre-metaphase stretch. Bivalents then recontract and progressively congress to the spindle equator.

The bipolar tension generated by the spindle fibers at the kinetochores of paired homologues does not initially lead to a separation of those homologues because this tension is resisted by the continued attraction between homologous chromatids in pairs. When this lapses, following congression, sister chromatids, which until now have maintained a close association except at chiasma sites, separate along their entire length except for the regions immediately adjacent to the centromeres where, in most organisms, association is retained. Coupled with this, the centromeres start a poleward movement (anaphase-1, Fig. 2.8*a*). Each anaphase-1 half bivalent thus consists of two divergent chromatids associated only in the regions proximal to the centromere. Telocentric half bivalents are thus two-armed and V-shaped, metacentrics are four armed and arranged in the form of a double V while acrocentrics have a double J shape.

At the end of first anaphase a haploid number of half bivalents, each consisting of two chromatids, are present at each spindle pole (Fig. 2.8*b*). According to the species involved this may, or may not, be followed by a division of the cytoplasm. The period between the first and second meiotic division (*interkinesis*) is also variable, both in extent and character, between species. In some, chromosomes do not even decondense before the formation of the second division spindle. In others there is a well defined interphase followed by a second prophase (Fig. 2.9*a*, *b*). In either event, the half bivalents, whose two chromatids lie splayed, orient on the second division spindle in such a way that each chromatid is directed to an opposite pole (Fig. 2.9*c*, *d*). Since the precise disposition of the kinetochore on the chromosome surface is the key factor determining orientation, this means that, in the transition from the first to the second meiotic division, the two kinetochores of each half bivalent must change from the adjacent position they occupy at first anaphase to the opposite position they adopt at metaphase-2 (Fig. 2.10). Consequently, when the centromere regions finally separate at anaphase-2, single chromatids from each half bivalent pass to opposite poles (Fig. 2.11) so that four haploid nuclei result from each meiocyte that entered meiosis. Two different

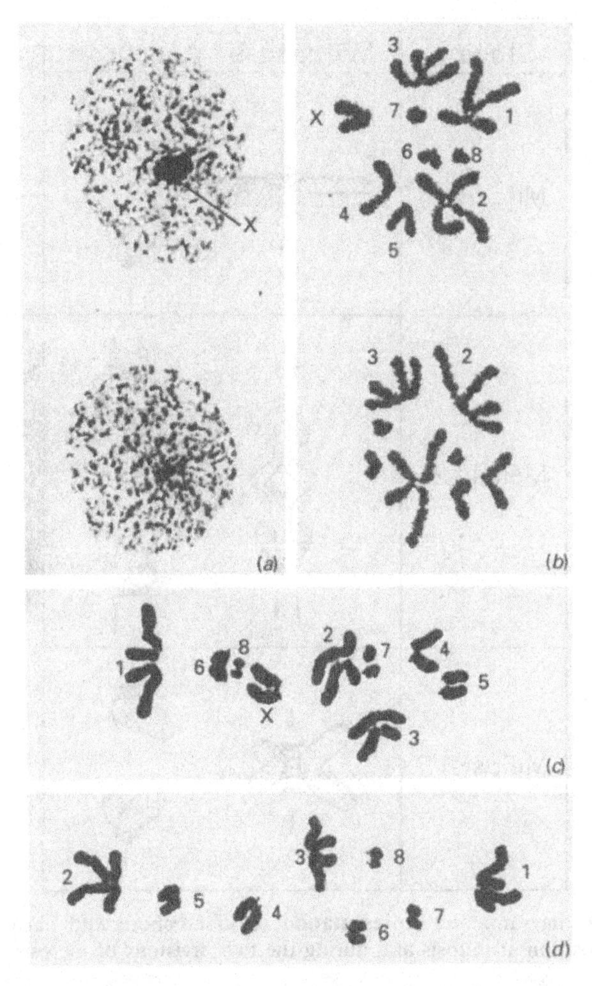

Fig. 2.9. (*a*) Interkinesis, (*b*) prophase-2 and (*c*) and (*d*) metaphase-2 in male meiocytes of *Chorthippus parallelus*. The movement of the X univalent to only one pole at anaphase-1 leads to the formation of two classes of interkinetic nuclei, with (upper) and without (lower) the X-chromosome, which is visible as a condensed body. Second division meiocytes thus have either nine (+X) or eight (−X) chromosomes. The chromatids of each autosomal half bivalent are widely splayed at prophase-2, a relationship they retain at metaphase-2.

arguments have been proposed to explain why sister chromatids proceed to the same pole at first anaphase but separate to opposite poles at second anaphase. The first assumes that they share a common kinetochore. The second proposes that each has its own kinetochore but that they are constrained in some way to orient to the same pole at metaphase-1 whereas no such constraint operates at metaphase-2. On both arguments, sister centromeres behave as a unit during the first phase of spindle activity

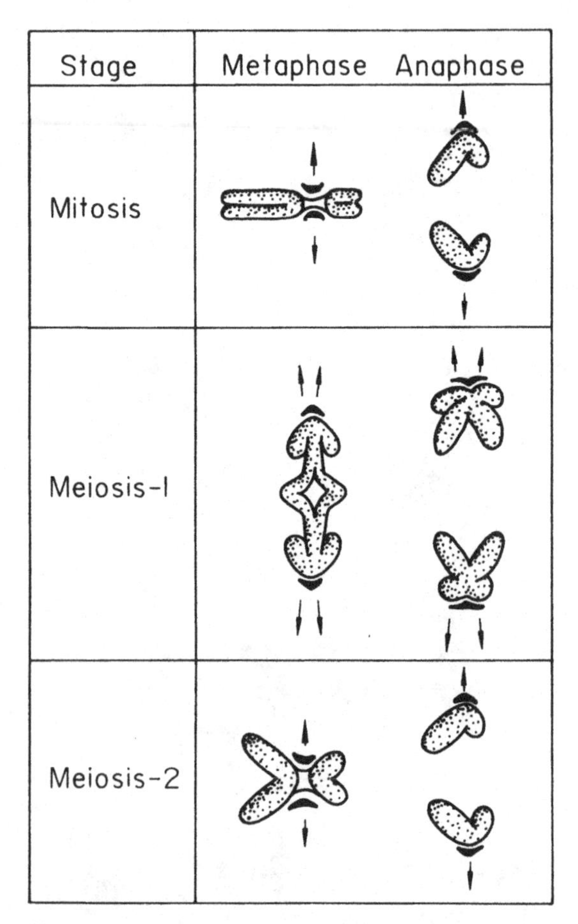

Fig. 2.10. A diagrammatic representation of kinetochore and chromatid disposition at mitosis and during the two divisions of meiosis.

but behave independently at the second division. Correct second division segregation is dependent on this transformation.

In most monocentric chromosomes, the association of sister chromatids until anaphase-2 is maintained by specialized regions which lie proximal to the centromere (Lima-de-Faria, 1956). However, in some flea beetles it is the distal tips (the *telomeres*), or in some cases specialized interstitial regions, which function in this capacity. Virrki (1989a), has applied the term *collochore* to all these regions. This extends the use of the term beyond that originally proposed by Cooper (1964) who used it specifically to define those interstitial regions by which the non-chiasmate sex chromosomes of male *Drosophila melanogaster* were associated at metaphase-1 of meiosis (see Chapter 5B.1.2).

Contrary to the statements in many textbooks, meiosis-2 is not a

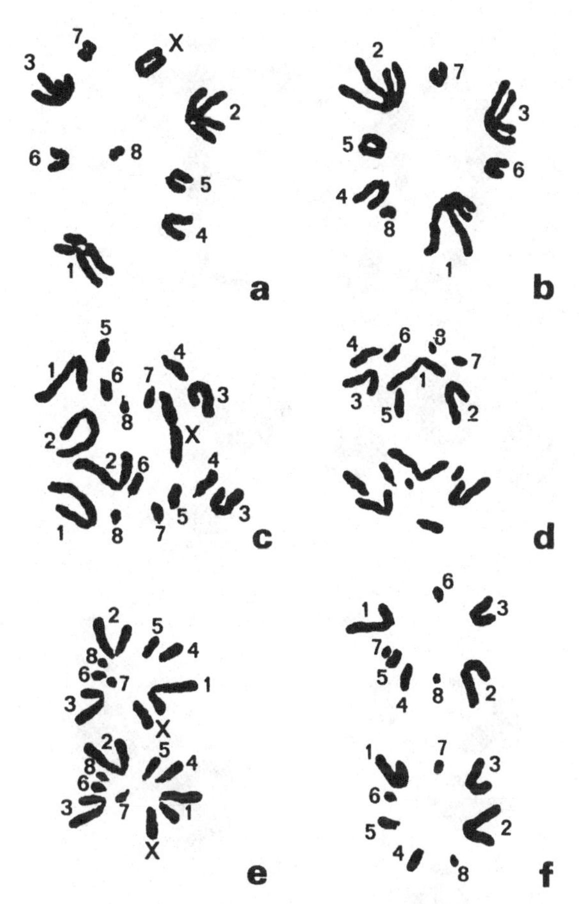

Fig. 2.11. (*a*) and (*b*) Metaphase-2 and (*c*) to (*f*) anaphase-2 in male meiocytes of *Chorthippus parallelus*. The left hand series (*a*, *c* and *e*) includes an X-chromosome whereas the right hand series (*b*, *d* and *f*) does not.

conventional mitosis though, mechanically, it is certainly reminiscent of a mitosis in terms of the orientation of half bivalents. It differs from it in two, quite fundamental, respects. First, except in cases where distal collochores are present, the partner chromatids of each half bivalent are widely splayed except at the regions adjacent to the centromere. This disposition reflects the lapse of association which occurred between the two chromatids of each half bivalent in the transition from metaphase-1 to anaphase-1. Second, since one or both of the partner chromatids of each half bivalent may have exchanged sections of their length at chiasma formation (see Chapter 2.A.2) these chromatids are no longer necessarily identical in origin, as they invariably are at mitosis. Moreover, in respect of DNA value, the number of chromatids and also allelic segregation, it

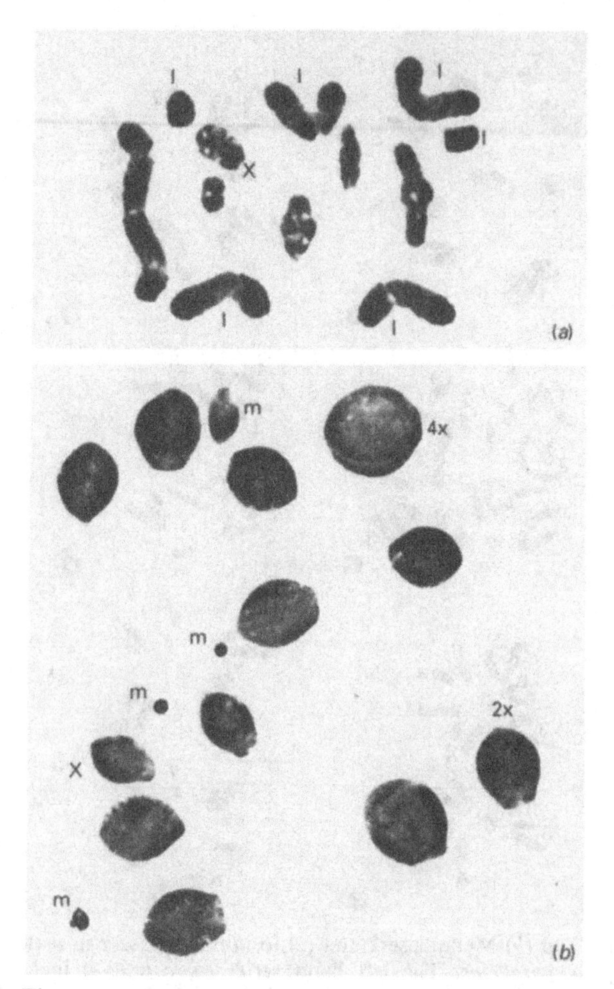

Fig. 2.12. The consequences of univalent formation in an asynaptic male individual of the grasshopper *Chorthippus dorsata* (2n = 2x = 17, X0). In this individual from 6 to 16 autosomal univalents were present at first metaphase with a majority of cells having 10–12 of them. In the cell illustrated (*a*) there are six (I). By lagging during either, or both, of the meiotic divisions these univalents sometimes form microspermatids (m) of various sizes depending on the size and number of chromosomes involved. (*b*) Alternatively univalent lagging may inhibit cytokinesis at either, or both, division leading to the production of diploid (2x) or tetraploid (4x) macrospermatids, respectively.

is clear that meiosis-2 is an integral part of the reduction sequence. Indeed, as will become apparent later (see Chapter 7.B), reduction can be avoided by the suppression of the second meiotic division. In sum, the two successive nuclear divisions that constitute a meiosis partition the four chromatids of each bivalent, one to each of the four haploid nuclei, which

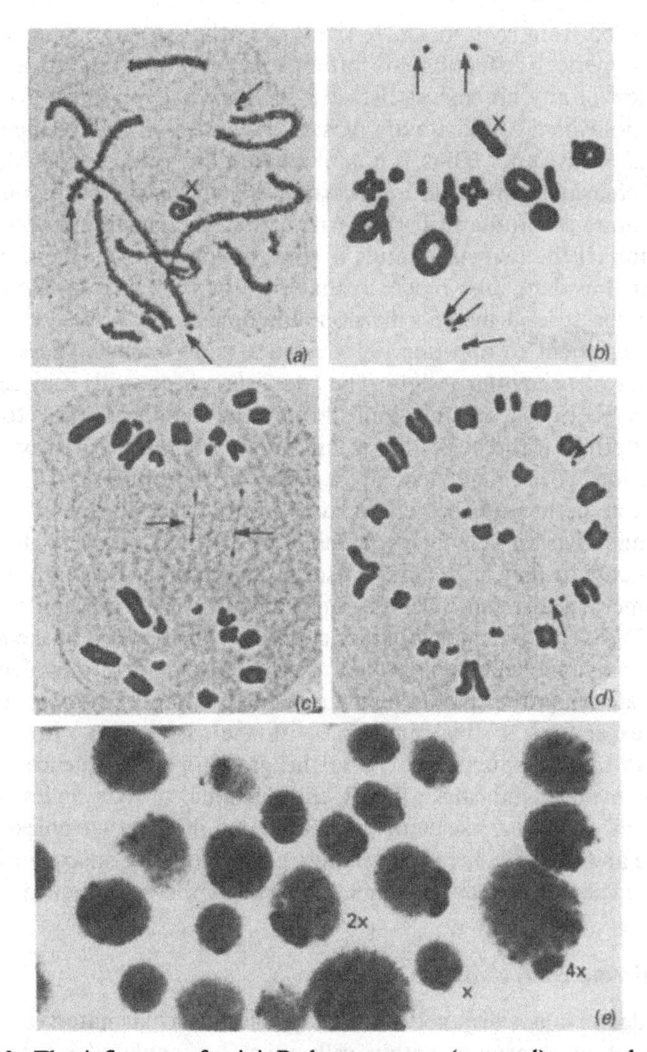

Fig. 2.13. The influence of mini B-chromosomes (arrowed) on male meiotic cytokinesis in the grasshopper *Buforania* sp. 6 (2n = 23♂, X0). Individuals of this species may carry from 1–5 small additional or supernumerary chromosomes. The pachytene meiocyte (*a*) illustrated has three, while the metaphase-1 meiocyte (*b*) has five. At anaphase-1 the Bs frequently lag (*c*) and may or may not divide. In either event they commonly lead to a diploid restitution nucleus at second division and in the cell illustrated (*d*) there is one divided and one undivided B. Division of such a diploid restitution nucleus leads to the formation of a diploid (2x) macrospermatid. Alternatively, lagging of an undivided supernumerary in a diploid second division anaphase may give rise to a tetraploid (4x) macrospermatid. Both classes are illustrated in the sample of spermatids (*e*) taken from an individual with 5Bs. The large, deep-staining structures in these spermatids are X-chromosomes.

are produced from each meiocyte that enters the sequence. This leads to both a numerical reduction in chromosome number and to a genetic segregation of any differences between the parental genomes involved.

In species where chiasmata are present they are essential for the orderly behaviour of bivalents. Thus, if homologues fail to pair (*asynapsis*), or fail to form chiasmata after pairing (*desynapsis*), the unpaired or univalent chromosomes so produced behave irregularly both at orientation and at segregation. If the two chromatids of a univalent orient to the same pole then that univalent may move to that pole at first anaphase and then divide at the second meiotic division. In some cases, however, the two chromatids orient to opposite poles. Under these circumstances, one of two modes of behaviour follow. The two sister chromatids may separate at anaphase-1 or else the univalent stalls on the spindle and fails to move. In the latter event there is either a failure of cytoplasmic division or else an exclusion of the univalent as a separate micronucleus (Fig. 2.12). Comparable patterns of behaviour may occur at second division. Where cytoplasmic division fails the products of meiosis will either be diploid, if failure occurs at first or second division, or tetraploid if it fails at both. Supernumerary univalent chromosomes sometimes produce similar results (Fig. 2.13). Even if normal cytokinesis occurs, a proportion of the meiotic products may still be aneuploid because the movement of univalents and half univalents on the spindle tends to occur randomly, though there are notable exceptions to this randomness in the case of sex chromosome univalents (see Chapter 5B.1.2). As the grasshopper sequence used to illustrate meiosis indicates, the X-chromosome of X0 males is also exceptional since it consistently passes to one of the two poles at first anaphase and then divides equationally at anaphase-2. Consequently, half the male meiotic products transmit the X while the other half do not.

A.2 Chiasma characteristics

For regular bivalent disjunction at anaphase-1 in chiasmate systems the distribution of chiasmata within cells must be regulated so that each bivalent forms at least one chiasma. Where this requirement holds it follows that not only must the minimum chiasma frequency equal the haploid chromosome number but, additionally, there must be a well defined relationship between mean cell chiasma frequency and the distribution of chiasmata within cells and between bivalents. If chiasmata were necessary only for chromosome disjunction there is no rational reason why they should not always be restricted to terminal regions. While this is certainly the case in some species it does not hold for most and, for this reason, it is important to analyse the factors which regulate chiasma frequency and chiasma distribution.

Table 2.2. *Mean cell chiasma frequencies in six species with proximal chiasma localization*

Species	Mean cell Xa frequency	Bivalent Number	Bivalent Type[a]	Reference
Plants (PMCs)				
Trillium kamtschaticum	8.71	5	M	Dyer, 1964
Allium fistulosum	16.81	8	M	Levan, 1933
Fritillaria acmopetala	23.00	12	M	Frankel, 1940
Animals (male meiocytes)				
Stephophyma grossum	11.24	11	T	Shaw, 1971a
Bryodema tuberculata	11.50	11	T	Henderson, 1969
Parapleurus alliaceus	11.50	8	3M+5T	White, 1954

[a] M = metacentric, T = telocentric.

A.2.1 Chiasma frequency and distribution

When size differences exist between the chromosomes in a given diploid set, while all bivalents form one chiasma, regardless of size, larger bivalents may form more than one. Where this situation occurs, the variation in the distribution of chiasmata between bivalents within a cell is partly dependent on chromosome length. Even so, relatively few species have cell chiasma frequencies in excess of twice the haploid number so that, since no bivalents lack chiasmata, those with chiasma numbers in excess of 2–3 are usually uncommon. This is particularly so in the case of acro- or telocentric bivalents. Higher bivalent chiasma frequencies involve metacentric chromosomes predominantly. In the absence of any restriction, chiasma frequency and distribution is expected to be random relative to chromosome length. In no case, however, is this so, for chiasmata do not occur with equal frequency in all bivalent regions. In particular, there is a disproportionately lower chiasma frequency per unit length in longer chromosomes. Nor, usually, are the positions of two or more successive chiasmata within the same bivalent independent of one another.

The most overt form of nonrandomness is seen in species where chiasmata show a marked preference for occurring at particular sites along a bivalent. The mean cell chiasma frequency is then a function of the pattern of localization and the number of meta- and acro–telocentric bivalents present in a species. There are two major forms of localization, though in no case is this localization absolute since chiasmata may also form at uncommon sites in a small number of bivalents. In the first of these two classes a majority of chiasmata are proximally sited and so lie close to the centromere (Table 2.2). This gives rise to an unusual bivalent

Table 2.3. *Chiasma (Xa) distribution in two species with proximal chiasma localization*

Species	Xta per bivalent 0	1	2	3	4	% Bi-Xte bivalents	Xa site[a] P	I	D	% Proximal Xte
Trillium kamtschaticum[b]										
Clone 1	3	26	216	5	0	87	466	5	2	98
Clone 2	7	54	163	24	2	65	407	33	20	89
Allium fistulosum[c]	0	6	953	40	0	95	2056	41	40	96

[a] P = proximal, I = interstitial, D = distal chiasmata (Xta).
[b] After Jones, 1974 (data of Dyer, 1964) with the permission of *Heredity*.
[c] After Maeda, 1937.

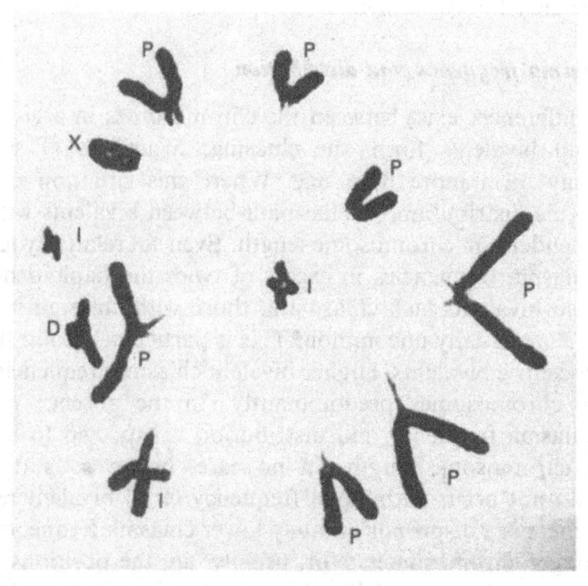

Fig. 2.14. Chiasma distribution at metaphase-1 of meiosis in the male grasshopper *Stethophyma grossum* (2n = 23♂, X0). Here seven of the 11 telocentric bivalents have proximally localized (P) chiasmata. In three of the four remaining bivalents the single chiasma is interstitial (I) while in the fourth it is distal (D).

configuration at meiosis. The situation has been particularly well characterized in *Allium fistulosum* where 95 % of all bivalents have two chiasmata and where 98 % of such bichiasmate bivalents have both sited proximally (Table 2.3). A similar situation occurs in *Trillium kamtschaticum*. Here 87 % of all bivalents are bichiasmate and 98 % of all chiasmata are proximally sited. In *Stethophyma grossum* (2n = 23, X0♂,

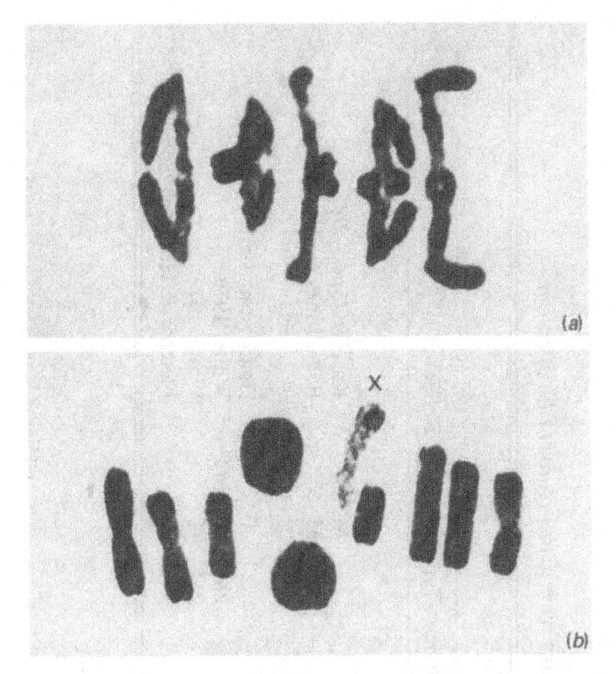

Fig. 2.15. Distal chiasma distribution in a PMC of the plant *Paeonia lutea* (*a*) with 2n = 10 and in a male meiocyte of the grasshopper *Propseudneura peninsularis* (*b*) with 2n = 19♂, X0). In *Paeonia lutea* all the chromosomes are biarmed. In *Propseudneura peninsularis* they are all uniarmed. Here seven of the telocentric bivalents have a single terminal association while the two ring bivalents have one proximal and one distal association.

all telocentric) the eight longest autosomal bivalents of the male invariably form a single proximally localized chiasma (Fig. 2.14). In some bivalents, this chiasma lies very close to the centromere; in others it is less so but in no case does the distance of the chiasma from the centromere exceed the length of the smallest bivalent. This behaviour is associated with an incomplete development of the SC, which is also confined to the proximal region. The three remaining, shortest bivalents show complete SC formation and this parallels the non-localized pattern of chiasma formation which they exhibit (Fletcher, 1978; Wallace & Jones, 1978; Jones & Wallace, 1980). Thus, here, a clear correlation exists between the pairing behaviour of the autosomes at the ultrastructural level and the extent to which chiasmata are localized within them.

No less impressive are the cases of extreme distal chiasma localization (Fig. 2.15) which are also more numerous in occurrence (Table 2.4). A particularly striking example is found in the mosquito *Aedes aegypti* where some 99 % of the chiasmata that do form are distally located (Table 2.5).

Table 2.4. *Mean cell chiasma (Xa) frequencies in six species with distal chiasma localization*

Species	Mean cell Xa frequency	Bivalent		Reference
		Number	Type[a]	
Plants (PMCs)				
Lolium perenne	11.8	7	M	Karp & Jones, 1982
Secale cereale	14.9	7	M	R. N. Jones & Rees, 1967
Animals (male meiocytes)				
Tetrix ceperoi	6.4	6	T	Henderson, 1961
Culmacris archaica	8.8	8	2M + 5T + MT	White, 1979
Spectriforma gracillicollis	10.0	8	2M + 6T	White, 1977
Euthystira brachyptera	13.8	8	3M + 5T	Fletcher & Hewitt, 1980a

[a] M = metacentric, T = telocentric.

Table 2.5. *Chiasma distribution (I = interstitial, D = distal) in three stocks of the yellow fever mosquito (Aedes aegypti)*

	Bivalent type				
Stock	ID	1I 1D	2D	2D1I	Percent distal chiasmata
Ganga	204	4	152	0	99%
Gkep	368	5	225	2	99%
Rock-red	205	1	211	3	99%

After Ved Brat & Rai, 1973.

Using a hypotonic bursting technique coupled with electron microscope analysis, Stack & Soulliere (1984) showed that in the plant *Rhoeo spathacea* (2n = 12) both synapsis and SC formation were confined to distal segments although, occasionally, one or two synaptic failures were found. Significantly, in this structurally heterozygous species, a ring of 12 multiple most commonly forms at meiosis (see Chapter 2A.4) and, in this multiple configuration, successive members are associated terminally. In newts, on the other hand, chiasma distribution is not related to the sequential characters of synapsis. Here pairing starts at the chromosome ends of both sexes regardless of whether the chiasmata that form are, or are not, localized (Callan & Perry, 1977) so that there is certainly no simple or consistent relationship between pairing behaviour and chiasma localization.

In outbred material of the perennial rye grass, *Lolium perenne*, chiasmata are predominantly (90%) distally localized at the ends of the chromosome arms in PMCs giving rise either to bichiasmate ring bivalents or monochiasmate rod bivalents. Proximal chiasmata are rare (0.8%) and only 9% of the chiasmata are interstitial (Karp & Jones, 1982). This pattern of chiasma distribution is uniform both between different PMCs and between bivalents within individual PMCs. Inbreeding leads to a decrease in mean chiasma frequency ranging from 10.56 to 6.76 and is accompanied by the production of univalents (Karp & Jones, 1983). Coupled with this, there is a decrease in both cell and bivalent variances and a much higher proportion of chiasmata are found in interstitial or proximal positions (Fig. 2.16).

Inbred lines of rye, which have been used extensively for chiasma studies, also show significant differences in both mean cell chiasma frequency and between-bivalent variance. However, while chiasma frequency is affected by inbreeding there is little effect on chiasma distribution and a majority of chiasmata are distally localized in one or both arms of each bivalent in both PMCs (Jones, 1967) and EMCs (Davies & Jones, 1974) and, within a given line, no significant differences

Table 2.6. *Mean cell chiasma frequencies of PMCs and EMCs in five lines of rye, Secale cereale*

		Xta per bivalent				Mean Xa frequency	
Line	Sex	0	1	2	3	Per bivalent	Per cell
Self pollinated							
Stårag var. P_2	♂	2	33	131	9	1.840	12.880
	♀	0	17	50	3	1.800	12.600
Derived from interspecific crosses							
J93	♂	8	55	110	2	1.617	12.489
	♀	0	24	46	0	1.657	11.599
J33	♂	18	71	58	28	1.549	10.843
	♀	6	32	27	5	1.443	10.101
J115	♂	21	81	60	13	1.377	9.039
	♀	9	33	23	5	1.357	9.499
J113	♂	23	100	51	1	1.170	8.990
	♀	8	34	20	0	1.286	9.002

After Davies & Jones, 1974. In each case the data relate to 25 PMCs and 10 EMCs per line.

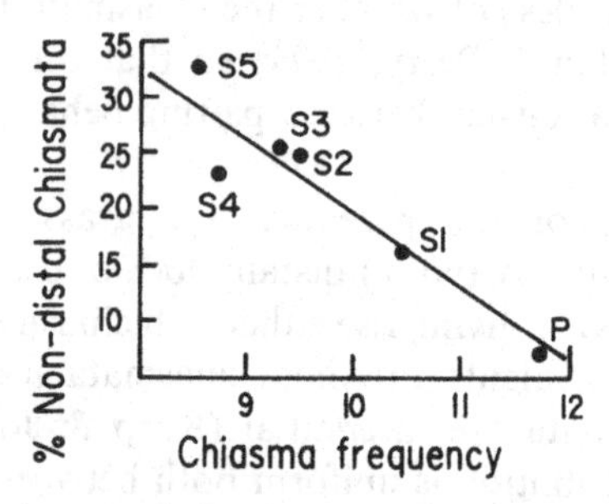

Fig. 2.16. The negative correlation between chiasma frequency and the percentage of non-distal chiasmata in the parental (P) and five inbred lines (S) of *Lolium perenne* (after Karp & Jones, 1982 and with the permission of Springer-Verlag).

exist between the two sexes (Table 2.6). An increase in the amount of phosphate available to a plant leads to an increase in mean cell chiasma frequency in PMCs of rye. Given that a relationship exists between chromosome length and chiasma frequency in species with non-localized systems of chiasma distribution it is possible that increasing the phosphate content may lead to an increase in bivalent length as it most certainly does to the length of root meristem chromosomes (Bennett & Rees, 1970).

By crossing the grasses *Secale dighoricum* and *Secale turkestanicum*, Jones (1967) succeeded in producing an F_2 segregant in which chiasmata

Table 2.7. *A comparison of the observed chiasma distribution in normal and abnormal rye genotypes with the expectations based on randomness*

	Genotypes			
	Normal		Abnormal	
Xa distribution	Observed	Expected	Observed	Expected
Distal	546	192	199	134
Interstitial	30	192	132	134
Proximal	0	192	71	134
Total	576		402	

After Jones, 1967.

regularly occurred at proximal and interstitial positions in addition to the conventional distal sites. Here, while the frequency of interstitial chiasmata agreed well with the expectation of randomness there was a deficiency of proximal, and an excess of distal, chiasmata (Table 2.7).

Some species, for example the grasshopper *Culmacris orientalis* (White, 1979) and species of the genus *Hastella* (White, 1981), show a combination of proximal and distal chiasma distribution within the same complement. Figure 2.17 illustrates two other instances of such a mixture, which imply that the control mechanism regulating chiasma distribution does not always affect all members of the same nucleus in the same way. This is particularly evident in *Chloealtis conspersa* where there is a sharp differentiation between the behaviour of the three metacentric bivalents and the five telocentric bivalents which make up the autosomal complement (Table 2.8).

The third major pattern of chiasma distribution is described as non-localized, with the chiasmata distributed in a more nearly random manner along the length of the chromosomes. Even here, however, while chiasmata are not restricted to specific regions, they nevertheless occur more commonly in some regions than in others. Mather (1937) was the first to attempt an explanation for the chromosome length–chiasma frequency relationship observed in species with non-localized distributions. This explanation was based on studies of the distribution of crossing-over in female *Drosophila melanogaster* where it had been recognized that successive crossovers were subject to interference so that one crossover event was prohibited from forming within a minimum distance from another. Such interference necessarily provides a major form of constraint on multiple crossover events since it leads to a narrower distribution of crossing-over per bivalent around the mean than would be predicted if all crossovers were at random. Mather therefore postulated that the position

Table 2.8. *The distribution of chiasmata in 70 male metaphase-1 meiocytes of Chloealtis conspersa, a grasshopper with six metacentric (M) and 11 telocentric (T) chromosomes (2n = 17, X0♂)*

	Chiasma distribution				
	2Xta		1Xa		
Bivalent category	2D	1D1P	P	I	D
M 1–3	70	—	—	—	—
T4	—	62	1	—	7
T5	—	4	56	—	10
T6	—	—	20	10	20
T7	—	—	54	2	14
T8	—	—	2	2	66

After John, 1987.

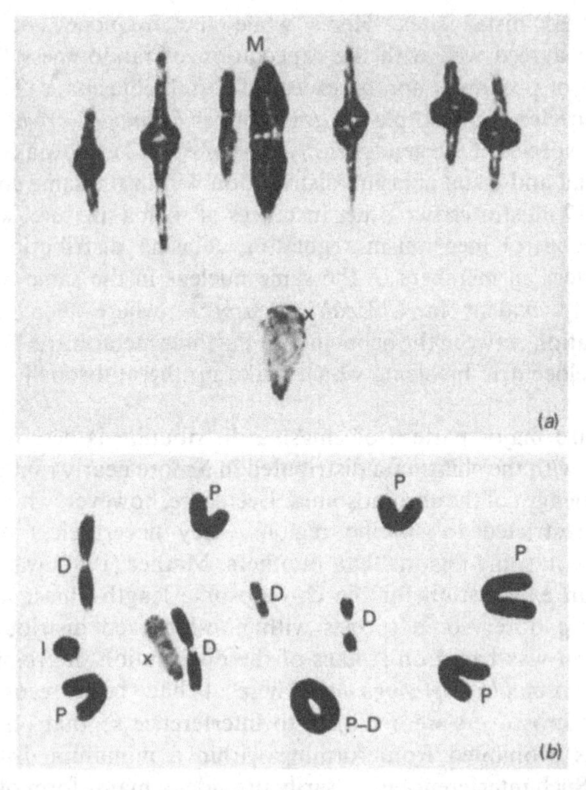

Fig. 2.17. Two species which combine different forms of chiasma distribution within the same male meiocyte. In the eumastacid grasshopper *Teichophrys* sp., (*a*) with 2n = 15♂, X0, the single metacentric (M) bivalent has two distal chiasmata. The other six bivalents are telocentric and five have a single interstitial chiasma. (*b*) In the grasshopper *Pezotettix giorni* with 2n = 23♂, X0, there is a mixture of proximal (P) and distal (D) chiasmata.

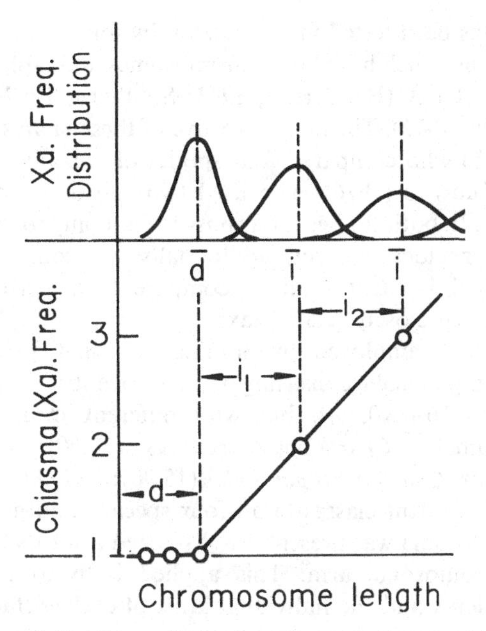

Fig. 2.18. The relationship of chiasma frequency and chiasma frequency distribution to chromosome length according to the model proposed by Mather (1937). The differential distance (d) represents the interval between the centromere and the first chiasma in a given chromosome arm so that $\bar{d}$ is the mean differential distance. The interference distance (i) defines the distance between successive chiasmata and $\bar{i}$ is the mean interference distance (after Mather, 1937).

of chiasmata within a bivalent depended on two parameters: (1) the mean cytological distance from the first formed chiasma to a terminal point which he suggested might be the centromere. This he referred to as the differential distance (d); and (2) the mean cytological distance between successive chiasmata, which represented the interference distance (i).

Mather argued that, whereas d was variable, depending on chromosome size, i was constant for all chromosomes within a given complement. In these terms, the relationship could be represented in a simple graphic form (Fig. 2.18). In this model the smallest members of a complement have a constant chiasma frequency of one, irrespective of their precise length, because a second chiasma can only form if the minimum i is less than the residual length of chromosome beyond the first chiasma. After this cut-off point, bivalents might then form one or more additional chiasmata depending on their length and on the value of i. The observed relationship between mitotic length and mean chiasma frequency per bivalent supports this predicted pattern (John & Henderson, 1962). The implication of this model is that non-localized chiasma formation is a sequential process in which the mean position of chiasmata can be described by the operation of interference not only between the first-formed and second chiasma but also between successive chiasmata.

This model has been tested in some detail by four studies all involving male orthopterans which have large chromosomes and diplotene bivalents of unparalleled clarity (Henderson, 1963; Southern, 1967b; Fox, 1973; Shaw & Knowles, 1976). The most objective of these analyses are those of Southern (1967b) who compared four species of truxaline grasshoppers, and Shaw & Knowles (1976), who used two subspecies of the *Caledia captiva* complex. In both studies it was possible not only to unambiguously define centromere location but, additionally, to consistently identify specific individual bivalents and so compare their performance both within and between closely related taxa.

The four species employed by Southern all had strikingly similar chromosome complements, consisting of six metacentric and 10 telocentric autosomes ($2n = 16+X0$, ♂), but with different mean cell chiasma frequencies, namely: *Chorthippus brunneus* (13.60); *Myrmeleotettix maculatus* (15.10); *Chorthippus parallelus* (15.90); and *Omocestus viridulis* (17.35). For all bivalent classes in all four species, Southern found that when only one chiasma was present it was located towards the more distal end of the chromosome arm. This applied both to the one-armed telocentric bivalents and the individual arms of each metacentric. When two or more chiasmata were present in the same arm there was an especially pronounced tendency for one chiasma to be restricted to the distal, non-centric, end with the other located proximally. That is, there was a marked change in chiasma pattern on switching from one to two chiasmata (Fig. 2.19). Little further change in this pattern occurred when three or more chiasmata were present because the additional chiasmata simply intercalated between the proximal and distal types. Thus, contrary to Mather's assumption, i was not constant but varied according to the number of chiasmata which formed in a given chromosome arm. Moreover, in all four species, higher mean cell chiasma frequencies were accompanied by a longer chromosome arm length in all chromosomes, as well as lower i values. Thus the interference distance in fact varied both within and between bivalents.

Southern suggested that the observed pattern of chiasma distribution was most easily explained on the assumption that chiasma formation begins in a distal position and then proceeds sequentially toward the centromere, the opposite of Mather's original claim. The principal controlling factor determining chiasma pattern was assumed to be a variable interference operating in relation to bivalent length so that multiple chiasmata tended to be equally spaced along the bivalent. Finally, and again in all four species, the centromere of the metacentric bivalents acted as a barrier to interference so that each arm of each metacentric behaved independently of the other arm with no evidence of any competitive interaction between either different arms of the same bivalent or between different bivalents.

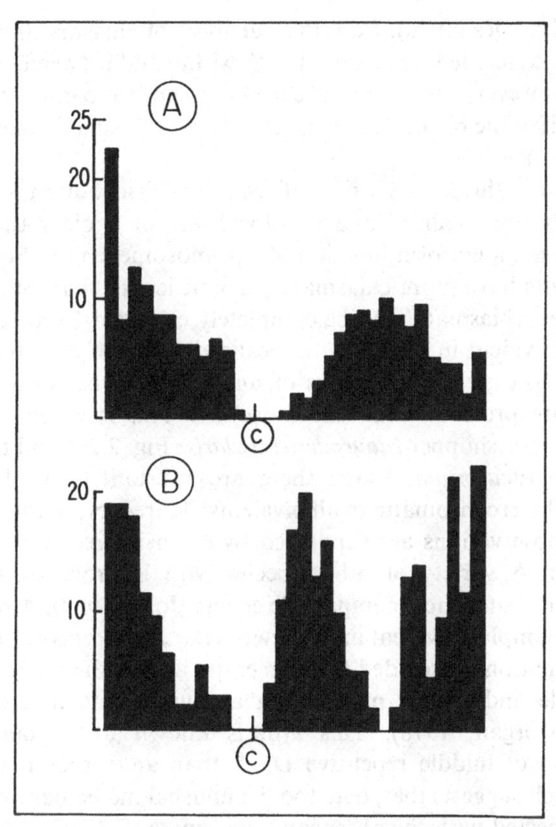

Fig. 2.19. The distribution of chiasmata in 389 metacentric L3 bivalents of the male grasshopper *Chorthippus brunneus* with one chiasma in each chromosome arm (A) and 94 L3 bivalents with one chiasma in the shortest arm and two in the longest arm (B). The position of the centromere is defined by C and each column of the histograms correspond to 5% of the long arm (after Jones, 1984).

In support of Southern's findings, Maudlin & Evans (1980) subsequently reported that, in a metacentric bivalent from a stock of mice homozygous for a single Robertsonian fusion, there was a consistent increase in chiasma frequency with increasing arm length in female oocytes. The mouse data also indicate the occurrence of a centromeric barrier to interference, the absence of competition between bivalents and the operation of interference as the principal factor governing chiasma distribution with chiasma formation proceeding sequentially from telomere to centromere. Thus, chiasma control appears to be bivalent, rather than cell-based (Hewitt & John, 1965). That is, cell chiasma frequency represents the sum of independently determined bivalent frequencies rather than bivalent chiasma frequencies resulting from the partitioning of a predetermined cell allocation between different bivalents.

Shaw & Knowles (1976), from their analysis of chiasma distribution in *Caledia*, also concluded that *i* varied both within and between bivalents. In this case, however, since single chiasma tend to concentrate in the proximal half of the bivalent, they assumed that chiasma formation begins at the centromere.

Thus, in all three categories of chiasma distribution – proximally localized, distally localized and non-localized – it is clear that chiasma formation is not a uniform function of chromosome length. Some regions of all bivalents have more chiasmata per unit length than others and, in extreme cases, chiasmata may be completely excluded from certain sites. This is most evident in species with localized forms of distribution but it applies also to cases where blocks of uninterrupted constitutive heterochromatin are present within chromosomes. This is especially clear in cases like the grasshopper *Stauroderus scalaris* (Fig. 2.20) and the tomato, *Lycopersicon esculentum*, where there are substantial blocks of pericentromeric heterochromatin in all bivalents. Moreover, in the tomato the cytological observations are supported by extensive recombination data (Rick, 1974). A variety of other species with heterochromatic blocks located at other sites show comparable effects (John, 1988). A particularly interesting example is evident in the newt *Triturus cristatus carnifex* where chiasma formation is excluded from an entire arm of the longest bivalent, at both male and female meiosis, on account of its heterochromatic character (Morgan, 1978). This arm is known to contain a larger concentration of middle repetitive DNA than any other region of the genome, which suggests that, here too, its unusual molecular organization may be connected with the absence of chiasmata.

On the other hand, chiasmata do occur in the C-band positive regions of Chinese hamster sex chromosomes. These regions differ from conventional constitutive heterochromatin both in respect of their differential staining (Murer-Orlando & Richter, 1983) and by virtue of the fact that they lack repetitive DNA (Arrighi *et al.*, 1974). Similarly, Hale & Greenhaum (1986) observed chiasma within the heterochromatic portion of the XY-bivalent in the deer mouse *Peromyscus sitkensis* where the heterochromatin involved is again not of the same type as that found in the autosomes. Chiasmata also occur in the facultatively heterochromatic X-autobivalents of *Carausius morosus* (see Chapter 7B) which are positively heteropycnotic at pachytene (Pijnacker & Ferwerda, 1986) as well as in the bivalents formed by the facultatively heterochromatic X chromosomes of autotetraploid grasshopper spermatocytes (Nur, 1978; Shaw & Wilkinson, 1978). In none of these facultative cases do the sex chromosomes consist of repetitive DNA.

In the female meiosis of *Drosophila melanogaster*, the pericentromeric heterochromatin, which *in toto* occupies some 30% of the genome, is known to be excluded from crossing-over. The basis of this exclusion is

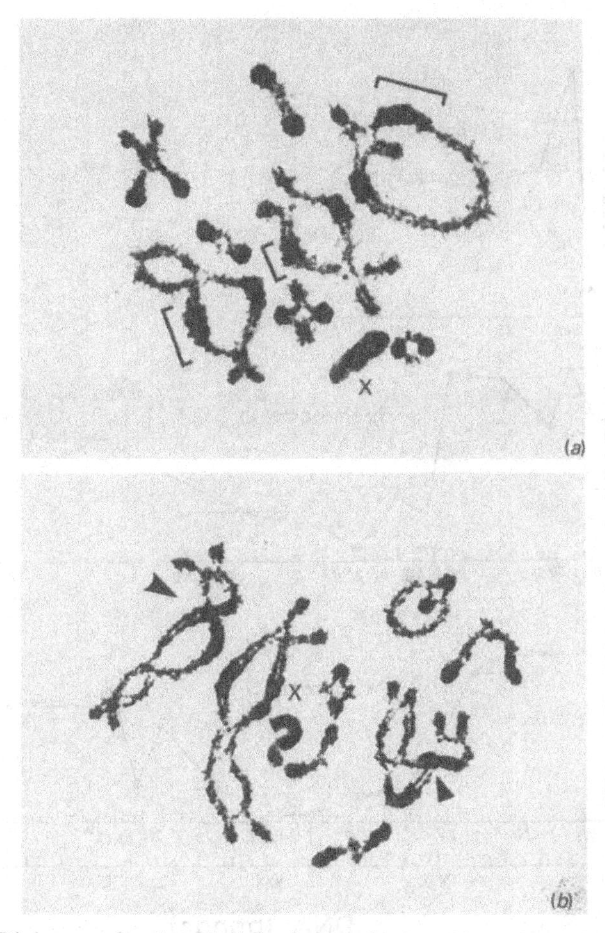

Fig. 2.20 Diplotene of male meiosis in the grasshopper *Stauroderus scalaris* (2n = 17♂, X0). In this species there are substantial blocks of condensed, and deeply staining, heterochromatin around the centromeres of all eight autosomal bivalents. In the three metacentric bivalents these blocks are interstitial in location (⊔) while in the five telocentric bivalents they are terminal (*a*). The arrowheads in (*b*) identify two regions where one homologue is relationally twisted over its partner. These regions are clearly distinguishable from genuine chiasmata.

evidently intrinsic to the heterochromatin because its capacity to exclude recombination is maintained when that heterochromatin is moved elsewhere in the genome either by inversion or by translocation (Szauter, 1984). Like the heterochromatic arm of the newt *Triturus cristatus carnifex* the pericentromeric heterochromatin of *Drosophila melanogaster* also has a distinctive molecular organization being composed predominantly of highly repeated DNA. Additionally, in *Drosophila* the centromere itself has been shown to suppress crossing-over in its

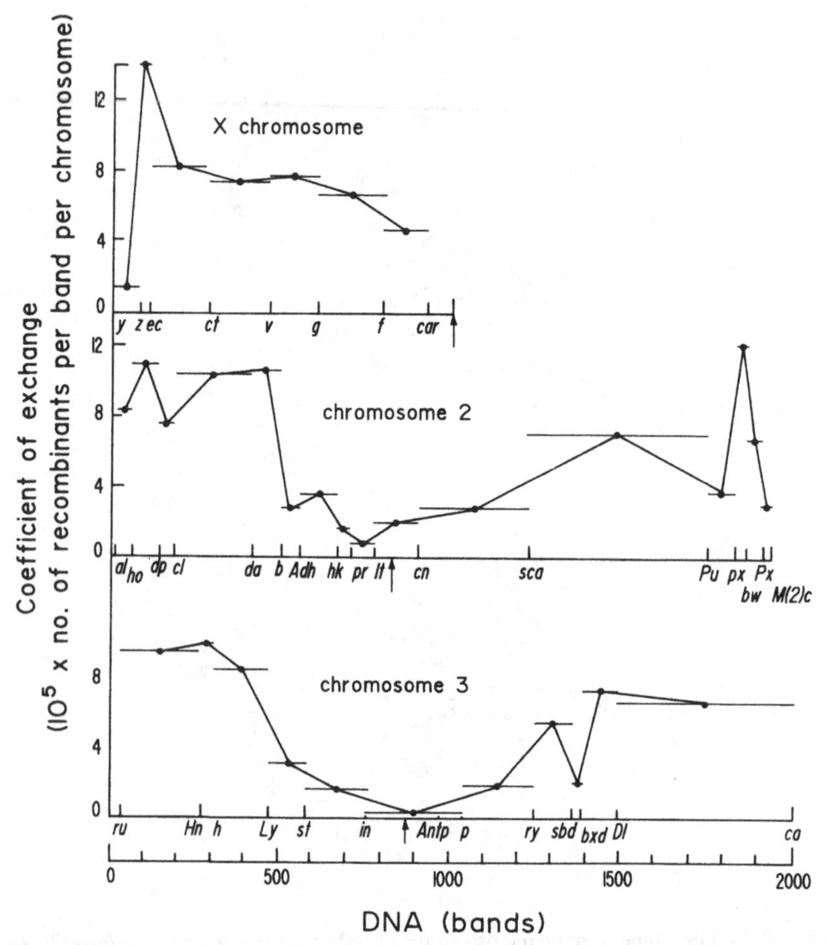

Fig. 2.21. The distribution of exchanges in the euchromatic arms of the three major chromosomes (X, 2 and 3) of female *Drosophila melanogaster* relative to the amount of DNA present in the polytene bands of those arms. The absissae also show the relative positions of the gene loci used to estimate the coefficients of exchange while the centromeres are marked by arrows. The individual points represent the mean coefficients of exchange over the chromosomal segments defined by the horizontal bars (after Lindsley & Sandler, 1977 and with the permission of the Royal Society of London).

immediate vicinity. This centromere effect is normally both masked and enhanced by the presence of pericentric heterochromatin but is revealed by deletion of that heterochromatin (Yamamoto & Miklos, 1978) as well as by structural rearrangements which move euchromatic segments next to the centromere (Beadle, 1932; Mather, 1939).

As estimated by Lindsley & Sandler (1977), the distribution of crossover exchanges per unit length of DNA in female *Drosophila* is, as expected, low in the euchromatic regions around the centromeric heterochromatin of all the autosomes but increases towards the chromosome ends, though it is also low at the actual telomeres. The X-chromosome behaves differently. Here the coefficient of exchange is low at both extreme distal and proximal regions but increases abruptly distally and then levels off towards the middle of the arm (Fig. 2.21). By comparing coefficients of exchange in specific regions, both in their normal location within the genome and when translocated elsewhere, Lindsley & Sandler also provide evidence that the level of exchange is an autonomous function of the local DNA sequence organization so that the amount of exchange in a given segment is modified when it is translocated elsewhere. Additionally, by comparing the coefficients of exchange in the wild type with mutations of genes normally functioning at pachytene (see Chapter 6A.2) they conclude that the mechanism governing the production of exchanges simultaneously promotes interference between adjacent exchanges.

Like female *Drosophila*, the centromeres of both the fungus *Neurospora* and the yeast *Saccharomyces cerevisiae* have an inhibitory effect on meiotic crossing-over. By means of transformation, Lambie & Roeder (1986) were able to relocate a 627 bp fragment of CEN III DNA in *S. cerevisiae* next to the HIS-4 locus (Fig. 2.22). This had two effects. It led to an increase in the frequency of crossing-over in the LEU-2 and PGK-1 regions, normally located adjacent to CEN-III, as well as to a decrease in the distal HIS26–LEU2 region.

The nucleolus organizing region (NOR), too, may act as a constraint to meiotic exchange. Moens (1969), for example, drew attention to the fact that in the migratory locust (*Locusta migratoria*) the nucleoli first appear in early prophase-1 prior to the formation of the axial cores of the SC. Consequently, the lateral elements of the SC are often separated at the site where they attach to the nucleoli and so fail to form a normal tripartite structure. This puts into clear perspective the early claim of Upcott (1936) that, in the plant *Eremurus spectabilis*, the nucleolus organizing bivalent has a reduced chiasma frequency with chiasmata commonly specifically excluded from the chromosome region which carries the terminal NOR (Fig. 2.23). In the silkmoth *Bombyx mori*, chromosome 2 includes a NOR, while chromosome 1 is characterized by a chromatin knob. Both of these structures have been shown to impede synapsis and SC formation (Rasmussen, 1986).

Not only are chiasmata absent within heterochromatic regions but such regions may also influence the occurrence of chiasmata in adjacent euchromatic segments and so lead to their redistribution within the chromosome. The grasshopper *Cryptobothrus chrysophorus* exists in two chromosome races. One of these, distributed in the north of Eastern Australia, has a series of fixed heterochromatic blocks located terminally on all five medium-sized bivalents, which are completely lacking in the

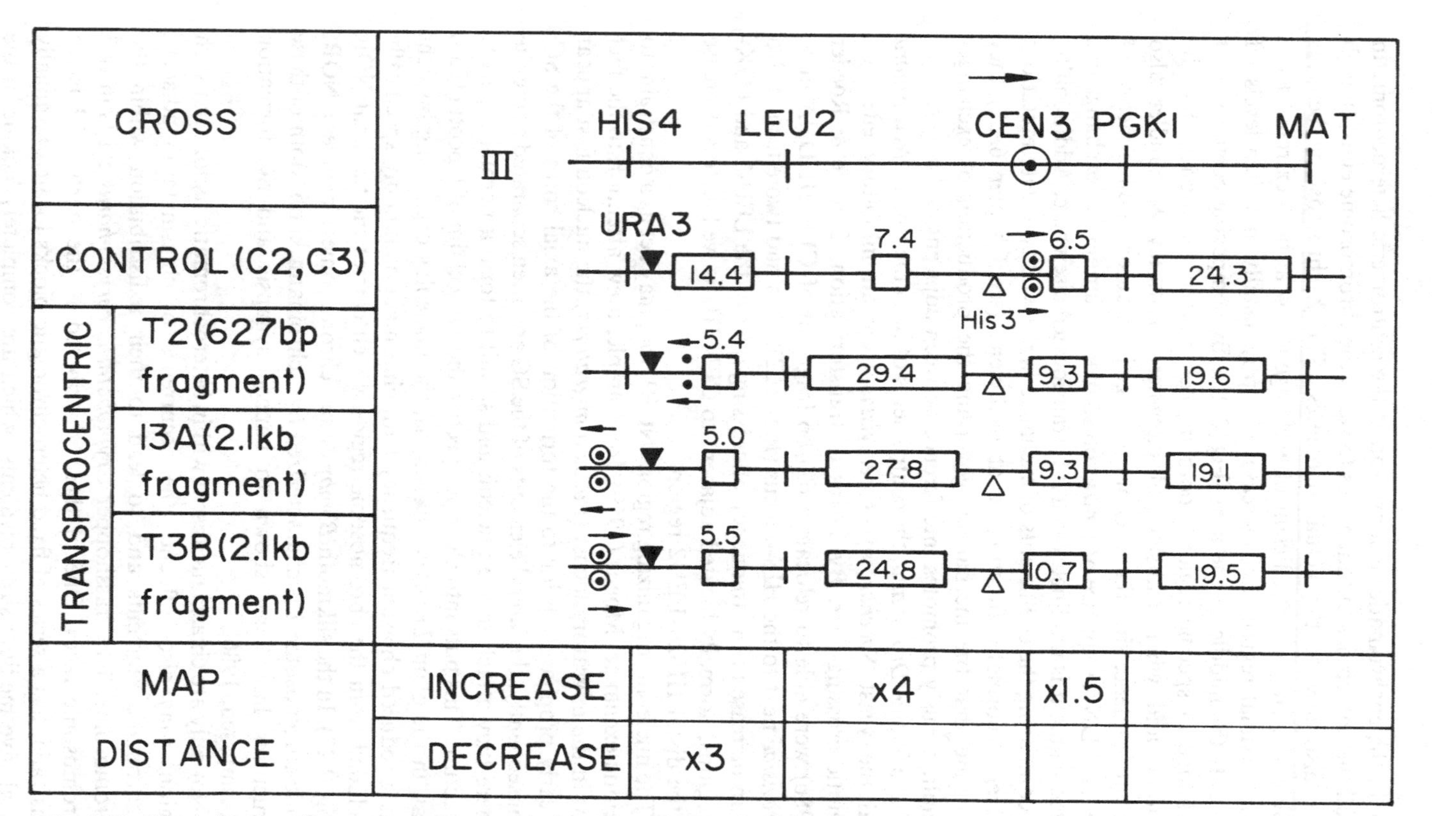

Fig. 2.22. The influence of centromere location (⊙) on map distance in chromosome 3 of the budding yeast *Saccharomyces cerevisiae* (after Lambie & Roeder, 1986 and with the permission of *Genetics*).

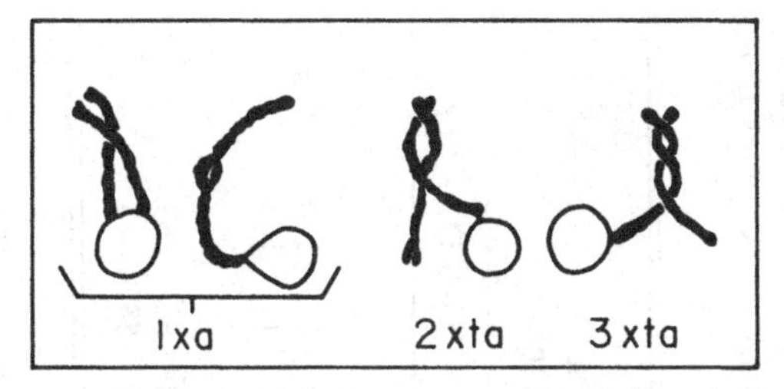

Fig. 2.23. Patterns of chiasma distribution in the nucleolus organizing chromosome of the plant *Eremurus spectabilis* (after Upcott, 1936).

southern race of this species. These two races differ strikingly in their chiasma distribution pattern with a dramatic increase in the number of proximally-sited chiasmata in the medium bivalents of the northern race (Table 2.9, Fig. 2.24). A similar effect has been noted in at least six other species where polymorphisms exist for heterochromatic supernumerary segments (John & King, 1985b). In all of these, the presence of a heterochromatic segment in a specific bivalent leads to a radical redistribution of chiasmata in that bivalent compared to its behaviour in the absence of the segment. Where the additional segment is distally located it leads to an increase in proximal chiasmata. Where the segment is proximally located it leads to a predominantly distal distribution. The precise basis for such an exclusion of chiasmata adjacent to entire blocks of constitutive heterochromatin is not known. Loidl (1987) suggests that it results from the fact that heterochromatic regions pair late and so do not complete synapsis until late pachytene, by which time crossing-over no longer occurs. Consequently, were such delayed pairing to extend into adjacent euchromatic regions it could account for the suppression of chiasmata in them too. This suggestion has yet to be supported by hard data.

Shaw (1971a) draws attention to an additional point of interest in relation to species with non-localized versus localized chiasma patterns. In the former, as exemplified by *Schistocerca gregaria*, cell chiasma frequencies are normally distributed whereas in the latter, as exemplified by *Stethophyma gracilipes*, they are Poissonian except in cases where supernumerary segments are present and produce an increased variance in chiasma frequency both within and between individuals. Here the distribution approximates to normality.

Less frequently, the presence of heterochromatic supernumerary segments may also influence mean cell chiasma frequency when compared to that of non-segmented individuals of the same species. This is true for *Chorthippus parallelus* (John & Hewitt, 1966a), *C. jucundus* (John, 1973) and *Tolgadia infirma* (John & Freeman, 1976). What applies to some

Table 2.9. *A comparison of chiasma distribution in 10 diplotene cells from each of two populations of the northern (N) and southern (S) chromosome races of the Australian grasshopper Crypotobothrus chrysophorus. The N-race carries a series of fixed distal blocks of heterochromatin on all five medium (M) pairs of chromosomes which are not present in the S-race. Both races are polymorphic for a variety of distal supernumerary heterochromatic blocks on the two small (S) pairs though these are more numerous in the N-race*

Population	Mean cell chiasma frequency	Chiasma distribution[a]											
		L(1–3)			M(5,6,8,9)			M(7)[b]			S(10–11)		
		P	I	D	P	I	D	P	I	D	P	I	D
S-race													
Mount Aggie	15.6	175	163	254	194	157	**325**	—	27	**74**	14	55	130
Forbes Creek	14.6	170	150	244	139	141	**324**	1	11	**88**	25	11	163
N-race													
Bolivia Hill	14.6	178	159	244	**328**	174	106	**73**	18	5	52	56	63
Glen Innes	14.3	164	200	232	**334**	152	100	**69**	23	9	51	58	43

[a] P = proximal, I = interstitial and D = distal chiasmata.
[b] The M7 is a megameric bivalent in this species and condenses differentially at prophase-1. It can, therefore, be readily distinguished at diplotene.
After John & King, 1980.

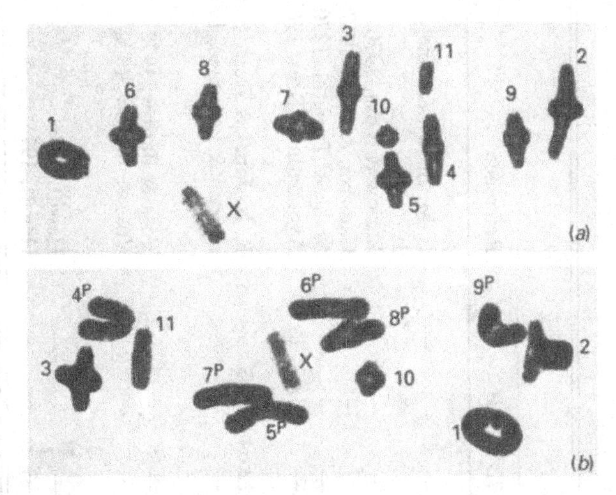

Fig. 2.24. Chiasma distribution in (*a*) southern and (*b*) northern chromosome races of the grasshopper *Cryptobothrus chrysophorus*. The presence of homozygous distal blocks of heterochromatin in the northern race (see Fig. 2.26) is correlated with the presence of proximal chiasmata (P) in the six medium-sized bivalents (numbers 4–9) whereas in the southern race, where these bivalents lack distal heterochromatic blocks, chiasmata are predominantly interstitial–distal with only occasional proximal chiasmata, as is evident in bivalent number 7 in the cell illustrated (*a*).

supernumerary segment systems applies more commonly to supernumerary or B-chromosomes (Table 2.10). Southern (1967b), who analysed the pattern of chiasma distribution in male individuals of *Myrmeleotettix maculatus* both with and without B-chromosomes, showed that in this case the effect of the B-chromosome is simply to increase the frequency of bivalents carrying higher numbers of chiasmata without any alteration in the basic pattern of distribution. In *Euthystira brachyptera*, on the other hand, the increase in mean cell chiasma frequency resulting from the presence of B-chromosomes is accompanied by a reduced chiasma localization pattern (Fletcher & Hewitt, 1980a). Even in the absence of any effect on mean cell chiasma frequency, B-chromosomes may still affect chiasma distribution. Thus, in Israeli populations of the plant *Triticum speltoides*, individuals carrying supernumerary chromosomes show an increased frequency of interstitial chiasmata and a compensatory decrease in the number of terminal associations (Zarchi *et al.*, 1972, 1974). Finally, in *Zea mays* both heterochromatic segments and B-chromosomes exert comparable, though not identical, effects (Table 2.11).

Compared to prokaryotes, which have very much smaller genomes, there are relatively few exchanges per genome in eukaryotes where the genomes are not only larger but usually contain a substantial repetitive DNA component which is not present in prokaryotes. Added to this, in

Table 2.10. *The effect of supernumerary (B) chromosomes on the mean cell chiasma frequency of the standard (A) chromosomes in eight eukaryote species*

Species	Number of A bivs	Mean cell chiasma frequency 0B	1B	2B	3B	4B	5B	6B	7B	8B	10B	Reference
Crepis capillaris	3	4.08	3.90	4.65	4.68	5.33	—	—	—	—	—	L. M. Brown & Jones, 1976
Puschkinia libanotica	5	9.07	10.82	11.70	11.62	13.80	—	—	—	—	—	Barlow & Vosa, 1970
Lolium perenne	7	11.93	11.21	10.00	—	—	—	—	—	—	—	Cameron & Rees, 1967
Festuca mairei	7	18.87	20.95	22.85	26.25	28.00	—	—	—	—	—	Malik & Tripathi, 1970
Zea mays	10	18.50	—	19.50	19.70	19.80	—	22.53	—	23.80	19.80	Ayonoadu & Rees, 1968
Secale cereale	7											
inbred		14.85	13.43	13.22	13.39	12.64	12.86	13.20	12.64	13.03	—	R. N. Jones & Rees, 1967
wild		13.20	15.12	16.57	17.78	18.40	—	—	—	—	—	Zečević & Paunović, 1969
Listera ovata	17											
PMC		26.9	28.9	28.2	30.3	29.1	—	—	—	—	—	Vosa & Barlow, 1972
EMC		30.3	32.6	31.3	32.5	32.7	—	—	—	—	—	Vosa & Barlow, 1972
Myrmeleotettix maculatus												
♂ (X0)	8	14.37	16.23	16.17	—	—	—	—	—	—	—	Hewitt, 1976
♀ (XX)	9	14.74	16.72	19.00	—	—	—	—	—	—	—	Hewitt, 1976

Table 2.11. *The influence of supernumerary (B) chromosomes, and of supernumerary segments (K = knobs) on chromosome 10, on crossing-over in two chromosomes of Zea mays*

	% change in recombination									
	Chromosome 5				Chromosome 9					
Comparison	A_2 – Bt		Bt – Pr		C – Sh		Sh – Wx		Wx – Gl	
2B versus OB	32.8	55.4	16.6	14.5	−18.6	9.5	1.6	−15.9	32.8	58.4
K10 versus k10	177.0	107.4	81.8	38.5	−2.3	2.4	4.8	14.2	6.0	18.0

After Nel, 1973 and with the permission of Springer-Verlag.

Table 2.12 *Relationship of recombination frequency to map length in 10 eukaryotes*

Species	Total map length	Average frequency of intergenic recombination (cM × kb^{-1})
Saccharomyces cerevisiae	3600	0.26
Schizosaccharomyces pombe	1300	0.08
Neurospora crassa	600	0.05
Aspergillus nidulans	1600	0.03
Caenorhabditis elegans	360	0.004
Drosophila melanogaster	275	0.0017
Homo sapiens	2600	0.0009
Mus musculus	1200	0.0006
Pisum sativum	1200	0.0002
Zea mays	1100	0.0001

After Thuriaux, 1977 and with the permission of *Nature*.

eukaryotes the amount of recombination per unit length of DNA decreases with increasing genome size so that recombination frequencies are much lower in large, as compared to small, genomes (Table 2.12). The lily genome, which is some 300-times larger than that of *Drosophila melanogaster*, has only 0.2 the amount of crossing-over per 100 kb of DNA (0.03 versus 0.17). As already mentioned, within a species, with the exception of the smallest chromosomes, a positive relationship exists between bivalent length, and hence DNA content, and the chiasma frequency per bivalent. Here too, however, longer chromosomes form fewer chiasmata per unit length of DNA. Similarly, when the chiasma frequency per picogram of DNA was plotted against total DNA content for a series of related plant species, in both *Lathyrus* (12 species) and *Lolium* (six species), there was a significant negative regression (Rees & Durrant, 1986; Rees & Narayan, 1988). This indicates that chiasma frequency per unit length of DNA also decreases with an increasing genome size, though it is worth pointing out that, whereas in *Lolium* there was a positive correlation between the chiasma frequency per PMC and the DNA content of different species, no such correlation was present in *Lathyrus*. There were also marked chiasma frequency differences between bivalents with similar DNA contents in different species belonging to the same genus (Table 2.13). It is possible, therefore, as Rees & Narayan (1988) point out, that the change in recombination with genome size reflects the alteration in DNA composition that accompanies this change. In both genera, the size increase is known to be accompanied by a disproportionate increase in the amount of repetitive DNA which may not be involved in chiasma formation (see Chapter 4D.2.2). However, factors

Table 2.13. *The comparative DNA contents of bivalents with means of 1.5, 2 and 3 chiasmata in four species of Lathyrus and two species of Lolium*

Genus and species	2C DNA (pg)	DNA content (pg) of bivalents with means of 1.5 Xta	2 Xta	3 Xta
Lathyrus				
L. clymenum	13.43	1.0	1.4	2.2
L. cicera	14.64	1.3	1.8	2.8
L. sativus	16.78	2.1	2.7	3.8
L. tingitanus	22.08	2.4	3.1	4.4
Lolium				
L. perenne	4.16	0.36	0.8	—
L. temulentum	6.23	0.51	1.2	—

After Rees & Durrant, 1986 and Reese & Narayan, 1988.

other than DNA composition and amount also affect recombination. Different mean cell chiasma frequencies may sometimes occur between the sexes within a species, where DNA differences are at best minimal, or even between the male and female meiocytes of hermaphrodites (see Chapter 7A.1), where no DNA differences of any kind are present.

A.2.2 The nature and behaviour of chiasmata

It was Rückert (1892), and especially Janssens (1909), who first noted that the homologues constituting a bivalent were associated at diplotene by nodes which Janssens termed *chiasma*. To quote his own words 'Nous appelons chiasma ou noeud l'endroit d'entrecroisement des deux chromosomes d'une dyade'. He also proposed that these chiasmata represented sites where two thread pairs changed partners. Again to quote him 'Aux chiasmas on voit souvent que l'un des deux filaments passe d'une anse à l'autre'.

This view, referred to subsequently as the *chiasmatype hypothesis* (Janssens, 1924), was championed and expanded by Darlington (1932) who argued that each chiasma was the result of a breakage and rejoin mechanism leading to a physical exchange between two non-sister chromatids, one from each of two homologues. The effect of such an exchange would, of course, be to recombine any genetic differences that existed between paired homologues. The fate of such recombinants would, however, then depend on the form of nuclear fusion that took place at reproduction. The fusion of gametic nuclei produced by different individuals (*amphimixis*) is expected to realize the greatest possible diversity of genotypes among progeny though, even here, 50% of the heterozygosity will be lost through segregation in every generation unless mechanisms exist to converse it. By contrast, *automixis* involves the fusion of nuclei descended from the same zygote. The most common form of automixis is *autogamy* involving the fusion of gametic nuclei produced by the same individual and this is expected to lead to increasing homozygosity among progeny. In some categories of automixis distinct gametic nuclei are not produced (see Chapter 7B) and here heterozygosity-enforcing mechanisms may operate so that automixis need not lead to complete homozygosity.

Darlington also proposed that the end-to-end associations commonly seen between homologues at first metaphase of meiosis represented chiasmata which had moved from their initial site of origin to progressively more distal positions coincident with the opening out of the diplotene bivalent. He called this event *terminalization* and argued that, in extreme cases, a chiasma could terminalize completely and be replaced by a purely end-to-end association of homologous chromatids. This terminalization, he held, was conditioned by the same forces that were responsible for the opening out of the bivalent.

The evidence which Darlington, and others, offered in support of terminalization, was based on comparisons between the frequency of interstitial chiasmata and the end associations present at successive stages of meiosis. These comparisons commonly identified more terminal associations at metaphase-1 than at diplotene and diakinesis. Coupled with this, in extreme cases, terminalization was assumed to lead to an actual reduction in chiasma frequency because two or more chiasmata could merge by terminalizing to the same chromosome end. There are, however, at least five possible explanations for these findings:

(1) Underscoring of metaphase-1 bivalents, stemming from their contraction, coupled with overscoring of diplotene cells because of the failure to adequately distinguish overlaps between chromatid pairs from genuine chiasmata (see Fig. 2.20). The most blatant example of this available in the literature is to be found in the case of the plant *Allium fistulosum* in which PMCs are characterized by extreme proximal chiasma localization at metaphase-1 but were illustrated by Levan (1933) as having what he initially interpreted to be large number of interstitial chiasmata at diplotene but which he later admitted were overlaps (Levan, 1935).

(2) Since cells enter meiosis in clusters this means that comparisons between different meiotic stages in the same individual involve different populations of meiocytes which may have different chiasma

Inbred line	 a	 b	c	d
Rheidol	5(6%)	20(23%)	39(46%)	22(26%)
J-46	6(8%)	22(31%)	31(43%)	13(18%)

Fig. 2.25. The frequencies of the four classes of C-banded bivalents observed in inbred line J-46 and in the Rheidol variety of the plant *Secale cereale* (after Jones, 1978 and with the permission of Springer-Verlag).

characteristics. Indeed, it is well known that, within individuals, chiasma frequencies differ significantly between different cells.

(3) There may be genuine terminalization. As we shall see subsequently (Chapter 2C), there is no doubt that terminalization occurs regularly in holocentric systems where it may well be a prerequisite for inverted meiosis. Added to this, in many monocentric systems some chiasmata form very close to a chromosome end so that the amount of movement required for terminalization in these cases is minimal.

(4) Constitutively heterochromatic regions characteristically give a positive staining reaction when treated with hot barium or calcium hydroxide prior to staining with Giemsa. Such C-banding serves to identify constitutively heterochromatic regions at condensed stages of the division cycle; that is, at a time when they are indistinguishable from euchromatin in their degree of contraction. In rye, *Secale cereale*, distal C-bands are present terminally on either the short or else on both arms of the seven biarmed bivalents that make up the diploid set. These correspond to blocks of constitutive heterochromatin, which appear in resting nuclei as chromocentres and are composed of repetitive DNA. What appear, from conventionally stained preparations, to be terminal associations of homologues are very frequent at metaphase-1. Unfortunately in rye, as indeed in many plants, diplotene–diakinesis stages are unsuitable for defining the exact location of chiasmata though there is no doubt that they are distally localized. The C-banding technique, however, enables one to examine the nature of the metaphase-1 associations and, using this approach, Jones (1978) identified four categories of metaphase-1 bivalents (Fig. 2.25). The arrangement of banded and unbanded material in the vicinity of genuine chiasmata (Fig. 2.25*a*) suggested to Jones that the exchanges leading to chiasma formation occurred in the unbanded

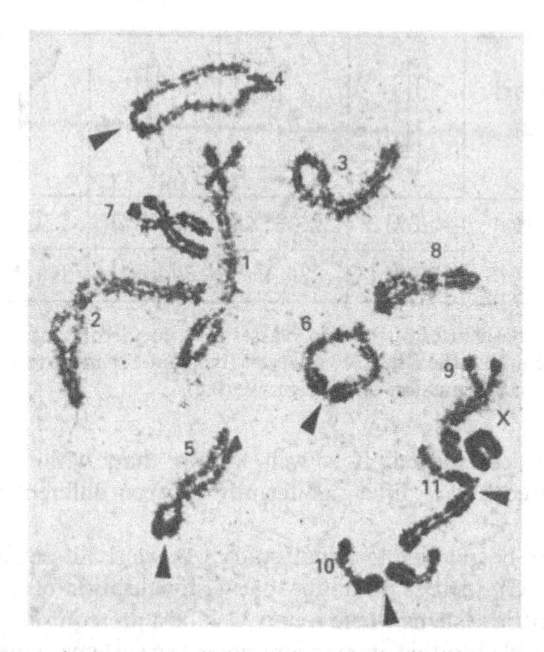

Fig. 2.26. Late pachytene–early diplotene in an individual from the northern race of *Cryptobothrus chrysophorus* in which the S10 bivalent is homozygous for distal supernumerary blocks of heterochromatin while the S11 bivalent is heterozygous for a distal, and homozygous for a proximal, supernumerary block. In both of these chromosomes, the two homologues are associated solely by a non-chiasmate association between the homozygous blocks of heterochromatin (arrowheads). Equivalent heterochromatic associations are also present in three of the six medium-sized bivalents (arrowheads, bivalents numbers 4, 5 & 6) which are homozygous for fixed distal heterochromatic blocks but which also form chiasmata in their euchromatic portions.

regions immediately adjacent to the C-bands and that as a result of conformational changes associated with metaphase contraction the C-bands became absorbed into the width of the bivalent giving rise to what Jones referred to as *pseudoterminalization*. The strongest evidence for this was the presence of a high frequency of bivalents in which the banded material of the associated arms was penetrated by unbanded material (Fig. 2.25*c*). If this interpretation is correct, it implies that most of the assumed terminal associations in this case are, in reality, sub-terminal and result not from terminalization but rather from the distorting effects of bivalent contraction and stretching. In some cases of terminal homologue association, however, there is a clear space evident between the pairing partners (see for example Figs. 2.3*a*, bivalent number 6 and Fig. 2.4*g*). Such situations are not easily interpreted in terms of pseudoterminalization.

(5) There remains the possibility that some categories of terminal association are non-chiasmate. This was a view held by Janssens himself. More recently, it has received support from observations involving the behaviour of chromosome ends which, like those of rye, are heterochromatic in character. The most convincing of these cases concerns the segregation of homologous partners which otherwise lack chiasmata. Examples of this behaviour have been described in several grasshopper species including the northern race of *Cryptobothrus chrysophorus* (John & King, 1985a and see Fig. 2.26). Additionally, similar associations between the ends of non-homologous chromosomes have been shown to lead to persistent pseudo-multiple formation in *Metrioptera brachyptera* (Southern, 1967a, 1968) and *Heteropternis obscurella* (John & King, 1982, 1985a and see Fig. 2.27).

As far as the chiasmatype hypothesis itself was concerned, however, the initial evidence for it, while convincing, was also indirect (Darlington, 1931; Mather, 1938). The first direct validation of the hypothesis came from a demonstration of label segregation following the application of tritiated thymidine at the penultimate S phase prior to meiosis. This provided good evidence for a general correlation between the location of chiasmata at diplotene–diakinesis and the sites of label exchange as inferred from anaphase-1 and metaphase-2 label segregants (Taylor, 1965; Church & Wimber, 1969; Craig-Cameron & Jones, 1970; Peacock, 1971). This correlation was especially convincing in the case of *Stethophyma grossum* where chiasmata are proximally localized and where, as expected, half of all chiasmata produced generate a proximal label exchange (Jones, 1971 and see Fig. 2.28). Even more convincing evidence was subsequently provided using bromodeoxyuridine (BrdU)-labelling. BrdU is a thymidine analogue and, following exposure to this compound during replication, the thymine of DNA is replaced by BrdU. After two rounds of replication in the presence of BrdU, chromosomes contain one chromatid in which the DNA is unifilarly substituted with BrdU and one which is bifilarly substituted. These two chromatids stain differentially either with Giemsa alone, with the fluorescent compound 33258H, with acridine orange or with fluorescent dyes followed by Giemsa (FPG). More specifically, in chromatids which contain BrdU in place of T, both strands of the double helix stain more weakly than chromatids with BrdU in only one strand.

Using BrdU-labelling coupled with FPG staining to differentially label the two pairs of sister chromatids in a given bivalent, Tease (1978) was able, for the first time, to directly analyse individual bivalents at both diplotene and metaphase-1. On the assumption that chiasma formation

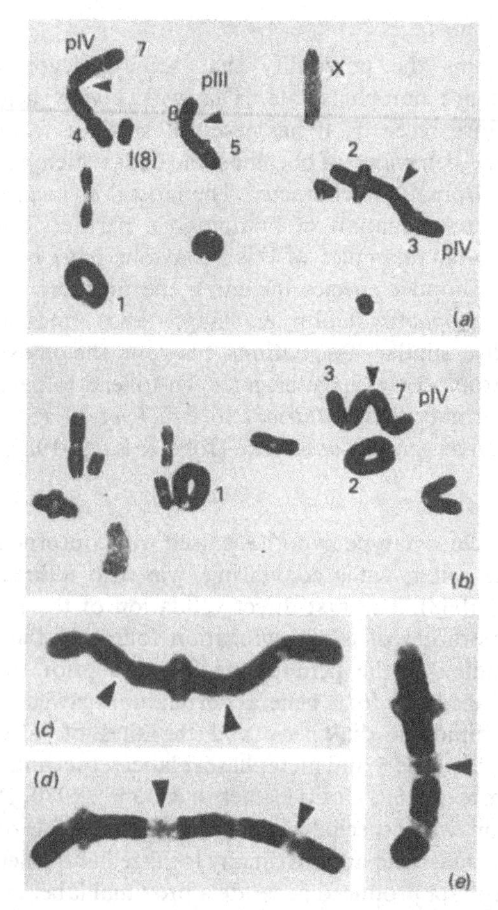

Fig. 2.27. Metaphase-1 pseudomultiples in male F_1 hybrids of the grasshopper *Heteropternis obscurella* (2n = 23, X0). In (*a*) there are two pseudomultiples of four (pIV, arrows) and one of three (pIII) together with an accompanying univalent (I). In (*b*) there is a single pseudomultiple of four (pIV) and in (*c*) a pseudomultiple of six. When suitably stained these pseudomultiples can often be seen to arise from terminal, non-chiasmate and heterochromatic associations between non-homologues, as is evident (arrowheads) in the pIV of (*e*) and the pV of (*d*).

involves an exchange between two non-sister chromatids of a given pair of homologues, and that this event is random with respect to the four chromatids of that bivalent, half of the observed chiasmata should be associated with an exchange between differentially stained chromatids, i.e. to what Tease & Jones (1978) have termed a 'visible' exchange. The other half would be between identically stained chromatids and so be 'hidden'. In an analysis of 295 monochiasmate bivalents of *Locusta migratoria*,

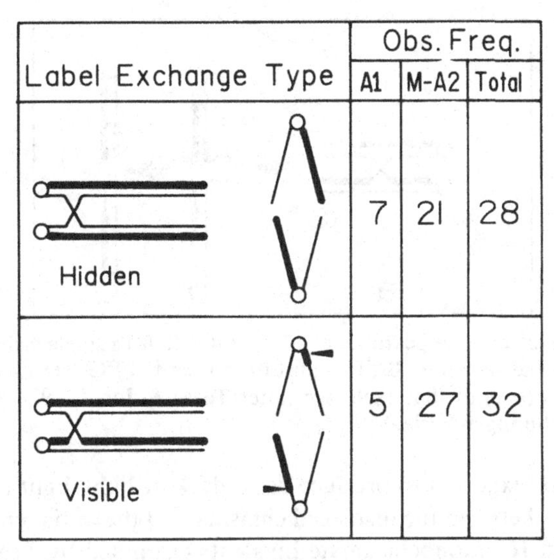

Label Exchange Type	Obs. Freq.		
	A1	M-A2	Total
Hidden	7	21	28
Visible	5	27	32

Fig. 2.28. The observed incidence of hidden and visible non-sister chromatid exchanges in bivalents of the male grasshopper *Stethophyma grossum*, with a single proximal chiasma, following pre-treatment with tritiated thymidine (after Jones, 1971 and with permission of Springer-Verlag).

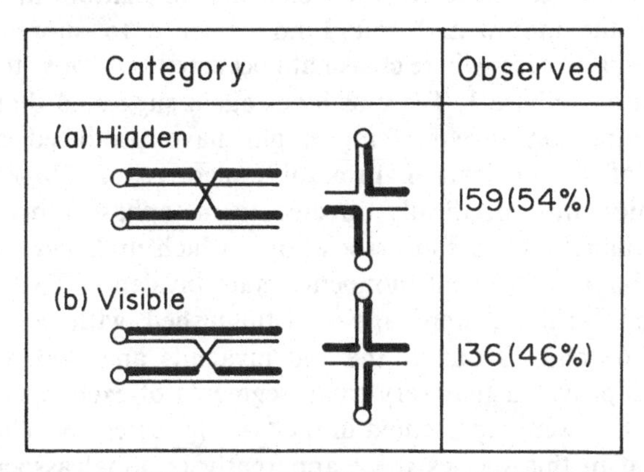

Category	Observed
(a) Hidden	159 (54%)
(b) Visible	136 (46%)

Fig. 2.29. The observed frequency of hidden and visible non-sister chromatid exchanges in bivalents of male *Locusta migratoria*, with a single interstitial chiasma, following BrdU substitution and FPG staining (after Tease & Jones, 1978 and with permission of Springer-Verlag).

using this approach, Tease & Jones (1978) found no significant departure from equality (Fig. 2.29).

The BrdU technique also offers an opportunity to objectively examine the question of chiasma terminalization since this should lead to a displacement of a chiasma relative to the actual site of exchange. This, in

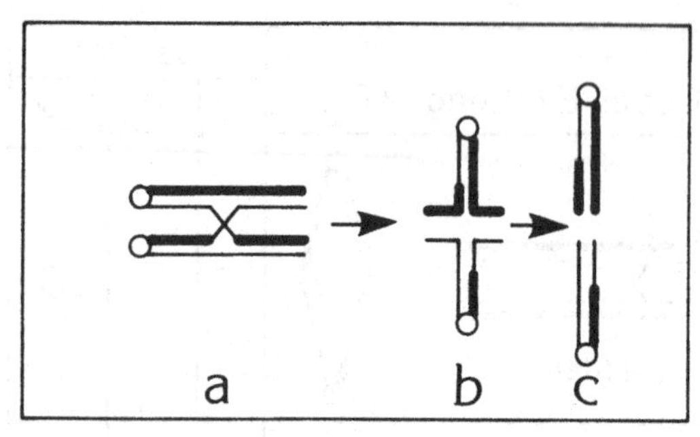

Fig. 2.30. The expected appearance of a bivalent with (*a*) a single visible non-sister chromatid exchange after BrdU substitution and FPG staining and with terminalization (*b*–*c*) of that exchange (after Tease & Jones, 1978 and with the permission of Springer-Verlag).

turn, would be expected to produce dark–dark or light–light associations of chromatids between the displaced chiasma and the initial crossover site (Fig. 2.30). In 167 monochiasmate bivalents examined by Tease & Jones (1978) no such regions were seen and all the observed chiasmata coincided precisely with the exchange sites identified by a switch of BrdU-labelling. Using a similar approach, Kanda & Kato (1980) also observed that visible crossovers coincided with sites of chiasma formation in autosomal bivalents of the mouse and found no evidence to support chiasma terminalization. In cases where chiasmata occurred very close to the end of a bivalent at metaphase-1, Tease & Jones again suggested that increased bivalent contraction or stretching might have produced a spurious appearance of terminalization. In a subsequent report (Jones & Tease, 1984) attention was specifically directed to metaphase-1 bivalents with apparent terminal end-to-end associations, which in *Locusta migratoria* constitute about 16% of all monochiasmate bivalents. Two categories, 'hidden' and 'visible', were again distinguished with respective frequencies of 63 and 70 out of the 140 bivalents analysed. Convincing evidence was provided that very small segments of exchanged dark and light chromatids were represented in the 'visible' category. This evidence indicates that in this species these apparently terminal associations are indeed of chiasmate origin. In XY bivalents of the male mouse all associations were between lightly-and darkly-stained chromatids and appeared to be terminal (Kanda & Kato, 1980). However, because of the analytical limits of the BrdU technique it was not possible in this case to exclude the occurrence of crossing-over in very small, near terminal, segments of the kind identified in *Locusta migratoria*. On the other hand, in the Armenian hamster, *Cricetulus migratorius*, Allen (1979) found that the single chiasma in the large XY bivalent did not always coincide with the point of visible exchange but showed clear evidence of terminalization (see his Fig. 66 right). Unfortunately he offered no quantitative data on the point so it is not possible to decide whether terminalization is a common or an uncommon event in this case.

A.3 Spontaneous U-type meiotic chromatid exchange

Meiotic crossing-over leading to chiasma formation is best considered as part of a wider category of exchange events which operate both within and between chromosomes. Such events include the conventionally recognized categories of structural chromosome rearrangements (inversions, interchanges and fusions) as well as mitotic recombination events (see Chapter 4D.3). Viewed in these terms, crossing-over is an X-type exchange between non-sister chromatids of homologous chromosomes. Equivalent U-type exchanges are also known to occur at meiosis though they are usually found with any frequency only in genetically unbalanced types where they form part of a syndrome of errors reflecting defective control of meiosis. Each exchange again involves only two of the four chromatids within a bivalent and is usually complete. The two chromatids concerned may be sisters (sister union, SU) or non-sisters (non-sister union, NSU).

That complete reciprocal U-type exchanges can occur between sister chromatids in a meiotic bivalent was first recognized in plants by Matsuura (1950) in *Trillium* and Haga (1953) in *Paris*. Such spontaneous exchanges lead to the production of asymmetrical bivalents at metaphase-1 and subsequently to dicentric bridges and acentric fragments at anaphase-1 (Fig. 2.31). A number of studies have subsequently drawn attention to the parallelism between such spontaneous U-type exchange and the X-type exchanges between non-sister chromatids that lead to chiasma formation. Thus:

(1) Variation in chiasma frequency is accompanied by a parallel variation in the frequency of bridges and fragments resulting from U-type exchange (Rees, 1962).

(2) The distribution of U-type exchanges parallels the distribution of chiasmata. In *Paeonia* (Walters, 1956) and *Secale* (Rees & Thompson, 1955), where chiasmata are distally localized, so too are exchange events and only small acentric fragments result. In *Paris*, where chiasmata are proximal, much larger anaphase fragments form. By comparing different lines of rye, which show striking differences in their chiasma distribution patterns, Jones (1968) demonstrated a consistent correlation between the distribution of chiasmata and the distribution of fragment sizes.

(3) In rye, chromosome ends sometimes show active (neocentric) mobility on the meiotic spindle. Jones (1969) described one such a case where

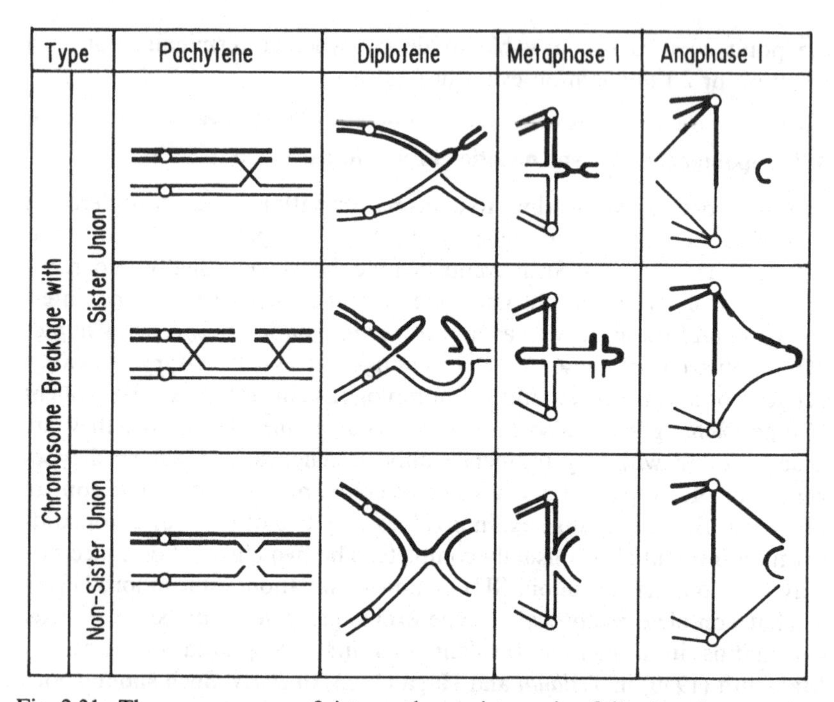

Fig. 2.31. The consequences of sister and non-sister union following chromosome breakage at meiosis (after Lewis & John, 1963).

the 'neocentromere' appeared as a small satellited region in a specific bivalent arm giving rise to a distinctive heterozygous pattern of expression. Where a chiasma occurs in the neocentric arm, the neocentric chromatids separate equationally at anaphase-1, one passing to each pole. Where no chiasma forms in this arm both neocentric chromatids pass to the same pole. A non-sister U-type exchange within the short neocentric arm produces a detached fragment in which the neocentromere remains active despite the absence of the main centromere and so serves as a distinctive marker of U-type exchange in that arm. Using this marker, Jones was again able to confirm a significant correlation between the occurrence of U-type exchanges in this bivalent and the observed distribution of chiasmata (Table 2.14).

Because meiotic U-type exchanges were entirely of the chromatid category and showed such good correlations with both chiasma frequency and chiasma distribution, Lewis & John (1966) suggested that they arose as errors in chiasma formation. In keeping with this hypothesis, Jones & Brumpton (1971) reported a marked excess of NSU- compared to SU-types in rye.

Table 2.14. *Chiasma frequency and U-type exchange frequency in the neocentric bivalent and the six other bivalents of the rye complement*

Exchange category	Neocentric bivalent Long arm	Short arm	Total	$\chi^2_{(1)}$	Other six bivalents	Total	$\chi^2_{(1)}$
Chiasmate	53	17	70	0.678	428	498	0.926
U-type	53	12	65	$P > 0.3$	352	392	$P > 0.3$

After Jones, 1969.

A.4 Multivalents and multiples

The rules of meiotic synapsis and chiasma formation which operate in diploids are based on homology between pairs of chromosomes and lead to consistent bivalent formation. Structural homology is thus one of the conditions for regular segregation. When more than two homologues, or near homologues (usually referred to as *homoeologues*) exist within a meiocyte, or when one or more chromosomes share a partial homology with two others, a situation exists in which multivalent or multiple configurations respectively may arise at meiosis. Multivalents represent associations involving three or more chromosomes which are either completely homologous or else homoeologous. They are present in individuals which are either polyploid or polysomic, the size of the multivalent depending on the level of polyploidy or polysomy and the chiasma frequency of the chromosomes involved in the multivalents (Fig. 2.32). Thus, *chain* trivalents arise in triploids ($2n = 3x$) and trisomics ($2n = 2x+1$) while, maximally, chain or *ring* quadrivalents arise in tetraploids ($2n = 4x$) or tetrasomics ($2n = 2x+2$) and so on. *Multiples*, on the other hand, represent associations between three, or more, partial homologues and arise in individuals which are diploid but heterozygous for either a centric fusion, producing a chain multiple of three, or else for one or more serial interchanges, maximally producing chain or ring multiples of four (one interchange), six (two serial interchanges) eight (three serial interchanges) and so on (Fig. 2.33).

Polyploids are of different kinds (Table 2.15). There is no strict requirement for any of them to form multivalents. Whether they do so or not depends on the pairing properties of the individual chromosomes concerned and their chiasma frequency relationships. Thus, for a multivalent to form, each homologue, or homoeologue, must pair and form a chiasma with at least one of the other homologues or homoeologues. Comparative analyses indicate that regular bivalent

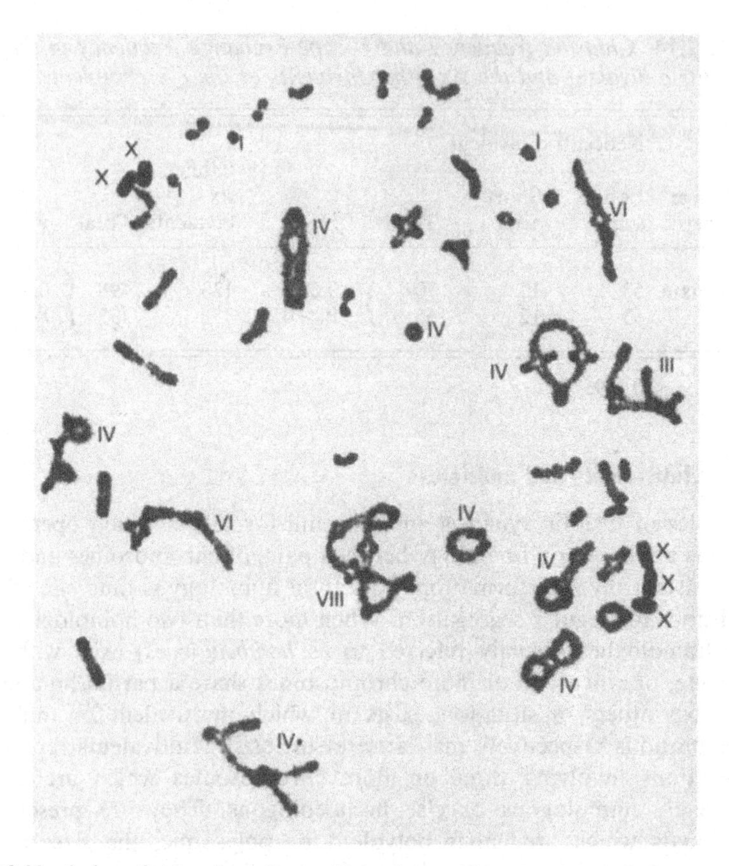

Fig. 2.32. A decaploid male meiocyte from the grasshopper *Urnisa guttulosa*. This species normally has a diploid count of 2n = 23♂, X0 and forms 11 bivalents plus an X univalent. The polyploid cell illustrated includes 1.VIII + 2.VI + 8.IV + 1(III + I) + 2I + 25II + 5X = 110 + 5X.

formation in autotetraploids can be achieved by at least three different mechanisms (Rasmussen, 1987):

(1) By initial recognition and SC formation being restricted to a single region in each pair of homologues. This occurs regularly in autotetraploids of the nematodes *Heterodera glycines* (Goldstein & Triantaphyllou, 1980) and *Meloidogyne hapla* (Goldstein & Triantaphyllou, 1981) where each SC is anchored to the nuclear envelope at one end only. In autotetraploid rye, too, there is evidence that pairing does not involve all the four members of a given homologous set. Here, pachytene associations form exclusively between pairs of homologues in about one of the seven sets of homologues in every PMC (Timmis & Rees, 1971). The net result is that bivalents outnumber multivalents.

Table 2.15. *Categories of polyploids*

Type	Structural differentiation of chromosome sets	Origin
Autoploids		
Strict, AAAA	None	Non-hybrid
Inter-racial, $A_1A_1A_2A_2$	Partial	Hybrid
Amphiploids		
Segmental, $A_sA_sA_1A_1$	Intermediate	
Genomic, AABB	Strong	
Autoallopolyploids, AAAABB		

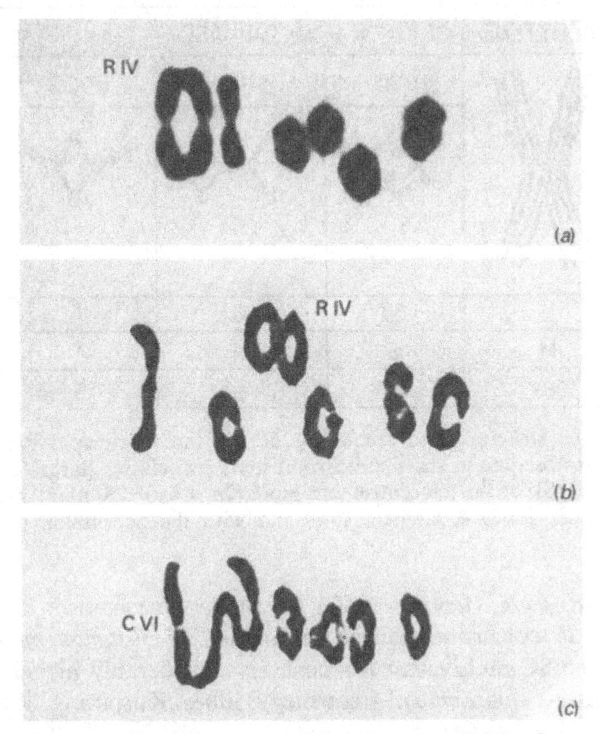

Fig. 2.33. Multiple chromosome configurations in PMCs from interchange heterozygotes of the plant *Secale cereale*. This species normally forms seven bivalents at meiosis (2n = 2x = 14) but the three cells illustrated include a single ring of four multiple (RIV) in (*a*) adjacent and (*b*) alternate orientation and a chain of six multiple (CVI) in adjacent orientation (*c*).

(2) By strict control of chiasma number and distribution. In autotetraploid *Allium porrum*, for example, whereas quadrivalent pairing configurations are present at pachytene, only bivalents occur at first metaphase and each of these has two chiasmata, one on each side of

Table 2.16. *Pairing behaviour of the different size classes of chromosomes in 33 metaphase-1 PMCs of the autotetraploid Tradescantia micrantha (2n = 4x = 24)*

Chromosome type	Metaphase-1 association				Mean Xta per arm pair
	2I	II	III+I	IV	
Long (8)	—	66	—	33	1.88
Medium (4)	1	48	1	8	1.22
Short (12)	1	189	1	3	1.01

After K. Jones & Colden, 1968 and with the permission of Springer-Verlag.

PACHYTENE	METAPHASE-I		
	2P+2D	2P+ID	IP+2D
L	57	3	II
M	6	4	2
S	-	-	5

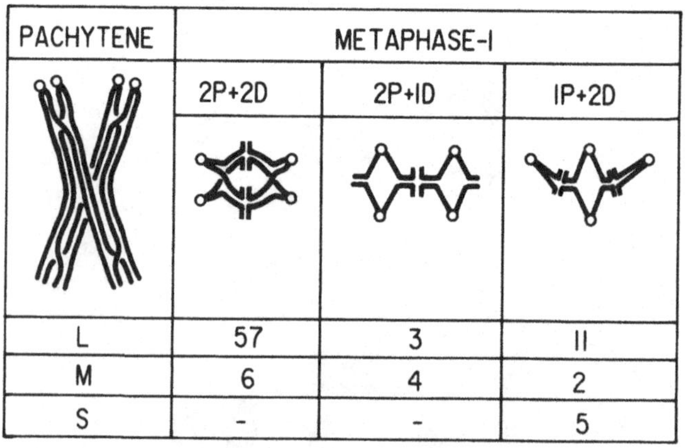

Fig. 2.34. The structure and frequency of the three principal categories of quadrivalents observed in the three chromosome size classes (large (L), medium (M) and small (S)) of the telocentric tetraploid (2n = 4x = 28) plant *Tradescantia micrantha* (after Jones & Colden, 1968 and with the permission of Springer-Verlag).

the centromere (Levan, 1940). A similar mechanism appears to operate in colchicine-induced tetraploids of *Triticum monococcum* where the SC multivalent frequency is considerably higher than the metaphase-1 quadrivalent frequency (Gillies, Kuspira & Bhambhani, 1987).

(3) In experimentally-produced autotetraploid females of the achiasmate moth *Bombyx mori*, the quadrivalents formed during zygotene are eliminated at early pachytene, in the absence of chiasmata, by a pairing correction mechanism (see Chapter 4C.5).

The morphology of the chromosomes involved in a multivalent plays a role in determining its appearance. The plant *Tradescantia micrantha* is an

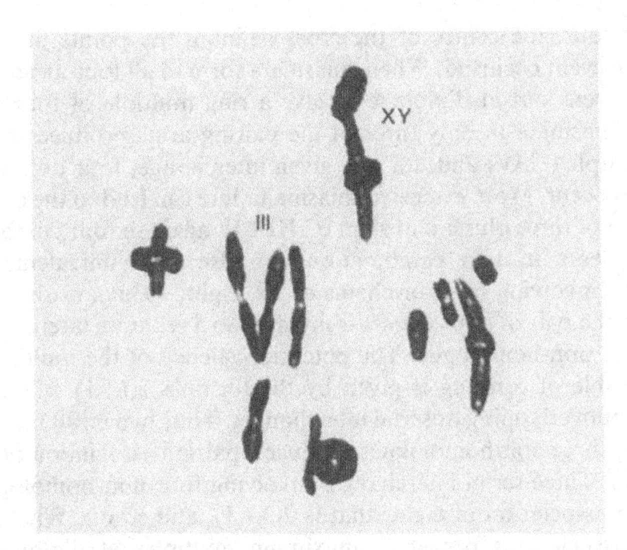

Fig. 2.35. A metaphase-1 male meiocyte from an individual of *Oedaleonotus enigma* heterozygous for a centric fusion between two non homologous telocentrics and with 2n = 1.III+8.II+XY.II (compare with Fig. 1.9).

autotetraploid with 2n = 4x = 24 and has eight large (L), four medium (M) and 12 small (S) chromosomes all of which are telocentric. From 0–3 quadrivalents are present per meiocyte and their frequency is correlated with chromosome length and chiasma frequency, which are themselves inter-related (Table 2.16). Both ring and chain quadrivalents occur, but the rings are structurally different from those which form in metacentric chromosomes with a similar number of chiasmata, simply because all four centromeres cannot be equidistantly spaced in the quadrivalent (K. Jones & Colden, 1968 and see Fig. 2.34).

Centric fusions result from whole arm exchanges between homologues where the exchange sites are either within, or very close to, the centromere. When heterozygous, a fusion system includes one metacentric chromosome and two acro- or telocentric equivalents. These, at synapsis, give rise to a chain multiple of three (Fig. 2.35). The metacentric involved in such multiples may be monocentric, with a single centromere homologous with that of one of the two acrocentrics, or else dicentric, with an homologous counterpart to each of the acrocentrics (John & Freeman, 1975). However, since the two centromeres in such dicentrics are located close to one another, they behave effectively as one. Interchanges arise from the mutual exchange of segments between two or more non-homologues. When heterozygous, such interchanges disturb the neat pair-wise homology of chromosomes. In the case of a single interchange it creates a situation in which four chromosomes each share a partial homology with two others. Consequently, at zygotene, all four associate as a pairing cross of four, the centre of the cross defining the points at which the exchange event occurred, When chiasmata form in all four arms, the cross of four opens out at diplotene to give a ring multiple of four (R.IV or ⊙IV). Chiasmata in only three of the pairing arms produces a chain of four multiple (C.IV) and, for any given interchange, four different chain multiples occur. More extensive chiasma failure can lead to the production of a chain of three plus a univalent (C.III + I), again in four combinations, two bivalents in two combinations or else four univalents. Larger multiples, involving rings or chains of six, eight, 10 etc., arise when each member of a pair of homologues is involved in a separate interchange with a different non-homologue. The potential valency of the multiple that is then capable of forming is given by the formula $2(i+1)$ where i is the number of overlapping or serial interchanges. Thus two serial interchanges involving three non-homologues produce a pairing association of six, that is $2(2+1)$. Three serial interchanges involving four non-homologues give a pairing association of eight, that is $2(3+1)$, and so on. Whether these pairing associations persist as maximum multiples at diplotene again depends on chiasma frequency and chiasma location.

The precise form of orientation adopted by multivalents and multiples at metaphase-1 determines the pattern of disjunction of the chromosomes involved in a given configuration. To produce meiotic products with a normal complement of chromosomes, or chromosome segments, disjunction has to be both numerically equal and genetically balanced. This is most easily attained in structural autoploids where numerically equal segregation is sufficient to guarantee genetically-balanced products. In other cases, however, only preferred patterns of orientation will guarantee genetic balance. Thus, chromosome translocations, whether involving interchanges or fusions, necessarily alter random assortment when present in a heterozygous condition, because they extend linkage beyond the limits of a single chromosome, and require preferential segregation to ensure the consistent production of genetically-balanced gametes. In interchange multiples, for example, only those orientations in which alternate centromeres throughout the multiple face the same spindle pole will produce balanced gametes (Fig. 2.36). Orientations in which adjacent centromeres, at one or more positions in a multiple, face the same pole invariably lead to the production of a proportion of unbalanced gametes. Alternate orientation in a ring multiple involves each centromere being co-oriented in relation to the two centromeres on either side of it. This means that, in large rings, such co-orientation must involve an increasing number of non-homologous centromeres.

The number of chromosomes in a multiple and the relative frequencies of ring versus chain configurations are major factors affecting the orientation behaviour of that multiple (Sybenga & Rickards, 1988). Alternate orientation is expected to predominate when the chromosomes

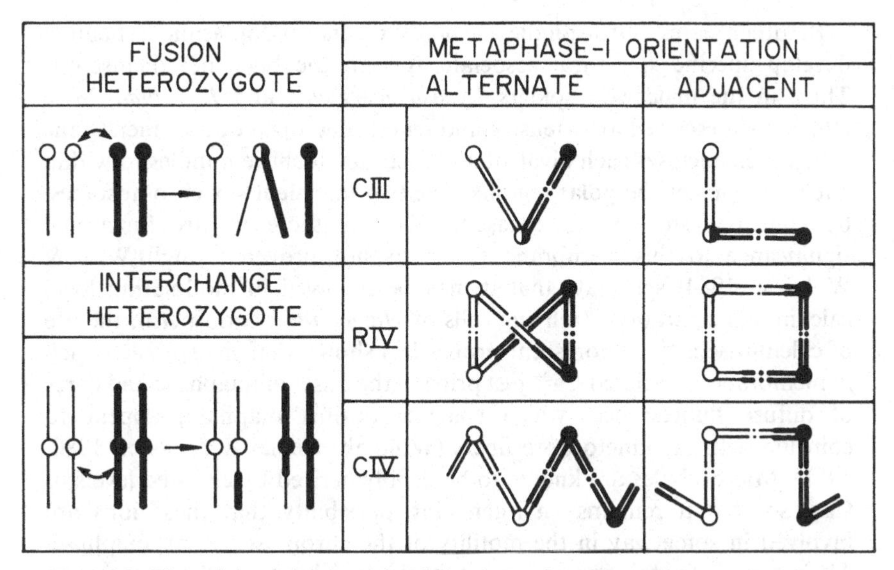

Fig. 2.36. The origin and orientation behaviour of single translocation multiples.

involved in an interchange are similar in size and isobrachial and when the chiasmata that are formed in them occupy distal positions in the multiple. Even so, those features of multiple morphology which can be expected to enhance alternate orientation are not always sufficient to guarantee it. Equally, multiples whose characteristics appear not to facilitate alternate orientation do not always exclude it. Here, undefined genetic factors also play a role (Rickards, 1983).

A.5 Membranes in meiosis

When the nuclear envelope breaks down not all the membranes of the cytoplasm disappear (Hepler & Wolniak, 1984). Indeed some may even become enhanced. In *Drosophila melanogaster* spermatocytes, for example, there is an elaboration of the membranes of the endoplasmic reticulum which accumulate as a multilayered system around the nuclear membrane at pre-meiosis (Ito, 1969). This system persists into meiosis. The fate of the original nuclear membrane has not been definitively settled. It probably does break down and is then effectively replaced by the multiple lamellae formed at pre-meiosis. However, it is not yet excluded that it may simply become the innermost layer of the lamellae. At prometaphase-1, microtubules radiate from extranuclear asters and are located in spaces between the double membrane layers. The perinuclear membranes become invaginated at the polar regions and microtubules extend into the nucleus through the membrane folds (Church & Lin, 1982).

In other cases, intraspindle elements of the endoplasmic reticulum develop specific structural associations with the bivalents themselves. Thus, in the male wolf spiders *Lycosa georgicola* and *L. rabida*, Wise (1984) has described an extensive and fenestrated open double membrane tube which encloses each bivalent and their kinetochore bundles to within a few microns of the polar regions. The two univalent sex chromosomes, by contrast, are not so encased. Whether there is any functional significance to this membrane system is not proven though Wise & Wolniak (1984) speculate that it may be involved in modulating local calcium concentration. In living cells of *Haemanthus* endosperm, the use of calcium-sensitive fluorescent probes has shown that the spindle is rich in membrane-associated Ca^{2+} just prior to the onset of anaphase and cores of diffuse fluorescence, which coalesce during anaphase, appear to coincide with the kinetochore fibers (Wolniak, Hepler & Jackson, 1980, 1981). Microtubules are known to be depolymerized by elevated levels of Ca^{2+} so that it remains an interesting possibility that these ions are involved in some way in the motility of the chromosomes at anaphase. The rationale for this suggestion is that the membranes which are present in division spindles are able to concentrate cations whose release at the meta–anaphase transition may play a role in initiating anaphase separation. As Pickett-Heaps (1986) points out, the closed spindles of lower eukaryotes are devoid of intraspindle membranes so that here anaphase movement cannot require their presence.

A second distinctive membrane system has been identified by Rieder & Nowogradzki (1983) in the oocytes of the strepsipteran insect *Xenos peckii*. During late prophase each of the randomly arranged intranuclear bivalents become encased in a tubular membrane sheath which appears to arise from the nuclear envelope. All the bivalents and their associated membranes become aligned within the nucleus prior to the formation of the first division spindle which is characterized by diffuse poles. Microtubules subsequently develop in association with, and parallel to, the tubular membranes that surround each bivalent and which then determine the orientation of the spindle microtubules in the absence of distinct polar MTOCs. This ultrastructural study of Rieder & Nowogradzki contradicts the earlier conclusion from light microscopy that each bivalent organized its own miniature spindle (Hughes-Schrader, 1924). Rather, it appears that in this instance the oriented tubular membranes serve as a scaffold for generating spindle bipolarity.

B ACHIASMATE MEIOSIS

In theory, the simplest and most direct method of achieving meiotic segregation is not by chiasma formation but rather through a mechanism that simply maintains the parallel synapsis of homologues throughout the entire first prophase. This prolongation of the lateral association of

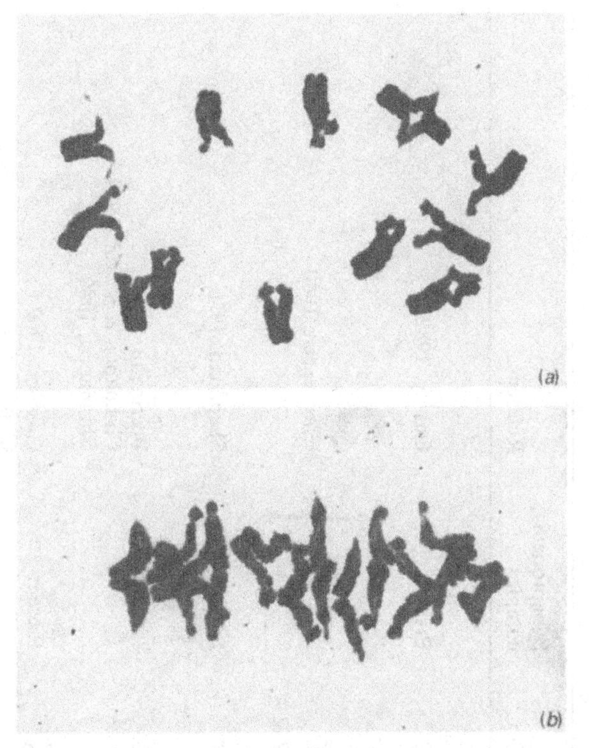

Fig. 2.37. (*a*) Metaphase-1 and (*b*) early anaphase-1 of the achiasmate meiosis in PMCs of *Fritillaria amabilis* (photographs kindly supplied by Dr S. Noda).

homologues is precisely what happens in a number of animals and in one plant genus. Here, when the first division spindle forms, the homologues, still in parallel alignment, orient so that sister centromere pairs are directed towards opposite spindle poles. Consequently, while there is a separation of the centromere regions of the two homologues, the remainder of the chromosome arms remain aligned. Homologues then separate progressively during anaphase-1 (Fig. 2.37). That the first meiotic prophase is somewhat telescoped in such achiasmate forms is confirmed by the finding that although there is a diffuse stage in chiasmate species of carabid beetles no such stage occurs in related achiasmate types (Serrano, 1981).

Superficially, achiasmate meiosis resembles a mitosis. It differs, of course, in having a haploid number of bivalents each composed of four chromatids arranged in two sister pairs. These, however, are not readily distinguishable from one another until the onset of first anaphase when associations of sister chromatids in each half bivalent lapse in some species, though not in others. It differs also by virtue of the fact that sister centromere pairs are oriented to the same spindle pole at metaphase-1.

As Table 2.17 indicates, achiasmate meiosis occurs sporadically in a

Table 2.17. *The occurrence of achiasmate meiosis in multicellular eukaryotes*

Group	Species	Sex	Synaptonemal complex	Reference
Animals				
Helminithes (⚥)	*Mesostoma ehrenbergii ehrenbergii*	♀	Present	Oakley, 1982
Annelida				
Enchytraeidae (⚥)	*Bucholzia*, all species *Fridericia*, all species	♀	?	Christensen, 1980
	Marionina subterranea and *M. sjaelandica*	♀ & ♂	?	Christensen, 1980
Arachnida				
Scorpionoidea	*Lychas marmoreus and L. variatus*	♂	Present	Shanahan, 1986
Crustacea				
Copepoda	*Cyclops strenuus*	♀	?	Beermann, 1977
	Mesocyclops edax			Chinnappa & Victor, 1979 Chinnappa, 1980
Insecta				
Dictyoptera	*Bolbe nigra*	♂	Present	Gassner, 1969
	Numerous other mantids			Gupta, 1966
Orthoptera	Several species of Thericleinae	♂	?	White, 1965
Mecoptera	*Panorpa communis*	♂	Present	Welsch, 1973
	Panorpa nuptialis			Gassner, 1967
Lepidoptera	*Bombyx mori*	♀	Present and	Rasmussen, 1977a

Lepidoptera	*Bombyx mori*	♀	Present and retained to M-1	Rasmussen, 1977a
	Probably present in all ♀ Lepidoptera			Suomalainen *et al.*, (1973)
Coleoptera	Some carabids	♂	?	Serrano, 1981
Diptera				
Phryneidae	*Phryne fenestralis*	♂	Absent	Meyer, 1964
Tipulidae	*Tipula caesia*	♂	Absent	Meyer, 1964
Simuliidae	*Cnephia dacotensis* and *C. ornithophilia*	♂	Absent	Procunier, 1975
	Also present in *Crozetia crozetia* *Eusimulium aureum* and *Prosimulium decemarticulatum*	♂		Rothfels & Mason, 1975
Drosophilidae	*Drosophila melanogaster*	♂	Absent	Meyer, 1960
Muscidae	*Glossina austeni*	♂	Absent	Craig-Cameron *et al.*, 1973
Plants				
Liliaceae	*Fritillaria japonica* group	♂	Present	Noda, 1975; Ito, Takegami & Noda, 1983

number of unrelated invertebrate animals. The only known bisexual groups which appear to be completely achiasmate are the Lepidoptera (Suomalainen, Cook & Turner, 1973) and the related Trichoptera (Suomalainen, 1966) where all females analysed to date lack chiasmata at meiosis. It is, however, especially common in enchytraeid worms (Annelida: Oligochaeta) where Christensen (1961) estimates that one-third of the species studied are characterized by achiasmate meiosis. In this group, which is hermaphrodite, it is sometimes, though not always, present in both male and female germ lines (Christensen, 1980), whereas in bisexual forms it is found only in one sex. It is also well represented in male Diptera where it characterizes both lower and higher forms.

Although seemingly simpler than chiasmate meiosis, there are good grounds for concluding that achiasmate meiosis has evolved secondarily from a chiasmate mode and has done so on many different occasions. This is especially clear in the Diptera, in mantids and in enchytraeids where only the advanced forms are achiasmate. The independent occurrence of achiasmate meiosis in phylogenetically remote groups points unmistakably to a multiple origin for this distinctive form of meiosis. Indeed, even within a given group it may sometimes be polyphyletic. This is the case, for instance, in blackflies (Diptera: Simulidae) where it occurs in occasional representatives of at least four different genera (Rothfels & Mason, 1975). Its multiple origin is also supported by the fact that at least three different forms of achiasmate meiosis are known. In some, SCs are formed and remain associated with each bivalent up to, and during, first metaphase. In the mantid *Bolbe nigra*, for example, electron micrographs indicate that the SC breaks transversely at the kinetochore during anaphase-1 so that, at first telophase, remnants of the SC are present either between, or at, the poles (Gassner, 1969). An even more spectacular behaviour has been reported in females of the silkmoth *Bombyx mori* (Rasmussen, 1977a). Here the oocyte is formed from one of a cluster of eight interconnected cells all of which develop SCs and proceed as far as pachytene. Then, in seven of them, the SCs are shed from the bivalents and develop into nurse cells. In the single survivor, which becomes the oocyte, the lateral elements of the SC grow in width and length. Both the chromatin and the modified SCs of the bivalents condense and, coupled with this, the bivalents associate end-to-end so that the modified SCs form a continuous sheet (Fig. 2.38). This is subsequently eliminated from the bivalents *en bloc* as homologues move to opposite poles at anaphase-1. Clearly, this must be a secondarily-derived form of meiosis. An earlier report of Sorsa & Suomalainen (1975) had described an essentially similar behaviour at the electron microscope level in the lepidopteran *Cidaria* but did not relate this behaviour to modifications in the SC. The Trichoptera, too, show an equivalent behaviour.

In a second category of achiasmate meiosis, the SC forms but does not

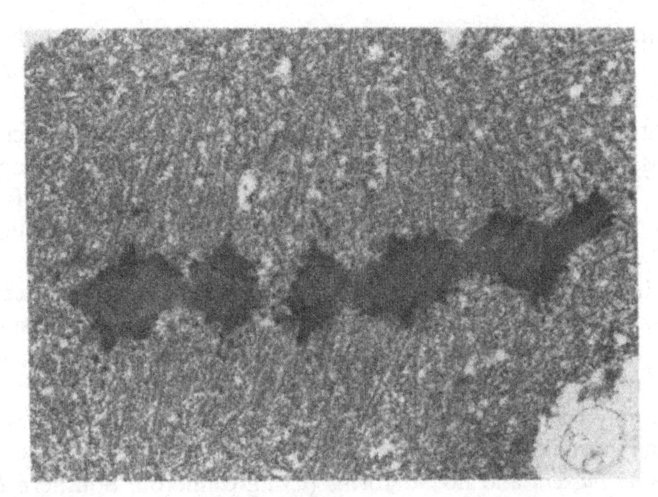

Fig. 2.38. An electron micrograph of first meiotic metaphase in a mature egg of the silkworm *Bombyx mori*. The modified and swollen synaptonemal complexes of the individual bivalents have fused to form a more or less continuous sheet which is easily distinguished from the more condensed chromatin of the two homologous chromosomes of each bivalent and which appears on opposite sides of the fused complexes (Photograph kindly supplied by Dr Søren Rasmussen).

persist. This is the case in oocytes of the flatworm *Mesostoma ehrenbergii ehrenbergii* (Oakley, 1982) where the SC is lost from the bivalents after pachytene. Here, therefore, some other mechanism must be responsible for the retention of parallel pairing.

In a third category, exemplified by the Diptera, SCs are absent altogether from the achiasmate male meiosis (Meyer, 1960; Procunier, 1975). In the female of *Drosophila melanogaster*, as in some other chiasmate organisms, precursor material of the central element of the SC has been identified in the nucleolus at early zygotene (Westergaard & von Wettstein, 1970). A similar material has also been identified in the nucleoli of primary spermatocytes of *D. melanogaster* though the exact stage has not been specified. Rasmussen (1973) therefore argues that the mechanism that normally functions in the transport of elements of the central region, from their site of primary assembly in the nucleolus to the pairing homologues, is defective and that this accounts for the absence of the SC in the male.

Almost all species of *Drosophila*, like *D. melanogaster* itself, have an achiasmate male meiosis with no SC and no crossing-over. *D. ananassae*, however, is a notable exception. Here, genetic data confirm that crossing-over does occur in the male though at much reduced frequencies compared to the female. The Tonga strain of *D. ananassae* shows the highest male recombination frequencies. Here crossovers occur with an average value of 10% compared to approximately 50% in the female.

Electron microscope sections of male leptotene chromosomes indicate the presence of axial filaments which, at zygotene, thicken and approach one another to some extent. An imperfect SC is formed with incomplete lateral and central elements (Moriwaki & Tsujita, 1974). The reduced crossing-over observed may, therefore, reflect the imperfect nature of synapsis. By contrast, Grell, Bank & Gassner (1972) failed to find any trace of a SC in a strain of *D. ananassae* with a reasonable amount of male crossing-over. Matsuda, Imai & Tobari (1983) using an air-drying technique, obtained a convincing cytological demonstration of X-type exchanges between non-sister chromatids in this species which they interpreted as chiasmata. A number of iso-site U-type chromatid exchanges were also seen in some 12% of the male meiocytes analysed.

What may be a similar situation to that found in the genus *Drosophila* has been reported in Tsetse flies. No chiasmata are present at male meiosis in *Glossina austeni* and no SC forms (Craig-Cameron, Southern & Pell, 1973). In line with this behaviour, B-chromosomes invariably behave as univalents in males irrespective of the number that are present. In chiasmate females, on the other hand, B-chromosomes form bivalents when two or more are present (Davies & Southern, 1977). A parallel, though curiously inverse, case has been reported in the Hemipteran genus *Saldula*. Here, the small m-chromosomes which characterize heteropterans behave differently in chiasmate and achiasmate forms. Where male meiosis is chiasmate, synapsis of the m-chromosomes is normal but they then undergo desynapsis. In the achiasmate forms, however, synapsis of the m-chromosomes is maintained (Nokkala & Nokkala, 1983).

By contrast with the situation in *Glossina austeni*, the related subspecies *G. morsitans* shows occasional chiasma-like configurations. These are not easily resolved because the chromatids within a bivalent are twisted and intertwined. Paralleling this, an SC of restricted length has been seen in this species although no unpaired axial cores have ever been observed. What is unusual is that autosomal homologues are fully paired when viewed with a light microscope at first prophase, yet the SC that forms is only of limited length. This, however, may simply reflect the fact that the somatic pairing that characterizes Dipterans is known in some cases to lead to close homologue association in prophase-1 of meiosis (see Chapter 4C.1) and so may, in some way, account for the prolonged pairing of homologues in the absence of any SC. This situation may also explain the cases of so-called cryptochiasmate meiosis identified by White (1965) in some mantids, as well as in grasshoppers of the family Thericleinae, where chiasmata, though assumed to be present, are believed to be concealed by the prolonged close synapsis of homologues. It would be worth determining whether partial SCs are present in this case also.

C INVERTED MEIOSIS

Meiosis in eukaryotes with monocentric chromosomes, whether chiasmate or achiasmate, involves a co-orientation of the localized centromere pairs in each bivalent at metaphase-1 followed by their reductional separation at anaphase-1. Half bivalents then auto-orient at metaphase-2 so that their sister centromeres separate equationally at anaphase-2. In a range of organisms, both plant and animal, the chromosomes are holocentric with non-localized centromeres. Here microtubules are attached at many points along each chromosome so that there is no localized kinetochore. Rather, there may be an extended kinetochore plate, as in the animal species *Rhodnius* (Buck, 1967), *Oncopeltus* (Comings & Okado, 1972a), *Tityus* (Benavente, 1982) and *Parascaris* (Goday, Ciofi-Luzzatto & Pimpinelli, 1985) or else multiple but discontinuous packages of MTs inserted into the chromatin, as in the plants *Luzula* and *Cyperus* (Braselton, 1981).

Holocentric systems are especially common in insects, where they are present in the Dermaptera, Mallophaga, Anoplura, Hemiptera, Lepidoptera, Trichoptera (Murakami & Imai, 1974), and the Acari (Oliver, 1967). They are also present in other arachnids, including scorpions and primitive spiders (Benavente, 1982), as well as in nematodes (Albertson & Thomson, 1982; Goday *et al.*, 1985) and heterotrichous ciliate Protozoa (Eichenlaub-Ritter & Ruthmann, 1982). In all these cases, chromosomes orient and subsequently move to the poles at mitotic division with their long axes parallel to the equatorial plate of the spindle. They also continue to behave in this way when spontaneously or experimentally fragmented.

Some, though not all, holocentric organisms are characterized by an inverted meiotic sequence in which the long axis of each bivalent lies in axial orientation, i.e. at right angles to the long axis of the spindle, and parallel to the equatorial plate, at metaphase-1. Following anaphase-1 the homologous chromatids of each half bivalent may remain held together at their ends or else may separate completely. In either event they re-pair prior to prophase-2 and then segregate at anaphase-2. The situation is best known in the plant *Luzula purpurea* (Juncaceae) and in homopteran bugs.

Luzula purpurea has only six chromosomes in the diploid set. The ultrastructural studies of Braselton (1971) indicate that multiple, though discontinuous, kinetochore regions are present within localized recesses along the poleward surfaces of the mitotic chromosomes at metaphase and anaphase. Bokhari & Godward (1980), on the other hand, find that in *L. nivea* a single diffuse kinetochore extends over most, but not quite all, of the chromosome length. Whether this is a genuine difference between the two species, or simply a result of technical treatment as Bokhari & Godward suggest, is not resolved.

Interstitial chiasmata are clearly visible in the three bivalents that form

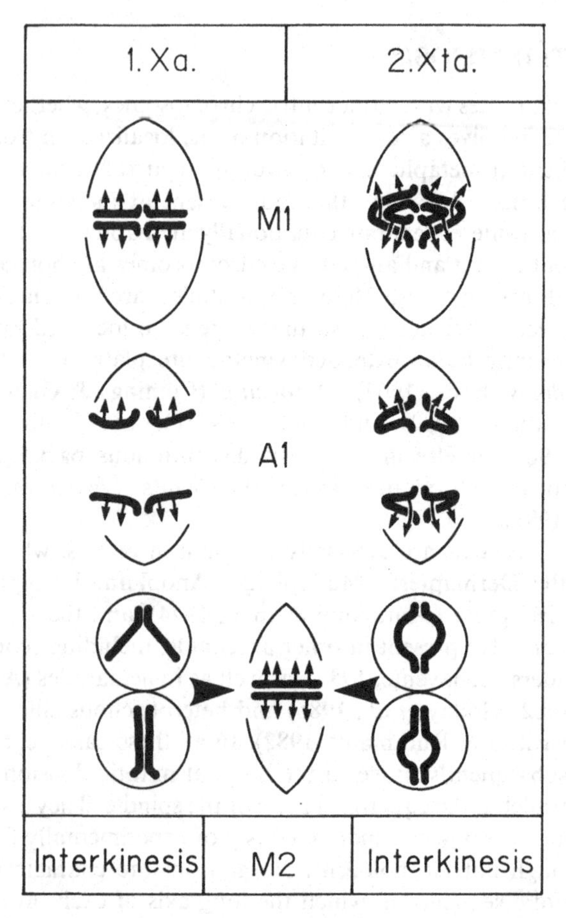

Fig. 2.39. Diagrammatic representation of meiosis in monochiasmate (1Xa) and bichiasmate (2Xta) bivalents of the holokinetic plant *Luzula purpurea* (after Nordenskiöld, 1962).

at meiosis in *L. purpurea*. These are more numerous at diplotene–diakinesis where 2–5 chiasmata have been reported per bivalent. By prometaphase-1, however, bivalents have at most only 1–2 subterminal chiasmata and a majority of the associations between homologues are terminal (Nordenskiöld, 1962). At metaphase-1, bivalent orientation is axial so that at anaphase-1 half bivalents take the form of either rings or else <-shaped rods in which the relicts of metaphase chiasmata may be visible. That is, where subterminal chiasmata are present at metaphase-1 they do not disappear during anaphase-1 but form relict half chiasmata (Fig. 2.39). In the late stages of the first division there is an active secondary pairing of homologous chromatids which is completed at interkinesis. Individual chromatids then segregate at anaphase-2.

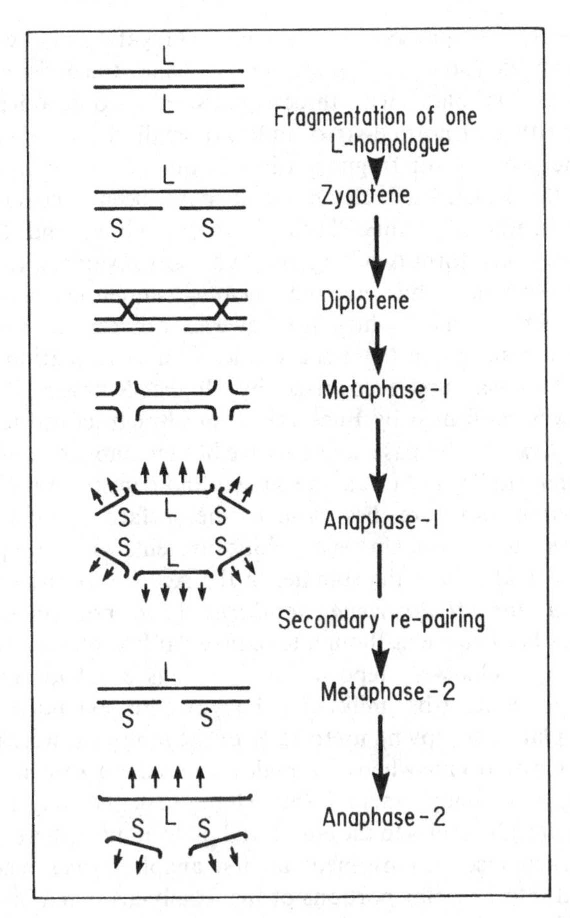

Fig. 2.40. The meiotic behaviour of a fission heterozygote (L, S, S) produced following irradiation of the holokinetic plant *Luzula purpurea* (after Nordenskiöld, 1963).

That it is the altered centromeric organization which is responsible for the inverted meiotic sequence of this holocentric type is confirmed by the behaviour of X-irradiated individuals in which one of the six chromosomes has been broken into two (Nordenskiöld, 1963). Here, one of two meiotic configurations is present at meta–anaphase-1 depending on whether pairing is complete or incomplete. In the former case an open association forms at metaphase-1 consisting of the long member, situated centrally, and two short members, located one at each end of the pseudo-trivalent (Fig. 2.40). In the latter case an unpaired fragment is present. At anaphase-1, mirror-image configurations are evident in which the long axes of the two sets of chromatids move parallel to one another while, at interkinesis, chromatids again re-pair according to homology and prior to their separation at anaphase-2. This mode of segregation can be confirmed after the mitotic divisions of the pollen tetrads where the division products contain two normal plates with three equal-sized chromosomes and two plates each with two normal-sized and two small chromosomes, which represent the products of fragmentation. Moreover, individuals homozygous for the fragmented chromosome have been recovered in the progeny of irradiated plants. These have four long and four short chromosomes which form two long and two short bivalents, respectively.

Both conventional and inverted meiotic sequences occur in the Hemiptera, all members of which are, however, holocentric. In chiasmate species of the Heteroptera there is a wholesale transformation of centric behaviour. This was first recognized by Hughes-Schrader & Schrader (1961) and was confirmed by Buck (1967) in ultrastructural terms using *Rhodnius prolixus*. In this case, an extensive but incomplete band of rather dense material, similar in both arrangement and density to the substance of a localized kinetochore, lies close to the surface of all the mitotic chromosomes but is missing in male meiotic bivalents which are positioned in the longitudinal axis of the spindle. In the Acari, too, the water mites *Eylais setosa* and *Hydrodroma despiciens* have holokinetic mitotic chromosomes but behave as though telokinetic at first metaphase (Oliver, 1967). Similar results were reported by Comings & Okada (1972a) for *Oncopeltus fasciatus* (the milkweed bug) where extended tripartite kinetochore plates, occupying up to 75% of the long axis, were present in each mitotic chromosome whereas at male meiosis there were no structures even vaguely resembling kinetochores. Here, also, the long axis of the bivalents lay at right angles to the equatorial plate at metaphase-1 (see Fig. 5.17*a*) and chromosome movement at first anaphase was mediated by MTs inserted into all polar portions of the axially-arranged bivalents. In cases like this, kinetic activity is concentrated at the end opposite the chiasma in the case of rod bivalents, while in rings such activity develops near the midpoint of the bivalent. A particularly convincing example of this behaviour was reported by Ueshima (1963) in F_1 hybrids between two species of the *Cimex pilosellus* complex. These hybrids combined 11A+X from the egg of one species with 10A+Y from the sperm of the other. At meiosis this produced 1 III+9 II+XY and eight of the nine bivalents were heteromorphic in size. Bivalents consistently had one chiasma and all of them were clearly interstitial at diakinesis. By metaphase-1, however, all chiasmata had terminalized with the two homologues lying end-to-end. The association of three always had two chiasmata at diakinesis and here terminalization produced a chain of three. At metaphase-1 all configurations were co-oriented and the products of this co-orientation were evident at the second meiotic division (Fig. 2.41).

Whether the holocentric system is a derived condition has never been

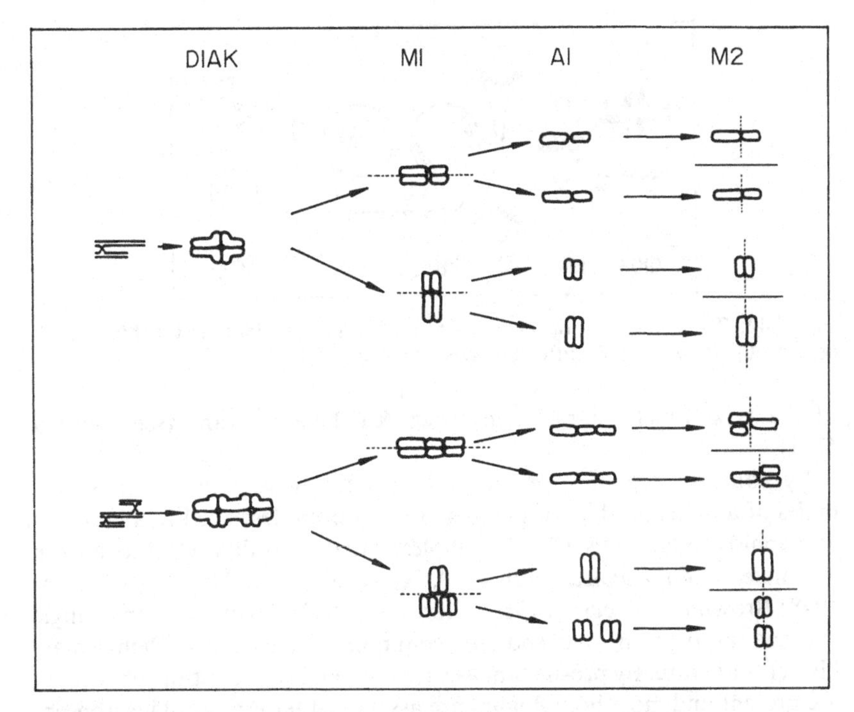

Fig. 2.41. The two theoretically possible modes of meiotic orientation (auto-rientation upper and co-orientation lower) which may be involved in the segregation of a heteromorphic bivalent and a trivalent of holokinetic chromosomes (after Ueshima, 1963 and with the permission of Springer-Verlag).

adequately resolved though most assume it is so. Sybenga (1981), however, suggests it is 'primitive', a view first expressed by Vaarama (1954). Sybenga argues for a gradual increase in specialization from a holocentric to a monocentric state. This, he believes, resulted first in increasing localization of function at meiosis, as is still evident in some hemipterans, followed subsequently by an equivalent localization at mitosis too.

The one exceptional form of heteropteran behaviour occurs in the achiasmate male meiosis of the genus *Nabis* (Nokkala & Nokkala, 1984). In *Nabis flavomarginatus*, for example, the earliest meiotic stage is diffuse except for the heteropycnotic X- and Y-chromosomes which may, or may not, be associated. When autosomal bivalents condense from this diffuse stage they each consist of two parallel homologues. At metaphase-1 these are equationally oriented, as too are the X- and Y-univalents, and move parallel to the equator at anaphase-1 in typical holocentric fashion. Also, unlike other heteropterans, there is an interkinesis and at second division the X and Y show distance pairing rather than the touch-and-go pairing

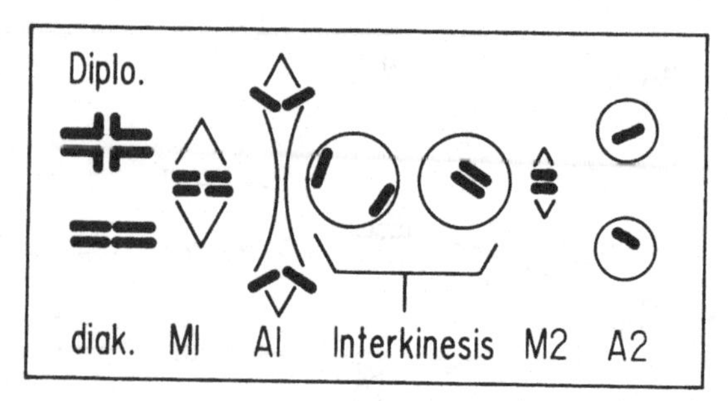

Fig. 2.42. The male meiotic behaviour of a single bivalent in the holokinetic coccid bug *Puto albicans* (after Hughes-Schrader, 1948).

of the type which characterizes most XY heteropterans (see Chapter 5B.1).

By contrast with heteropterans, inverted meiosis is common in the males of all advanced homopterans and has been described in all coccids and aphids so far analysed. The simplest situation is that which occurs in the male of *Puto albicans* (2n = 19, X0♂; 20, XX♀, Hughes-Schrader, 1948; Brown & Cleveland, 1968). Here (Fig. 2.42), bivalents with a single chiasma at or near one end are common at diakinesis. Bichiasmate bivalents are rare. By pro-metaphase-1, however, few interstitial chiasmata are present and most homologues are associated terminally. This appears to be a general feature of bugs with inverted meiosis; that is, only a single chiasma forms, and this is invariably completely terminalized, in the very contracted bivalents seen at metaphase-1. Indeed, this same behaviour often characterizes heteropterans (Fig. 2.43) which do not show inverted meiosis (John & Claridge, 1974). Presumably, chiasmata terminalize towards the nearest end, or if located centrally towards either end.

At metaphase-1 in *Puto* all bivalents are oriented with their long axes parallel to the spindle equator. It is not possible to resolve either the four chromatids of each bivalent or the two chromatids of the X univalent though the X is clearly double at first anaphase. In more advanced coccids, unlike the situation in *Puto*, homologous chromosomes do not pair at male meiosis. This lack of pairing is associated with the fact that the entire paternal set is facultatively heterochromatinized in early embryogeny and persists in this state within the germ line (Nur, 1980 and see Chapter 8). In the inverted sequence of homopterans the homologous chromatids of each half bivalent dissociate sometime during the first meiotic division, usually at anaphase-1 or early telophase-1, and then re-pair at the second meiotic division. This secondary pairing occurs even in those coccids which are regularly asynaptic at the first division.

In aphids (Homoptera: Aphididae), Blackman (1976, 1985) has described two novel forms of inverted male meiosis. Here the males are X0 and are produced by parthenogenesis, i.e. without fertilization, by XX-

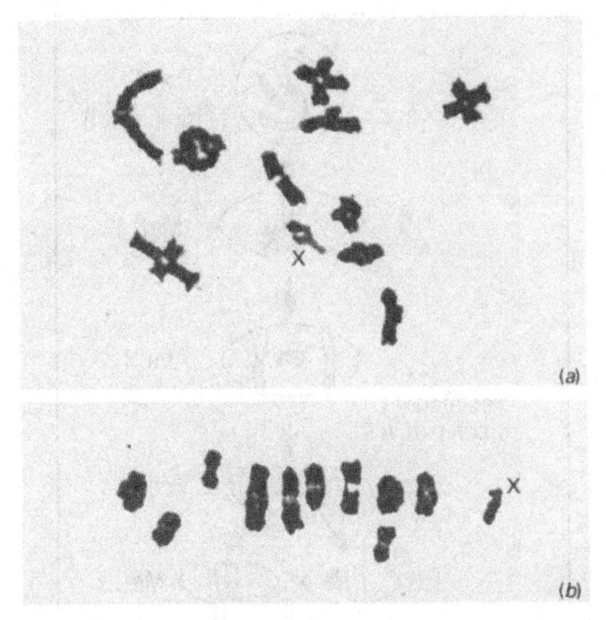

Fig. 2.43. (*a*) Diakinesis and (*b*) metaphase-1 of male meiosis in the holokinetic birch leafhopper *Oncopsis flavicollis* (2n = 21♂, X0). At diakinesis there are interstitial chiasmata in most of the bivalents but by first metaphase most of these have completely terminalized.

females (see Chapter 8). In parthenogenetic eggs destined to develop into males, the X undergoes a reductional division whereas the autosomes simply divide equationally so that both sexes end up with the same number of autosomes. At male meiosis in *Amphorophora tuberculata* (2n = 3 = AA + X0; Fig. 2.44) the two A homologues appear as largely unpaired threads at first prophase though they are associated end-to-end. They retain this association as they contract and then orient with their long axes parallel to the spindle equator. By contrast, the X-univalent orients with its long axis at right angles to the equator. Consequently, while the autosomal half bivalents move apart in parallel, and so maintain the end-to-end alignment of homologous chromatids, the univalent X remains at the equator giving rise to an abortive first anaphase in which the lagging X holds the two separating autosomal half bivalents in a loose association which does not completely dissociate until metaphase-2.

During second prophase the terminal association between the two chromatids of each half bivalent lapses and they become aligned side-by-side through secondary pairing. By metaphase-2 the undivided X has become associated with one of the two autosomal half bivalents and this combination now dissociates itself from the other half bivalent, the X orienting at right angles to the autosomal half bivalent with which it is

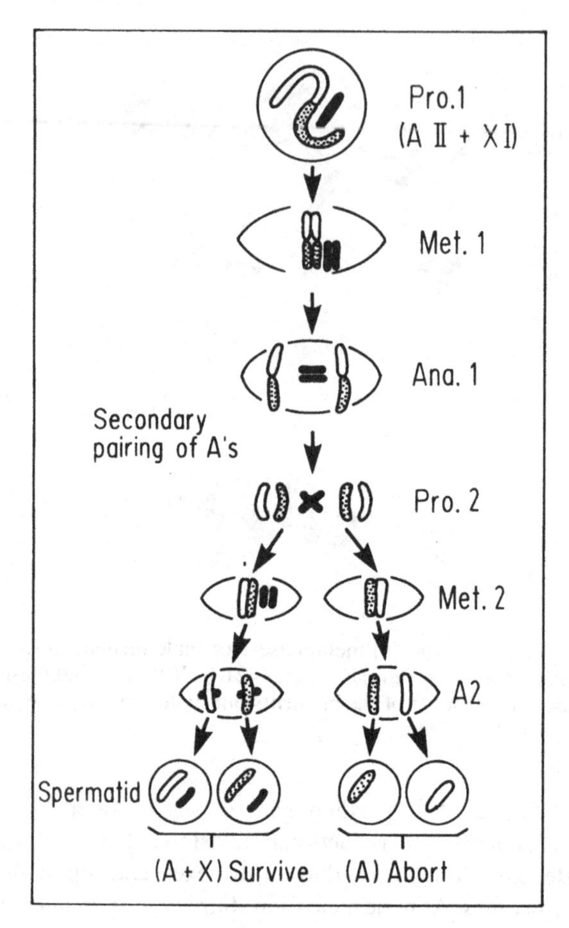

Fig. 2.44. Male meiosis in the aphid *Amphorophora tuberculata* (2n = 3, X0). The unpaired X-chromosome is shown solid. One of the two autosomal homologues is open, the other stippled (after Blackman, 1985 and with the permission of Springer-Verlag).

associated. In this same orientation, the two chromatids of each chromosome now move apart. The other, unassociated half bivalent also undergoes anaphase separation but the products degenerate so that only two spermatids form from each meiocyte. In other aphids, too, functional sperm are produced only from spermatids that carry an X-chromosome. Indeed, in some species, cells lacking an X-chromosome degenerate before second prophase. In others, degeneration occurs after an apparently normal second division, while in still others small nullo-X spermatids form and then regress.

A quite different situation occurs in birch aphids. Here two species are recognized, *Euraphis betulae* (2n = 10♀) and *E. punctipennis* (2n = 8♀).

The six shortest chromosomes are common to both species. Two of these, referred to as B-chromosomes, are strongly condensed at somatic interphase and prophase, show positive C-banding, and are often associated with one another at these stages. The four remaining short chromosomes are sex chromosomes, both species having an $X_1X_1X_2X_2$♀, X_1X_2♂ sex mechanism. The difference in diploid number between the two species is due to the presence of an exceptionally long chromosome (A-chromosome) in *E. punctipennis* which replaces two shorter chromosomes, A_1 and A_2, in *E. betulae*. Since the length of the single A-chromosome is the sum of A_1 and A_2 there would appear to be a simple Robertsonian relationship, involving either fusion ($A_1+A_2 \rightarrow A$) or fission ($A \rightarrow A_1+A_2$), between the two species. Consequently, male embryos of *E. betulae* have $2n = 8 = 2A_1+2A_2+B_1B_2+X_1X_2$ while in *E. punctipennis* male embryos have either $2n = 6 = 2A+B_1B_2+X_1X_2$ or $2n = 5 = 2A+B+X_1X_2$.

In *E. betulae* the homologues of both A_1 and A_2 show complete pairing, do not form chiasmata, and remain parallelly-aligned at first metaphase. By contrast the two X-chromosomes and both Bs remain unpaired and move, without dividing, into one of the two secondary spermatocytes at anaphase-1. These then transform into sperm without further division. Those cells which fail to receive an X degenerate so that all sperm cells have $n = 6$ ($A_1A_2+X_1X_2+B_1B_2$).

In *E. punctipennis*, all the chromosomes, including the very large As, are present as univalents at male meiosis though female meiosis is chiasmate. One of the A univalents in the male becomes positively heteropycnotic and by first metaphase it forms a condensed mass which takes no further part in division. The single meiotic division that does occur produces two sperm cells which have $n = 5 = A+X_1X_2+B_1B_2$ in $2n = 6$♂ and $n = 4 = A+X_1X_2+B$ in $2n = 5$♂.

Thus, the anomalous meiotic sequence of the birch aphid *E. punctipennis* is both inverted and asynaptic as well as lacking a second meiotic division.

An unusual pattern of behaviour has been described in holocentric F_1 hybrids between domestic, *Bombyx mori*, and wild, *B. mandarina*, silk moths (Murakami & Imai, 1974). These two taxa show a Robertsonian difference in diploid number so that F_1 hybrids have 26 II + 1 II in which the chain multiple of three is made up of a large M-chromosome from the wild form and two smaller, m_1 and m_2 chromosomes from the domestic form. The two chiasmata in the multiple of three are completely terminalized by first metaphase as too are all the other chiasmata in this hybrid. The bivalents orient axially at metaphase-1 but the chain of three, while also orienting predominantly in an alternate fashion, show half trivalent separation as a significant alternative.

In summary, in bivalent-forming, sexually-reproducing eukaryotes, the genome is rearranged during meiosis by the randomized allocation of one member of each pair of homologues to a given meiotic product. This may occur either with (chiasmate meiosis) or without (achiasmate meiosis) a recombination between the homologues themselves. The succession of two phases of nuclear division without an intervening phase of chromosome replication thus leads to both numerical reduction and to genetic segregation.

3

Occurrence and timing of meiosis

There must be a form of nuclear division in which the ancestral germ plasms contained in the nucleus are distributed to the daughter nuclei in such a way that each of them receives only half the number contained in the original nucleus.

August Weismann

A DIFFERENTIATION OF THE GERM LINE

In the late 1800s it was widely accepted that the germ cell line of animals was continuous from one generation to the next. Weismann (1892), however, proposed that the true continuity was provided not by the germ cells as such but by a nuclear substance handed down from parent germ cells to those of the offspring. He termed this substance *germ plasm*. In the early 1900s Hegner succeeded in tracing cytoplasmic granules from the pole plasm of insect oocytes of one generation to the germ cells of the next generation (Hegner, 1914). He referred to these granules as *germ line determinants* and argued that they were the visible expression of a specialized differentiation in the cytoplasm which controlled the production of the primordial germ cells. In Hegner's terms, therefore, the germ plasm was cytoplasmic in character and not nuclear, as Weismann had assumed. Time has confirmed the essential accuracy of Hegner's ideas and it is now clear that there is no difference in principle between the differentiation of the germ line and the differentiation of the various categories of somatic cells. Germ cells are simply cells specialized for meiosis and sexual heredity.

In most metazoans there is a clear division of labour between the cells responsible for sexual reproduction, the *germ line cells*, and the somatic lineages of the organism. The mode of origin of the germ cells is, however, varied and the differentiation between soma and germ line is not always abrupt. In coelenterates, for example, the interstitial cells serve as precursors for nerve cells, nematocysts and germ cells (David, 1983). Similarly, the neoblasts of flatworms can regenerate most somatic cell types as well as generating the germ cells (Lange, 1983). In both these cases, the germ cells arise as a specific cell lineage from pluripotent stem cells. Among chordate animals, the germ cells of Ascidians likewise originate from pluripotent stem cells, termed *hemoblasts*, which circulate in the blood.

Flowering plants, too, have an indeterminate pattern of development. While most cells of the very young embryo undergo mitosis, division in the vegetative sporophyte occurs principally in the localized meristems of the root and shoot apices and the leaf primordia. These meristems incorporate a stem cell population capable of giving rise to new tissues throughout the life of the plant. It is from the meristem of the shoot that the germ cell line is eventually set aside at a comparatively late stage of development. Here, then, the production of the germ line involves the repression of events which lead to somatic differentiation since the shoot apex produces leaves when vegetative and floral parts when reproductive.

In most multicellular animals, by contrast, the germ cells are set aside before the embryonic cells become committed to specific developmental pathways. In many, though not all, cases the gametes are derived solely from primordial germ cells whose production is specified, as Hegner first demonstrated, by the action of cytoplasmic determinants present at the time of fertilization. This situation has been especially well analysed in *Drosophila melanogaster* where primordial germ cell formation is triggered by the presence of specific cytoplasmic substances formed by the transcription of the maternal genome during oogenesis and hence present at fertilization. Here, the posterior tip of the growing oocyte contains a distinctive cytoplasm containing polar granules which are formed at a late stage of vitellogenesis. At the 8th–9th nuclear divisions following fertilization, a number of nuclei migrate from the syncytial blastoderm and enter the specialized polar plasm present at the posterior end of the egg, which houses the polar granules. These nuclei give rise to pole cells which are obligate precursors of the germ line. If pole cells fail to form, or are experimentally prevented from forming, no other cell type is able to differentiate into germ cells (Mahowald & Boswell, 1983).

A class of maternal effect, grandchildless, mutations are known in *Drosophila* which specifically affect the ability of the embryo to form pole cells. Consequently, while the first generation is normal except for the absence of germ cells, these mutations lead to second generation sterility. Of the eight known mutations of this kind (Konrad *et al.*, 1985), the most specific is tudor (*tud*). Here six alleles have been identified with a range of effects on the presence or size of the polar granules (Boswell & Mahowald, 1985). The most extreme of these have no detectable polar granules and never form pole cells. Another allele forms very small granules but still fails to produce pole cells. Thus, the gene product of the tud^{+} allele is required for the proper assembly of the germ plasm and hence for the formation of primordial germ cell precursors.

Equivalent germ plasm systems are known in a variety of invertebrate animals including rotifers, nematodes, annelids, molluscs, chaetognaths, crustaceans and a variety of insects other than dipterans (Eddy, 1975). In vertebrates, on the other hand, a cytoplasmic germ plasm is known only in the eggs of anuran amphibians. Here, it is initially located in the vegetal cortical region of the egg so that, by late cleavage, it has become restricted to a few cells on the floor of the blastocoel. These cells, and their clonal descendants, subsequently migrate to the genital ridges. At the ultrastructural level, the germ plasm of the frog is visually identical with that of *Drosophila* (Mahowald & Hennen, 1971) and, as in *Drosophila*, surgical removal of the germ plasm leads to sterility (L. D. Smith, Michael & Williams, 1983).

In mammals, the germ line forms as a cell population within the epiblast and by a determinative step that occurs only once during early development (Eddy & Hahnel, 1983). The migration of the primordial germ cells and their colonization of the genital ridges is similar in both male and female embryos. The subsequent phase of mitotic multiplication is also similar. From this point on, however, events diverge dramatically. In the mouse, for example, male germ cells enter mitotic arrest in a G_1-state and the next round of DNA synthesis is delayed until shortly after birth when the germ cells differentiate as spermatogonia. Successive waves of spermatogenic cells then leave the stem cell population and multiply by mitosis prior to subsequently entering meiosis. Whereas male meiosis is delayed until after birth, female germ cells embark on meiosis in the embryo and enter first prophase.

B MEIOTIC ARREST

There is a further difference in the character of male and female meiosis in animals. The stem cell population of male germ cells continues to proliferate throughout the life of an individual and spermatogenesis, once initiated, goes through to completion with the products of meiosis differentiating directly into spermatozoa. By contrast, in many species, oogenesis is a protracted process subject to one, or more, periods of temporary arrest.

In female mammals, for example, the primordial germ cells, which are set aside during early embryogeny, migrate to the genital ridges once the somatic gonad rudiments are established. Here, they become enclosed by somatic cells which form a single layer around each individual oocyte giving rise to a series of primordial follicles. The somatic follicle cells play an important role in the subsequent maturation of the oocyte. What actually triggers the primordial follicles to grow is unknown but, once initiated, they are committed to an irreversible sequence of events in which the follicle becomes multilayered and, in this form, is able to respond to gonadotrophins, and especially to the follicle stimulating hormone (FSH) produced by the pituitary.

Many thousands of primordial follicles are produced in the embryonic

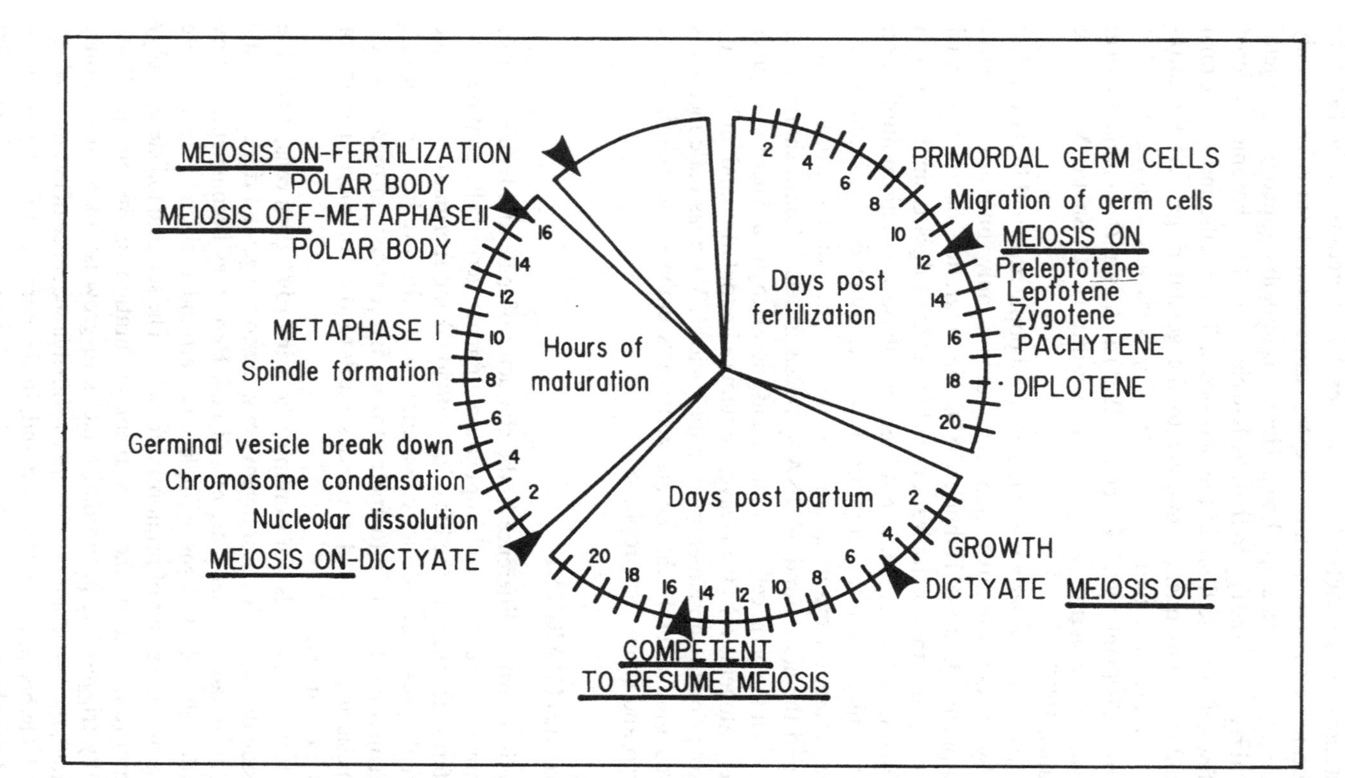

Fig. 3.1. Principal events in the oogenesis of the mouse (after Schuetz, 1985 and with the permission of Plenum Press).

ovary but most of them subsequently degenerate. It is within these primordial follicles that meiosis is initiated within the embryo. When female meiocytes reach diplotene, however, they enter a meiotic hold phase and are arrested. This inhibition is governed by a peptide produced by the granulosa cells of the follicle (Tsafriri & Pomerantz, 1984), an inhibition that is potentiated by cAMP (cyclical, 3′5′-adenosine monophosphate). During this meiotic arrest the oocyte persists in a diffuse diplotene stage referred to by mammalian cytologists as *dictyate*.

FSH not only stimulates the final phases of follicular growth, leading to the formation of the mature or *Graafian follicle*, but also induces the production of luteinizing hormone receptors in the follicle cells. In response to luteinizing hormone, the oocyte resumes meiosis in adult life following the pre-ovulatory surge of gonadotrophins produced shortly before the release of the oocyte from the follicle at ovulation. At this time, the nuclear membrane of the oocyte, the germinal vesicle, breaks down and this is followed by chromosome condensation, the development of the first meiotic spindle and the extrusion of the first polar body. By ovulation the chromosomes are again in a state of meiotic arrest, this time at metaphase-2, and require the activating stimulus provided by fertilization to complete meiosis (Fig. 3.1). The cytoplasm of oocytes arrested at second metaphase contains a specific 'cytostatic' factor which is responsible for the arrest and is activated by the increase in Ca^{2+} levels which occurs at sperm penetration (Masui, Lohka & Shibuya, 1984).

In amphibians, where female meiosis is also arrested at diplotene, the oocyte continues to grow because the chromosomes are especially active in transcription at this stage, often over an extended period of time. This transcription is not concerned with the metabolism of the oocyte itself, and hence with the process of meiosis, but rather with the production of a stockpile of maternal components required to support the early stages of development and at a time when there is little direct gene activity. After fertilization, amphibian eggs cleave rapidly to form about 4000 cells within only eight hours. Thus an adult cell of *Xenopus laevis* takes longer to complete a single cell cycle than to form 10000 cells in the early embryo. This is possible only because of the vast store of maternal products produced during oogenesis, as well as the uncoupling of chromosome replication from cell growth. An egg of *X. laevis*, for example, contains enough maternally produced histone to construct some 20000 nuclei (Woodland, 1980).

As a prerequisite for this phase of extensive transcription the chromosomes decondense and are transformed into a distinctive lampbrush form in which there is a periodic looping-out of DNA from the main chromosomal axes. This gives rise to several thousand transcriptively active and paired lateral loops from each of the homologues involved in a bivalent (Macgregor, 1980). At the end of transcription these loops

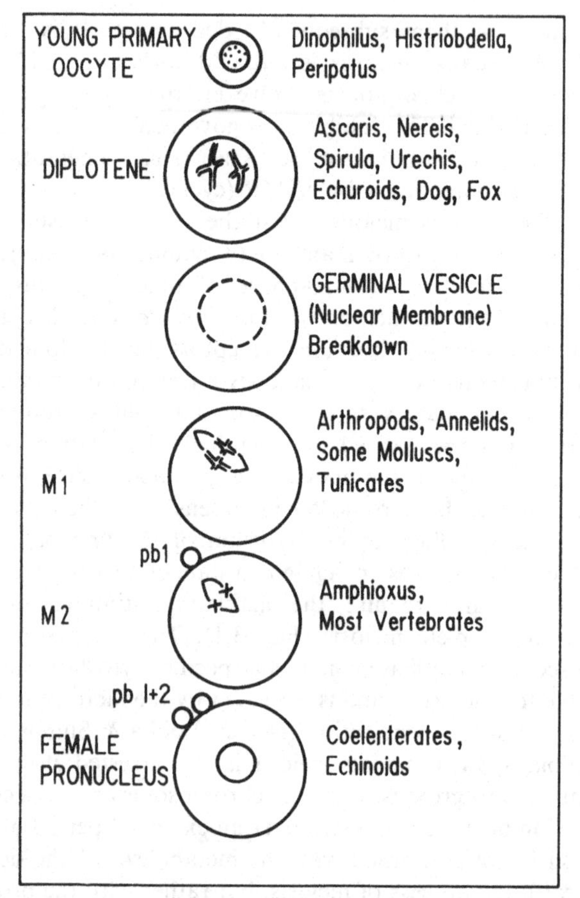

Fig. 3.2. The occurrence of meiotic arrest in animal oocytes.

retract and the chromosomes recondense. Male meiocytes too are active in RNA and protein synthesis throughout much of first prophase (Henderson, 1964), though in a far less spectacular guise. The products of these syntheses are subsequently involved in the transformation of spermatids into sperm.

A fertilization-dependent resumption of meiosis characterizes many other animals too (Fig. 3.2). In echuroids and some molluscs (*Urechis*) oocytes are fertilized at the germinal vesicle stage (dictyate) and only then do they resume meiosis. In most invertebrates, including arthropods, annelids, some molluscs and tunicates, oocytes are arrested at metaphase-1, whereas most vertebrates arrest at metaphase-2. In colenterates and echinoderms, on the other hand, oocytes complete meiosis and then are fertilized at the pro-nuclear stage. An equivalent phase of meiotic arrest and reactivation has been described in the orchid *Dactylorhiza*. Here

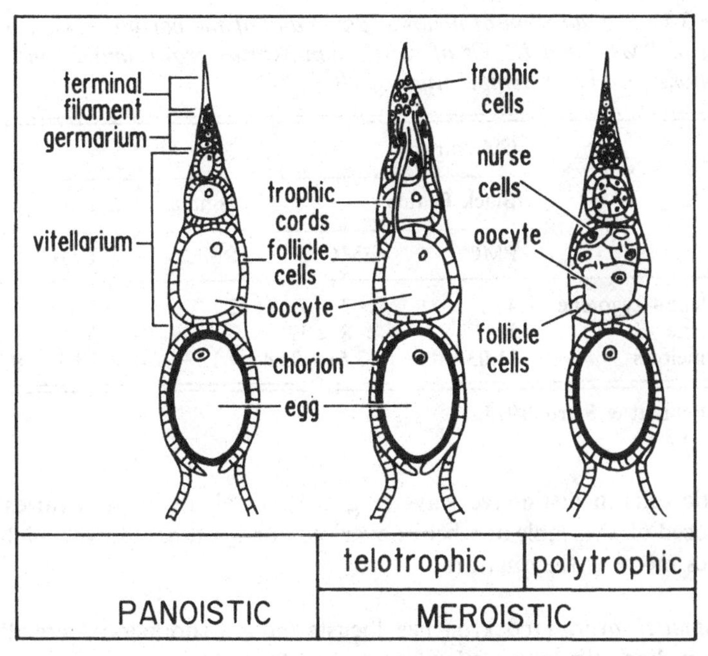

Fig. 3.3. The three categories of insect ovarioles. In all three, the oocyte is enclosed by a unilayered wall of follicle cells. Specialized nurse cells are present in meroistic, but not in panoistic, ovarioles. In telotrophic ovarioles these nurse cells are located terminally and are connected to the oocyte by trophic cords which increase in length as the oocyte moves down the ovariole. In polytrophic ovarioles the nurse cells are enclosed, with the oocyte, within the follicle (after Postlethwait & Giorgi, 1985 and with the permission of Plenum Press).

meiocytes in the ovule enter leptotene but require pollination for meiosis to resume (Heslop-Harrison, 1957).

C SOMATIC SUPPORT SYSTEMS

Germ cells are in constant interaction with the somatic tissues that constitute their immediate environment. Their migration in the embryo to the genital ridges is guided by a preordained somatic route. Their entry into meiosis and their mode of maturation depend on a series of somatic signals and triggers. This is especially evident in female germ cells which commonly become enclosed within follicles of somatic origin.

Particularly striking examples of the interaction between germ cells and somatic cells are evident in insect oogenesis. Here follicle cells synthesize yolk peptides, or vitellogenins, which are sequestered in the oocytes. Additionally, in the later stages of oogenesis, the follicle cells secrete a protective covering, the *chorion* or *egg shell.* Three different types of ovary are known in insects (Bowнes, 1983) which combine germ cells and

Table 3.1. *The duration of meiosis (days) and of the early prophase-1 stages in PMCs and EMCs of two Lilium hybrids grown under controlled conditions at a mean temperature of* 20 ± 1 *°C*

	Cultivar			
	Black Beauty		Sonata	
Stage	PMC	EMC	PMC	EMC
Leptotene + zygotene	4	$>3<7$	2	3
Pachytene	5	$>8<12$	3	5
Total meiosis	11.03 ± 0.51	17.61 ± 0.84	7.12 ± 0.24	10.32 ± 0.72

After Bennett & Stern, 1975.

somatic cells in distinctive ways (Fig. 3.3). In all three the ovaries are composed of a variable number of ovarioles, or egg tubes, down which the oocytes move as they mature:

(1) *Panoistic ovaries* (cockroaches, locusts and grasshoppers). Here all the germ line cells, present in the terminal filament or germarium of the ovariole, develop into oocytes. Mesodermal somatic cells form a sheath surrounding the entire ovariole as well as giving rise to the follicle cells that surround each oocyte. When the egg membranes are secreted the oocyte separates from the follicle cells and is released at the base of the ovariole.

(2) *Meroistic ovaries*. Here some germ line cells differentiate into oocytes while others form feeder or trophic cells which pass RNA, proteins and lipids into the oocyte, via cytoplasmic connections, and then degenerate. The precise location of these trophic cells, which are of germ line and not somatic origin, distinguishes two kinds of meroistic ovary:
   (a) *Telotrophic* (aphids and bugs). Each stem cell in the germarium divides to produce one oocyte and one trophocyte. All trophic cells remain at one end of the ovariole and are connected to a single oocyte by nutritive cords which increase in length as the oocyte moves down the ovariole.
   (b) *Polytrophic* (flies, bees, butterflies and moths). The trophic cells, commonly termed *nurse cells*, remain in close contact with the oocytes forming a cell cluster which moves as a unit along the ovariole. Each cluster is surrounded by follicle cells so forming an egg chamber. The nurse cells degenerate as the oocyte matures and the follicle cells eventually secrete the egg membranes.

Table 3.2. *The duration of meiosis (hours) and of the early prophase-1 stages in PMCs of five grass taxa*

Stage	Duration (h)				
	Hordeum vulgare var. Sultan 2n = 14	*Secale cereale* var. Petkus Spring 2n = 14	*Secale cereale* autotetraploid 2n = 4x = 28	*Triticum aestivum* var. Chinese Spring 2n = 6x = 42	*Triticale* amphiploid of wheat and rye 2n = 8x = 56
Leptotene	12.0	20.0	13.0	10.4	7.5
Zygotene	9.0	11.4	9.0	3.4	3.0
Pachytene	8.8	8.0	6.4	2.2	2.2
Remainder of meiosis	9.6	11.8	9.6	8.0	8.0
Total meiosis	39.4	51.2	38.0	24.0	20.7

After Bennett, Chapman & Riley, 1971; Bennett *et al.*, 1973a.

A not too dissimilar situation to that found in insects also occurs in amphibians where growing oocytes are supplied with maternally-derived vitellogenins whose secretion and uptake is regulated by oestrogen synthesized by the follicle cells. In mammals, the dialogue between germ cells and somatic cells differs from that in insects and lower vertebrates but is none the less indispensable. Somatic signals initiate meiosis during foetal life, arrest the meiotic cycle at the dictyate stage and determine which oocytes from the primordial pool eventually ovulate. These signals regulate oocyte maturation in three ways. They control the entry of all molecules into the oocyte. They provide energy, substrates and RNA precursors for the oocyte and they influence the pattern of proteins synthesized by the oocyte itself. In addition to the follicles which directly surround the oocyte, the liver and the oviduct of vertebrates also function as somatic sites for the synthesis of ooplasmic components. This is particularly evident in reptiles and birds where the oviduct provides massive amounts of the glycoproteins of egg white, namely ovalbumin, conalbumin, lysozyme and ovomucoid.

D THE DURATION OF MEIOSIS

The limited information available indicates that there is considerable variation in the length of meiosis in different organisms. Bennett (1973, 1976) reported that a positive correlation existed between DNA content and meiotic duration in a number of diploid plants. However, this relationship is complicated by the facts that:

(1) In some cases the duration of meiosis differs between EMCs and PMCs of the same species despite the fact that they contain the same amount of DNA. This is so in *Lilium* where meiosis is some 50% longer in EMCs (Table 3.1).

(2) In polyploid plants meiosis is shorter than in related diploids, despite the fact that polyploids have a higher DNA content. As in the case of the difference between EMCs and PMCs this relates largely to differences in the duration of the leptotene, zygotene and pachytene stages (Table 3.2).

Meiosis is also consistently longer in animals than in plants. This, presumably, relates to the fact that in both spermatocytes and oocytes there is a need for specialized syntheses required not for meiosis itself but for gamete development in the male and early embryonic development in the fertilized egg. In plants, by contrast, both the developing gametes and the developing embryo are supplied with all necessary molecular components by specialized trophic tissues.

4

Events and mechanisms of meiosis

> Models, of course, are highly seductive, especially where the realities of events are poorly known.
> *Herbert Stern*

A THE SWITCH TO MEIOSIS

The precise factors responsible for the transition from mitosis to meiosis are still largely unknown. In mammals, as we saw earlier, while differentiation of the gonad into a recognizable testis occurs in the embryo, male germ cells do not enter meiotic prophase before early puberty. Fetal mouse testes and ovaries, together with their urinogenital connections, can be cultured either singly, or in pairs, on nucleopore filters. When a fetal testis is cultured in combination with older ovaries containing germ cells at prophase-1 of meiosis, the primordial male germ cells are triggered to enter meiosis (Byskov & Saxén, 1976). The female gonad thus appears to secrete a meiosis-inducing substance which can trigger indifferent male germ cells to enter meiosis. In keeping with this, XY cells in fetal mouse chimaeras also embark on meiosis and McLaren (1984) has proposed that they are stimulated to do so by the surrounding XX cells.

The specific somatic cell system that appears to be essential for the differentiation of the mammalian fetal ovary is the *rete ovarii.* Both the onset of female meiosis and follicle formation seem to depend on this urinogenital connection, of mesonephric origin, in the early gonadal anlage. When fetal testes with developed testicular cords are cultured in combination with ovaries of the same age, but which contain germ cells already in meiosis, the oocytes are prevented from reaching diplotene. This implies that the male gonads are producing some kind of meiosis-inhibiting substance. Unfortunately no diffusible inducing or inhibiting substances have yet been isolated. Moreover, the application of different steroids has no influence on the onset of meiosis despite the fact that fully grown oocytes of the frog *Xenopus laevis*, arrested in prophase-1, can, when surgically removed, be triggered to resume meiosis using a variety of agents, including progesterone.

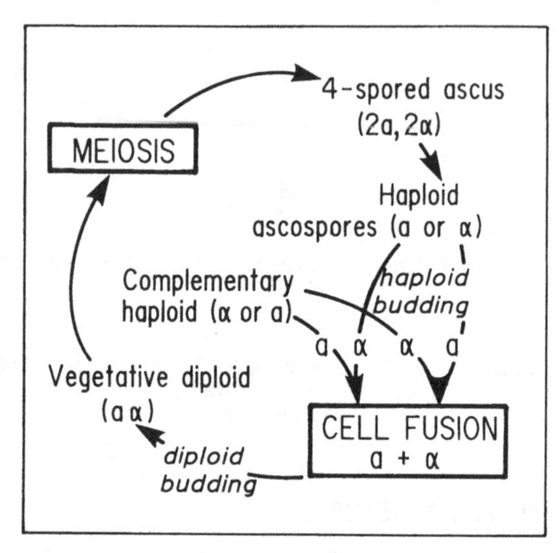

Fig. 4.1. The life cycle of the budding yeast *Saccharomyces cerevisiae*.

Experiments with the nematode *Caenorhabditis elegans* also identify somatic cells within the gonads that are critical for the entry of germ cells into meiosis (Kimble & White, 1981; Kimble, 1983). Here, the two distal tip cells of the somatic gonad are responsible for the local inhibition of germ cells from entering meiosis. This applies both to spermatogonia and oogonia. Destruction of these cells by laser microsurgery leads to the arrest of mitosis and the initiation of meiosis in the germ cells. How this inhibitory system functions is again unknown though, here too, it has been assumed to involve a diffusible factor of some kind. Both these cases thus suggest that entry into meiosis is dependent on some diffusible substance which originates outside the pre-meiotic cell itself, and an equivalent suggestion has been made for plants (Walters, 1985). Thus pre-meiotic anthers of the tomato plant can be induced to enter meiosis when the branches bearing them are grafted onto branches with buds already in meiosis (Gregory, 1940).

Considerable attention has been paid to the conditions necessary for the entry of the yeast unicell *Saccharomyces cerevisiae* into meiosis. Yeast grows vegetatively in either a haploid or a diploid phase. In the haploid phase there are two mating types, a and α, which by hybridization (≡ fertilization) give rise to diploid cells which divide mitotically, the so-called *vegetative diploids*. Such diploids can be induced to enter meiosis and subsequently produce an ascus containing four haploid ascospores. These, on germinating, produce vegetative haploids (Fig. 4.1). Thus, whereas fission yeast, *Schizosaccharomyces pombe*, remains haploid until transferred to a sporulation medium and then undergoes conjugation before pre-meiotic DNA replication, budding yeast, *S. cerevisiae*, enters meiosis from a vegetative diploid, i.e. in a manner equivalent to higher plants and animals.

Dawes (1983) lists three conditions that need to be met before a yeast cell culture can initiate meiosis and spore formation:

(1) Unsatisfactory nutritional conditions involving a depletion or limitation of either the nitrogen or the carbon source of the growth medium. This indicates that, at least under conditions of artificial culture, meiosis in yeast depends upon an inducible mechanism which facilitates escape from a vegetatively uncongenial environment

(2) Satisfactory completion of mitotic cell division in the sporulating medium. Meiosis is initiated only in the G_1-phase of the cell cycle

(3) Heterozygosity for mating type alleles, i.e. a/α or a/a/α. This represents an intrinsic form of control, determined by the genetic constitution of the cell. Thus, a/a or α/α diploids will not undergo premeiotic DNA synthesis under sporulating conditions. However, aneuhaploids disomic for chromosome III, which carries the mating type locus, and which are also a/α in constitution, can initiate a premeiotic S-phase. The mating locus is thus a master genetic switch for entry into meiosis in this organism

In *Schizosaccharomyces pombe*, too, sexual differentiation is governed both by cell type and by nutritional factors. Haploid cells propagate vegetatively as one of two mating types, h^+ or h^-. Cell type determination is controlled by a class of master regulatory, *mating genes*. Meiosis normally occurs only in h^+/h^- diploids under a state of nutritional stress. These conditions allow for the expression of the mating-type genes which, in turn, leads to the transcriptional induction of a dominant meiotic activator gene, *mei-3*$^+$. This gene encodes a transcript of 1300 nucleotides and a transcriptional product of 21K, both of which are detectable only during meiosis. In non-meiotic cells the *mei-3*$^+$ gene is transcriptionally silent.

The protein product of the *mei-3*$^+$ gene has recently been shown to act as an inhibitor of the protein kinase encoded by the *ran-1*$^+$ gene. This functions as a critical negative regulator of sexual conjugation, meiosis and sporulation. Inactivation of the *ran-1*$^+$ protein kinase thus results in the immediate onset of meiotic development. The interaction between these two meiotic regulators, *mei-3*$^+$ and *ran-1*$^+$, occurs at the protein–protein level (McLeod & Beach, 1988).

Gross overstimulation of cAMP-dependent protein kinase in *S. pombe* bypasses the requirement of the *ran-1*$^+$ gene for vegetative growth and also inhibits normal meiosis. Consequently, alterations in the activity of this

Table 4.1. *A comparison of cycle times in mitotic and premeiotic cells*

Species	Cell type	Duration (h)			Reference
		Cell cycle	Mitosis	S phase	
Triticum aestivum var. Chinese Spring	Root tip meristem	12.5	1.6	3	Bennett, 1973
	3rd mitosis before meiosis	25.0	1.9	—	
	2nd mitosis before meiosis	35.0	2.3	—	Bennett, 1976
	Pre-meiotic mitosis	55.0	2.6	12–15	
Lilium longiflorum	Root tip meristem	24.0	—	8	Bennett, 1976
	Pre-meiotic mitosis	> 72	—	49	Rasmussen & Holm, 1980
Saccharomyces cerevisiae	Vegetative cells	2.5	—	—	Simchen, Piñon & Salts, 1972
	Pre-meiotic cells	7–8	—	—	
Triturus vulgaris ♂	Somatic cells	—	—	24	Callan & Taylor, 1968
	Pre-meiotic cells	—	—	9–10 d	Callan, 1972

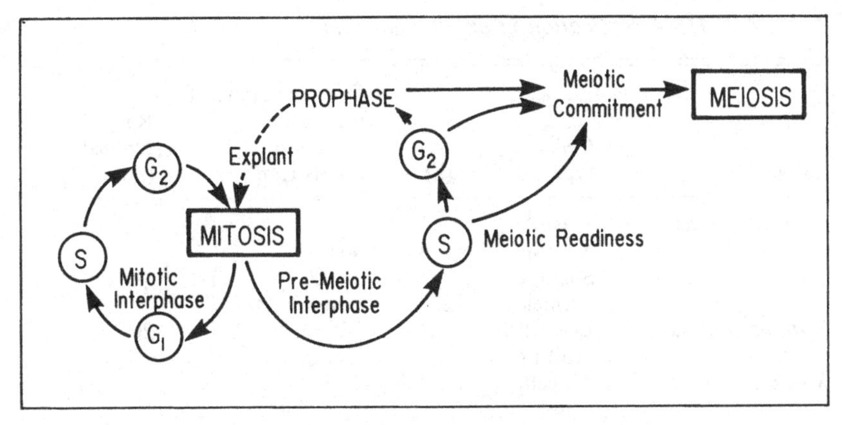

Fig. 4.2. Events preceding the switch to meiosis.

enzyme, too, may be involved in the switch to meiosis, possibly through an interaction of the pathways regulated by these two protein kinases. Whether homologues of the *ran-1*$^+$ and the *mei-3*$^+$ genes exist in other eukaryotes has yet to be determined.

It has been claimed that in several organisms, including mice and wheat, the transition to meiosis is achieved gradually. This is supported by a progressive increase in the duration of the cell cycle, and more particularly the S-phase (Table 4.1), as well as change in chromosome volume. Consequently, the S-phase preceding meiosis is significantly longer in duration than the S-phase of mitotic cycles from the same organism. Similarly, in *Drosophila melanogaster*, Carpenter (1981) reports a gradual lengthening of the S-period in oogonia as meiosis is approached.

B PRE-MEIOTIC SYNTHESES

The meiotic cycle, like that of mitosis, involves a pre-meiotic period with a replication or S-phase. Indeed, entry into pre-meiotic S is the most reliable signal that a cell is likely to proceed to meiosis. Even so, it is not an infallible signal. Microsporocytes of the plants *Lilium* and *Trillium* can be induced experimentally to revert to mitosis by explanting them from anthers and into a culture medium during pre-meiotic S. Such explanted cells do not show an irreversible commitment to meiosis until late G_2 or leptotene (Stern & Hotta, 1969; Ito & Takegami, 1982). Clearly, in these organisms, pre-meiotic S is not sufficiently different from mitotic S to preclude a cell from entering mitosis and there is a distinction to be made between meiotic readiness and meiotic commitment (Riley & Flavell, 1977 and see Fig. 4.2).

As in mitosis, DNA synthesis occurs by means of multiple replicons which, after replication, are joined to form full length molecules. In yeast

Table 4.2. *DNA replication in amphibian cells*

Species	Cell type	S (h)	Distance apart of replication initiation points (μm)	Rate of replication ($\mu m\ h^{-1}$)
Triturus vulgaris	Blastula	1		
	Neurula	4	40	6
	Somatic		> 100	12
	Premeiotic	200	Several 100	
Triturus cristatus carnifex	Liver cell culture	24	175	20
Xenopus laevis	Ab cell line		57.5	9

After Callan, 1972, 1973.

the size range of the meiotic replicons is essentially similar to that of mitotic chromosomes (Johnston *et al.*, 1982). Here the activation of replicon origins must be staggered throughout most of pre-meiotic S to account for the fact that its duration is at least 2–3 times that of mitotic diploids (Williamson *et al.*, 1983). In the newt *Triturus*, on the other hand, there are fewer initiation sites, and hence larger replicons, in meiotic chromosomes compared to mitosis. In *T. vulgaris* the S-phase takes one hour in the blastula and four hours in the neurula whereas pre-meiotic S occupies 200 h (Callan, 1972, 1973). Coupled with this there is a difference in the length of the replicons (Table 4.2). Additionally, the bidirectional pattern of DNA replication which operates at mitosis is here replaced at meiosis by a unidirectional form. Moreover, while the greater bulk of the DNA within a genome is replicated at pre-meiotic S, in a small fraction it is delayed until prophase-1 (see Chapter 4D.2.2). Thus, different organisms evidently employ different means of regulating the length of the S-phase. In amphibians, replicon spacing appears to be the prime determinant of the duration of replication. In mammals it is the rate of fork migration. In plants, according to Dubey & Raman (1987), neither replicon spacing nor fork migration vary significantly. Here they claim that modulation of the S-phase depends on variation in the time of initiation of replication. This contrasts with the opinion of Holm (1977b) that the increased length of pre-meiotic S in the lily, which is some six times that of mitotic S, is due to the low number of replication origins.

While the programming of germ line cells may sometimes begin several cell generations before they develop meiotic readiness it is clear that the events which are critical for a commitment to meiosis occupy the interval from pre-meiotic S to zygotene. Histone synthesis is uncoupled from DNA synthesis (Bogdanov *et al.*, 1968) and occurs during the whole of

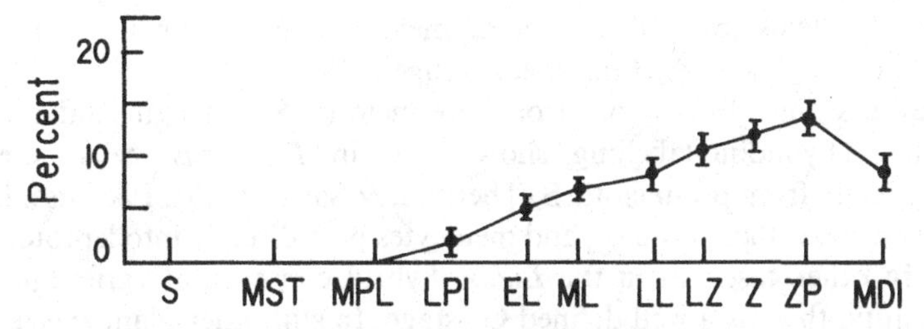

Fig. 4.3. The relative content of the specific meiotic histone fraction of *Lilium candidum* calculated as a percentage in excess of 100% of the somatic histone. The abscissa refers to the different stages of development with S = somatic tissue (leaves), MIT = mitoses in sporogeneous tissue, MPI = middle pre-meiotic interphase, LPI = late pre-meiotic interphase, EL = early leptotene, LL = late leptotene, LZ = leptotene–zygotene, Z = zygotene, ZP = zygotene–pachytene, MD1 = remainder of meiotic division-1 (after Strokov *et al.*, 1973 and with the permission of Springer-Verlag).

this period, though it is reduced after the onset of first prophase and it then involves the production of a specific meiotic histone (Liapunova & Babadjanian, 1973; Strokov, Bogdanov & Reznickova, 1973) which is formed in addition to the somatic histone fraction and not in place of it. The amount of additional material has been estimated as 15% in *Lilium* and 25% in the cricket. Its production declines after pachytene (Fig. 4.3) and, for this reason, it has been suggested that it serves for the formation of the lateral elements of the SC.

In male lepidopterans there are two pathways to meiosis. One is regular and is referred to as *eupyrene*. The other is irregular, leads to the production of anucleate sperm, and is termed *apyrene*. In apyrene meiosis no equatorial metaphase-1 plate forms. Instead, the chromosomes clump into irregular masses which remain connected by bridges at first ana–telophase. These subsequently break up into a series of chromatin fragments which move asynchronously to the poles and the unbalanced chromosome sets so generated are eventually discarded. The shift to apyrene meiosis is induced by a haemolymph factor which becomes functional towards pupation and then remains active throughout the pupal and adult stages. Consequently, whereas in young last instar male larvae meiosis is solely eupyrene, in old pupae it is exclusively apyrene, while in late last instar larvae and young pupae both categories occur together. Following the application of tritiated thymidine, Friedländer & Hauschteck-Jungen (1986) showed that in the coddling moth, *Cydia pomonella*, the pre-meiotic S-phase was shorter in apyrene cells. They suggest that this, in turn, leads to a shortened first prophase and so to a failure of the synthesis of meiosis-specific proteins and the consequent irregular behaviour of the chromosomes during the first meiotic division. In support of this they have been able to demonstrate that cytoplasmic lysine-rich proteins, present at regular meiosis, are lacking in apyrene meiocytes (Friedländer & Hauschteck-Jungen, 1982).

The events that follow on from pre-meiotic S differ in different organisms. Thymidine-labelling shows that in *Eremurus*, cells enter meiosis directly from pre-meiotic S (Therman & Sarto, 1977). Likewise, in *Triticum aestivum* there is no G_2 and meiocytes pass directly into leptotene from S. In other cases, as in the *Lilium* hybrid commercial variety (cv) Black Beauty, there is a well defined G_2-stage. In still other plant species, while there is a recognizable G_2, the chromosomes then enter a pre-leptotene contraction phase. This results in the production of a prochromosome condensation stage followed by a phase of despiralization prior to entry into leptotene. An equivalent pre-meiotic contraction phase is also known to occur in animals (Wilson, 1925; Church, 1972; John, 1976). In some such cases it is possible to see that the contracted chromosomes are clearly divided into two chromatids (Burns, 1972, figs 1*D* and 2). Walters (1978) argues that the pre-leptotene contraction phase represents a temporary reversion during an early stage of meiosis to a mitotic-like prophase state, a view reiterated by Bennett & Stern (1975). These authors find that pre-leptotene contraction occurs in both PMCs and EMCs of lily but that it is of much shorter duration in EMCs (0.7 d) compared to PMCs (1.16 d). Correlated with this there is less chromosome condensation.

There are a number of claims that either the synapsis of homologues is initiated at interphase preceding the onset of meiosis or else some association or alignment of homologues takes place prior to their eventual synapsis during meiosis and so acts to facilitate this process. This argument has been reinforced recently by the claim that chromosomes in somatic interphase nuclei are characterized by a mean spatial order determined by the juxtaposition of the most similar-sized chromosome arms and fixed about a single site of discontinuity (Bennett, 1982). Such a model represents a variation of an earlier suggestion by Shchapova (1971) that non-homologous chromosomes are associated at their telomeres during interphase so as to produce the best approximation to a constant intercentromeric distance. Either of these arrangements, if present in pre-meiotic nuclei, might be expected to influence meiotic pairing. Callow (1985) has questioned both the method of analysis and the evidence used in support of the Bennett model so that, like a number of its precursors, its relevance to meiotic pairing remains unconvincing. Dover & Riley (1977) had earlier suggested that the occurrence of pre-alignment during the early stages of meiotic development may be unique to those plant polyploids in which there is the additional problem of preventing homoeologues from pairing. Subsequent electron microscope studies have provided convincing evidence to discount even this suggestion (see Chapter 4C.5).

Table 4.3. *Effect of colchicine treatment on lily meiocytes*

Meiocyte category	Numbers of		Chiasmata/ bivalent			
	Bivalents	Univalents	1	2	3	> 3
Control	480	0	26	83	81	50
Colchicine treated	210	270	44	39	17	5

After Shepherd *et al.*, 1974.

At one time it was proposed that crossing-over leading to chiasma formation also occurred at pre-meiosis and in conjunction with DNA replication. The fact that colchicine treatment of wheat affected chiasma formation only if applied prior to the onset of meiosis was also initially assumed to support the concept of pre-synaptic homologue alignment (Bennett *et al.*, 1973b). In lily meiosis, however, Shepherd, Boothroyd & Stern (1974) were able to demonstrate that chiasma frequency could be affected by colchicine treatment after cells had entered meiosis. While such treatment abolished chiasmata completely in some bivalents, in others it only reduced them (Table 4.3), suggesting that drug treatment influenced zygotene synapsis. In line with this, only a small fraction of bivalents had complete SCs when treatment was carried out at leptotene. By contrast the SC was not affected in cells exposed during mid-zygotene to late pachytene. An even more direct demonstration that recombination does indeed occur at pachytene, as inferred indirectly from classical cytological observations and confirmed more recently by electron microscope studies, has come from the work of Borts *et al.* (1984). They used a pair of restriction endonuclease recognition-site polymorphisms to detect physically-recombined chromosome regions during meiosis.

Many of the arguments which have been used in support of the pre-meiotic pairing of homologues are based on associations involving heterochromatic chromosome regions. In relation to these arguments it is important to note the variable behaviour of such regions. Thus:

(1) While heterochromatic regions do indeed associate at pre-meiotic interphase in some organisms the evidence that these associations are homologous is not always convincing. Maguire (1967), for example, reported that homologous heterochromatic regions of maize were significantly nearer to one another than to equivalent heterologous regions at pre-meiotic interphase, tapetal mitotic interphase and root tip interphase. The regions involved were, however, all very small and not readily distinguishable so that the claim for homologous association was not convincing. By contrast, Palmer (1971) failed to

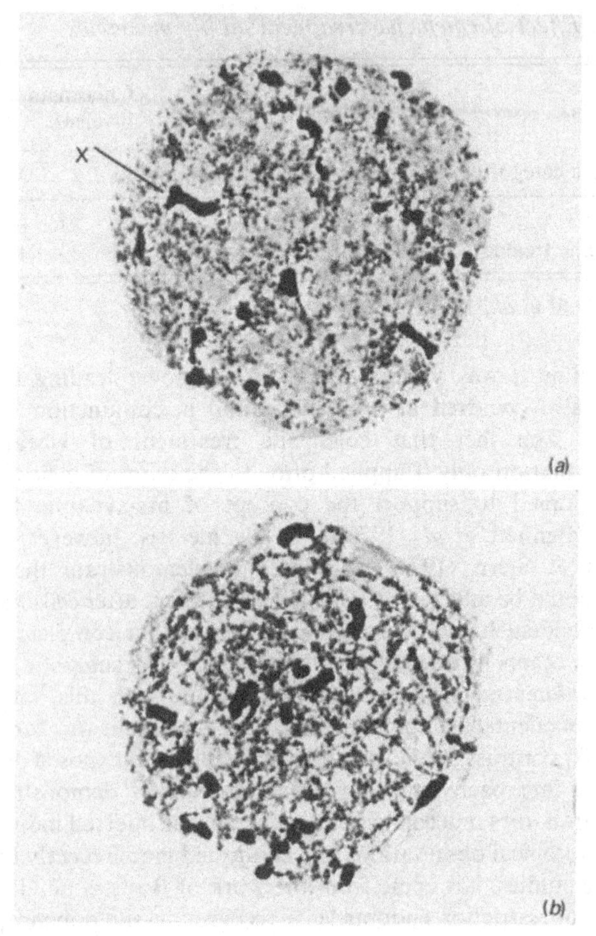

Figs 4.4. and 4.5. Premeiosis and early prophase-1 of male meiosis in the grasshopper *Stauroderus scalaris* (2n = 17♂, X0). In this species there are substantial blocks of procentric heterochromatin on all the autosomes (see Fig. 2.20). At pre-meiosis (Fig. 4.4*a*) and leptotene (Fig. 4.4*b*) all 16 autosomal heterochromatic blocks are unpaired and the single X-chromosome is also condensed. During zygotene the autosomal blocks of heterochromatin come into homologous alignment but do so only after homologous euchromatic regions have completed their pairing (Fig. 4.5*a*) so that by the end of zygotene most of them lie in parallel pairs (upper arrowhead, Fig. 4.5*b*). In the cell illustrated the heterochromatic regions of three of the smaller autosomal bivalents are also associated non-homologously at the base of the zygotene bouquet (lower arrowhead, Fig. 4.5*b*).

observe any tendency for the largest pair of homologues, chromosome number 1, or the satellited number 6 homologues, both of which are readily identified, to show any somatic association in either pre-meiotic mitosis of normal individuals or in the division of homozygous ameiotic mutants where meiosis is replaced by a synchronized mitosis.

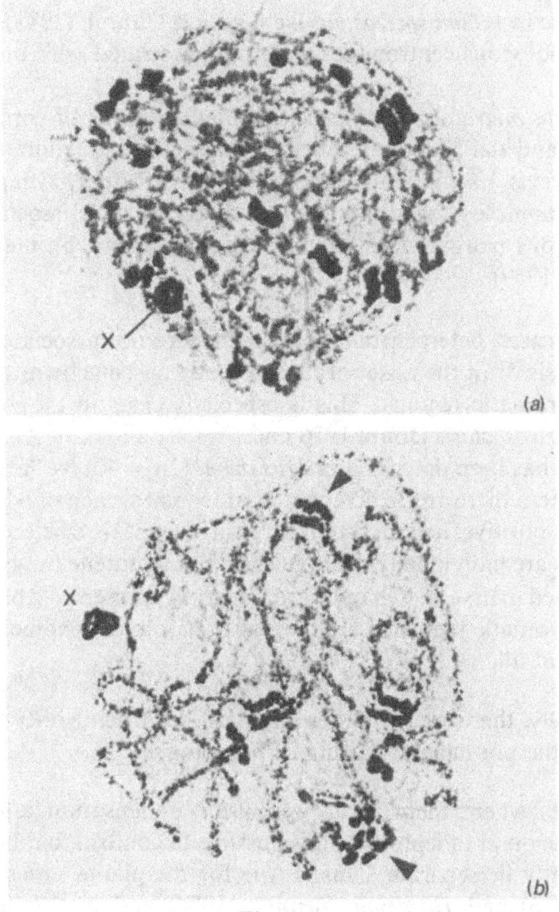

Fig. 4.5.

(2) While chromocentres involving heterochromatic regions of homologues are indeed present at pre-meiosis in some cases, as in the plant *Salvia* (Chauan & Abel, 1968), they lapse before the onset of meiosis. In the plant *Plantago ovata*, too, homologues are paired somatically in both the anther wall and the tapetum, as well as in young, unidentifiable, stages of the sporogenous cells and in the prophase preceding the first meiotic division. These prochromosomes, however, disappear late in pre-meiotic interphase as the cells increase in size and the chromosomes become diffuse (Stack & Brown, 1969). Likewise in *Allium fistulosum*, three-dimensional EM-reconstructions of centromere behaviour indicate that pre-meiotic associations of centromeres are resolved prior to meiotic synapsis and there is no indication that these associations are involved in any form of pre-zygotene alignment of homologues (Church & Moens, 1976). Similarly, in an electron

microscope study of centromere positions at mid- and late premeiotic interphase in *Lilium speciosum*, Del Fosse & Church (1981) concluded that homologous centromeres were not associated with one another.

(3) In *Locusta migratoria* the heterochromatic regions of chromosomes-1,3,4,11 and the X show a mutual nonspecific attraction which does not, however, facilitate synapsis at meiosis. Rather, synapsis occurs between homologues despite the occurrence of such associations, and homologous pairing is initiated outside the heterochromatic regions (Moens, 1969).

(4) In other cases, heterochromatic segments are not associated at either pre-meiosis or at the onset of meiosis and can be shown to pair after the euchromatic regions. This is especially clear in the grasshopper *Stauroderus scalaris* (John, 1976 and Figs 4.4 and 4.5). An equivalent situation has been described in *Mus dunni* (2n = 40) by Pathak & Hsu (1976). Here there are 38 acrocentric autosomes, each of which carries a C-band positive, heterochromatic, short arm. All 38 heterochromatic segments are individually distinguishable at leptotene but by zygotene 19 coupled pairs of C-bands are generally present. Moreover, the heterochromatic segments are the last to pair and, in some cases, may not pair at all.

Additionally, there are two other especially convincing lines of evidence that negate the pre-meiotic pairing of homologues:

(1) In species where there is a 'prophase' condensation after G_2 and before the onset of leptotene, it is possible to confirm that homologues consistently lie separate. This is true for the plants *Lilium* (Walters, 1970, 1976) and *Hordeum* (Bennett, 1984) and in the grasshopper *Melanoplus femur-rubrum* (Church, 1972).

(2) In the zygotic meiosis of the fungi *Neottiella* (Rosen & Westergaard, 1966) and *Coprinus* (Lu & Jeng, 1975), karyogamy, and hence meiosis, is delayed. Here the diploid state is restored only immediately prior to entry into meiosis so there can be no conceivable opportunity for any form of pre-meiotic pairing (von Wettstein, 1984).

C SYNAPTIC SPECIFICITY

C.1 The synaptonemal complex

The key event in meiosis, on which all other events subsequently depend, is the pairing of homologues or of homologous regions. In conventional diploid organisms this resolves the two parental genomes into a haploid number of bivalents. With the exception of some categories of achiasmate meiosis, pairing, as we have seen, is a complex process involving the elaboration of a tripartite intranuclear structure closely associated with the DNA of the chromosomes and referred to as the *synaptonemal complex* (SC) or, less frequently, as the *synapton* (Gillies, Rasmussen & von Wettstein, 1974).

Two rather different ultrastructural techniques have been employed to study the SC. The initial approach, which dates back to its independent discovery by Moses and Fawcett in 1956, employed three-dimensional reconstruction from serial sections. In 1973 Counce & Meyer introduced a simpler, and more rapid, microspreading technique. If animal meiocytes are spread on the surface of a saline solution, fixed with formaldehyde and then treated with phosphotungstic acid, the SC is selectively stained because of its proteinaceous nature. The centromere region is also visible as a lightly stained ball of fine fibrils through which the SC passes uninterrupted. Otherwise the chromatin associated with the SC is dispersed and not visible. More recently, microspreading has been coupled with silver staining (Fig. 4.6) and techniques have been developed for spreading plant meiocytes by hypotonic bursting, which overcomes the difficulty imposed by their rigid cell wall (Stack & Anderson, 1986a). While the reconstruction of serial sections involving whole nuclei has the advantage of preserving structures *in situ*, it is a laborious, exacting and time consuming process. The microspreading method is both rapid and direct and so allows for the analysis of many more nuclei, though it has the inherent disadvantage of producing a one-dimensional end product from a three-dimensional structure.

Despite the fact that it has now been studied over a 30-year period, there is still controversy over several aspects of the formation and function of the SC. In its definitive form the complex consists of two lateral elements, *c.* 40 nm thick, separated by a medial space, 100–300 nm wide, which contains a central element, 10–20 nm thick, coupled to each lateral element by a series of transverse filaments each 2 nm wide (Fig. 4.7). The precise chemical composition of the SC is still not well defined. Since the SC remains intact after DNase digestion (Fig. 4.7), but is completely destroyed by trypsin, there are good grounds for concluding it is predominantly proteinaceous (Comings & Okada, 1970, 1972b). Moreover, with the exception of the alignment of the lateral elements, each step in the synthesis and assembly of the complex in *Coprinus lagopus* has been shown to be preceded by protein synthesis (Lu, 1984). When explanted lily meiocytes were cultured in the presence of the protein inhibitor cyclohexamide during the synaptic phase of prophase-1 the bulk of the lateral elements failed to form. In explanted cells treated in early zygotene, SCs reformed once meiocytes were removed from cyclohexamide but failed to do so if removed at late zygotene or early pachytene. The

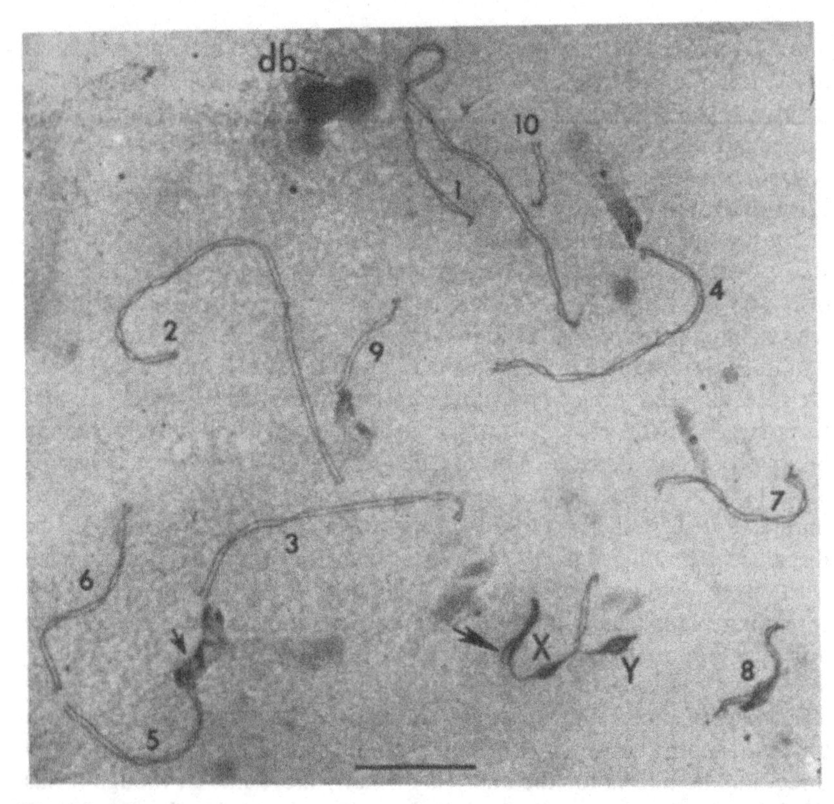

Fig. 4.6. Electron micrograph of a silver-stained, surface spread, spermatocyte of the Chinese hamster *Cricetulus griseus* at mid-pachytene of meiosis-1. The individual autosomal synaptonemal complexes are numbered according to ranked length and in decreasing order of size (1–10). Note the thickening of the unpaired portions of the synaptonemal complexes in the XY bivalent and the nucleoli associated with the ends of the complexes in bivalents 3, 4, 5, 8 and 9 (photograph kindly supplied by Dr Montrose Moses).

chromosomes then desynapsed and no chiasmata formed (Roth & Parchman, 1971). Evidence from selective staining, coupled with differential extraction following enzyme treatment, indicates that the SC is composed predominantly of arginine-rich, non-histone protein but, as yet, no unique protein, characteristic of the SC, has been identified (Dresser, 1987). This, however, may soon be rectified since monoclonal antibodies have now been developed which recognize specific SC-proteins (Moens *et al.*, 1987). Monoclonal antibody 1152F10, for example, identifies two major components of the lateral elements of the SC (Heyting *et al.*, 1988).

Using anti-smooth muscle myosin and anti-actin antibodies, De Martino *et al.* (1980) claimed to have demonstrated that part of the SC protein included actin and myosin. Specifically, actin was identified at attachment points of the SC to the inner nuclear membrane while myosin

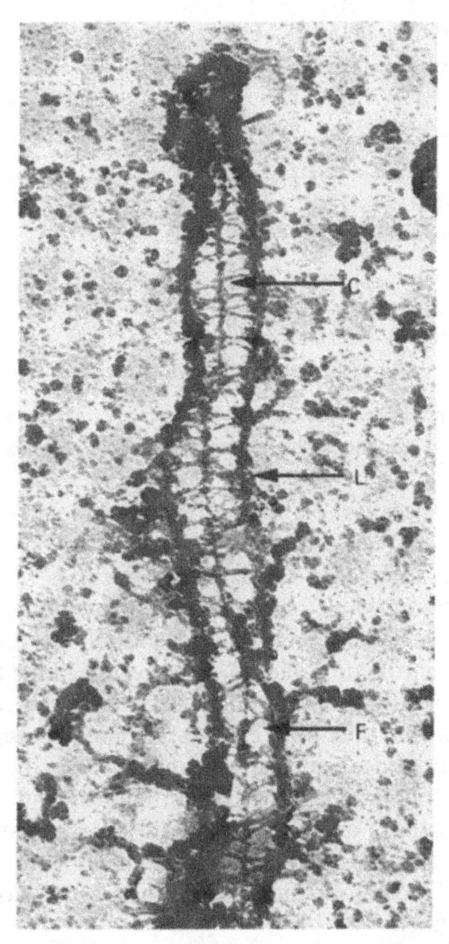

Fig. 4.7. Electron micrograph of a DNase-treated whole mount of a synaptonemal complex from a spermatocyte of the quail. The treatment has removed essentially all of the chromatin, exposing the lateral and central elements of the complex as well as the transverse fibers that interconnect them (photograph kindly supplied by Dr David Comings).

was reported as filaments along the pairing portions of the homologues. Spyropoulos & Moens (1984), however, were unable to confirm this claim in mouse SCs where the actin-specific stain, phallacidin, failed to produce any fluorescence in microspread pachytene nuclei.

Although leptotene chromosomes most commonly appear single-stranded with light microscopy they must each consist of two chromatids since they are in a post-replicative state. In the fungus *Neottiella* the strongly condensed chromatin of the unpaired leptotene chromosomes makes it possible to distinguish the two sister chromatids as discrete entities in the electron microscope whereas this is not possible in most

Table 4.4. *Relationship between SC length and DNA content of the genome*

Species	SC length (μm)	1C-DNA content (mm)	% SC/DNA
Neurospora crassa	50	16	0.3
Drosophila melanogaster	110	61	0.2
Bombyx mori	212	170	0.12
Physarum DK2	191	20	0.09
Zea mays	353	2500	0.015
Cricetulus griseus	130	1070	0.012
Homo sapiens	231	2300	0.01
Lilium	3700	60000	0.006

After Lie & Laane, 1982 and with the permission of *Hereditas*.

other organisms because of the diffuse nature of their chromatin (Westergaard & von Wettstein, 1972). Comings & Okada (1971b) were, however, able to show that the lateral element is composed of two filaments after treating whole-mount preparations of quail meiocytes with DNase. At leptotene, the lateral components of the future SC appear between the sister chromatids of each chromosome in the form of an axial core. If, as is generally claimed, only one axial element is present per leptotene chromosome then both chromatids must be associated with it. Wahrman (1981) provides evidence that, at pachytene, the lateral elements are sometimes visibly double, especially in microspread cells. In regions where homologous have undergone synapsis the chromatin fibers are no longer symmetrically disposed around the axial core as they are at leptotene. Rather, the pairing faces, where the SCs abut one another, are relatively free of fibers.

At the earliest leptotene stage the SCs are discontinuous and appear as a large number of individual pieces (Holm, 1977a). Precisely where and how these pieces originate is not clear. King (1968) and von Wettstein (1977) proposed that the formation of the SC depended on a special class of polynucleotide sequences, scattered along each chromatid, which were then brought into linear alignment by the folding of the DNA, but this remains nothing more than speculation. It has recently been discounted by Heyting *et al.* (1988) who conclude that the lateral elements of SCs are newly synthesized during zygotene and are not assembled by a rearrangement of pre-existing components. What is clear is that, while the SC normally joins homologues along their entire length, the length ratios of SC/DNA content indicate that only a small amount of chromosomal DNA is associated with each lateral element (Table 4.4). This means that the mutual accessibility of homologous DNA stretches must be quite severely restricted. For this reason Stern, Westergaard & von Wettstein

(1975) have argued that there must be some form of preselection of the DNA stretches that ultimately become incorporated into the SC. On the assumption that crossing-over occurs within and across the SC this implies there must also be a preselection of crossover sites well before pairing so that specific homologous DNA stretches become trapped within the SC at synapsis (see also pp. 133–187).

The pairing of homologues which leads to zygotene bivalent formation generally occurs in two steps. First, there is a gross pairing phase which brings homologues into rough alignment and to within 300 nm of one another. The second step involves a more precise alignment of homologues to within 100 nm following the assembly of the SC. From observations on chromosome pairing in the grasshopper *Chloealtis conspersa*, Moens (1985) theorized that specific elements, which he referred to as *synaptic units*, are present prior to SC formation. Subsequently, Rasmussen (1986) showed that in microspread spermatocytes of the silkmoth *Bombyx mori* the primary recognition between homologues was effected by short, subterminal sites located some 1–3 μm from the telomeres. This was followed by the formation of the SC itself which began from the closely associated telomeric attachment plaques of two homologues, already matched by such subtelomeric associations at one or both ends, and then developed progressively to the nearest association site. After a short delay at that site it then proceeded to completion unless impeded mechanically in some way. Rasmussen's observations imply that homologue recognition, the acquisition of pairing competence and SC formation are separate events of synapsis. This means that the specific pairing of homologues required for SC formation does not depend on any absolute requirement for homology but is a consequence of short subterminal recognition sites.

To what extent random chromosome movement plays a part in achieving initial synaptic alignment is not clear. In many cases this step appears to be related to telomere recognition and movement on the inner side of the nuclear membrane. This tendency is most pronounced in organisms where all the telomeres group together at a site on the nuclear envelope adjacent to the centrosome, where one is present. *Rabl orientation* of the chromosomes is a constant feature of telophase organization. It develops as a consequence of the orientation of the centromeres to the spindle poles, with the telomeres lying indifferently in the other half of the nucleus (Rabl, 1885). At first meiotic prophase, however, there is commonly an abrupt change in the orientation of centromeres and telomeres. This may be achieved either by a rotation of the nucleus after the last gonial division or else by the migration of the centrosome from the centromeric to the telomeric area of the nucleus. Either of these events results in a bouquet polarization in which the telomeres lie in a restricted area of the nucleus. In the case of acro- or telocentric chromosomes, the

Table 4.5. *Location of synaptonemal complex (SC) initiation sites*

EM technique	Species	Bivalent type[a]	SC initiation sites per bivalent	Reference
Microspreading	*Peromyscus maniculatus* and *P. sitkensis*	M/A	Single, distal	Greenbaum, Hale & Fuxa, 1986
	Cricetulus griseus	M/A	Single, distal	Moses, 1977a
	Locusta migratoria and *Schistocerca gregaria*	A/T	Single, proximal or distal, some interstitial	Jones & Croft, 1986
	Secale cereale	M	Multiple	Gillies, 1985
	Tradescantia sp.	M	Multiple	Hasenkampf, 1984
	Aedes aegypti	M	Multiple, 9 per bivalent	Wandall & Svendsen, 1985
Sectioning	*Bombyx mori*	holokinetic	1–2, distal	Rasmussen, 1976
	Homo sapiens	M/A	1–2, distal or interstitial	Bojko, 1983
	Pales ferruginea	M	Multiple	Fuge, 1979
	Mesostoma ehrenbergii ehrenbergii	M/A	Multiple	Oakley, 1982
	Lilium longiflorum	M	Multiple (4–36 per bivalent)	Holm, 1977a
	Zea mays	M	Multiple	Gillies, 1975

[a] M = metacentric, A = acrocentric, T = telocentric.

centric end appears also to behave like a telomere so that in some cases, e.g. the pig (Schwarzacher, Mayr & Schweizer, 1984), the bouquet includes a mixture of centric and telomeric ends. In rye, *Secale cereale*, there are substantial terminal blocks of heterochromatin on most of the seven pairs of meta–submetacentric chromosomes. These blocks give positive C-banding and, using C-banded preparations, Thomas & Kaltsikes (1976) were able to show that at pre-meiotic interphase these terminal C-bands all lay in one half of the nucleus, reflecting a Rabl orientation. A similar arrangement was evident at the onset of meiosis but this was followed during leptotene by a pronounced aggregation of the heterochromatic telomeres. This led to the production of a single large chromocentre which formed in relation to the *polar determinant*, or *organizing centre*, located in the cytoplasm of the PMC adjacent to the *tapetum*, which functions as an attachment site for the telomeres. The production of this chromocentre can be inhibited by colchicine pre-treatment but normally it persists well into zygotene, eventually breaking up at the onset of pachytene. At diplotene there are numerous end-to-end associations of bivalents. These involve both heterochromatic and euchromatic ends implying that non-heterochromatic telomeres must also be included in the original chromocentre. An alternative type of configuration, associated with the early stages of meiosis in many plants is the *synizetic knot*, a dense aggregation of chromatin at one side of the nucleus. This may represent an alternative and more cryptic form of expression of the bouquet arrangement found in many animals.

While the bouquet may facilitate synapsis it cannot be a prerequisite for it since it is absent, for example, in *Caenorhabditis elegans* (Goldstein, 1984). Added to this, while SC initiation sites are single and terminal in some species there are multiple initiation sites, including interstitial regions, in others (Table 4.5). Interstitial SC segments are occasionally present in the grasshopper *Chloealtis conspersa* where the subtelomeric initiation of synapsis is the general rule (Moens, 1985). Rasmussen (1986) interprets this to mean that interstitial initiation sites for synapsis exist in this species but are not usually expressed.

A second situation in which no bouquet stage develops is in species which are characterized by intense somatic association. In the dipteran *Aedes aegypti*, for example, visible leptotene and zygotene stages are absent from male meiosis. Rather, homologues retain the intimate somatic pairing of the previous mitotic division, which characterizes all mitoses in this organism (Akstein, 1962). Moreover, the two chromatids of each homologue are not discernible at any point during the entire first meiotic prophase and the earliest stage observed, which resembles pachytene in other eukaryotes, has been referred to simply as pre-pachytene. Even so, an SC develops despite the fact that no axial cores appear prior to pachytene. Here, the SC is formed gradually by the simultaneous production, between pairing faces, of both lateral elements and transverse filaments (Wandall & Svendsen, 1985). A similar behaviour characterizes other dipterans including *Glossina* (Craig-Cameron *et al.*, 1973), *Drosophila* (Rasmussen, 1974) and *Pales* (Fuge, 1979).

Prior to SC formation the two sister chromatids of a homologue rotate relative to the axial core so that it is relocated lateral to them. The core is thus exposed on one surface forming a potential pairing face. In this condition it is free to bind with the precursor material of the central region, which also becomes available at this time. The association of the two lateral elements with the central element then produces the characteristic tripartite complex in which the two lateral elements are now aligned in register with one another. The transverse filaments thus serve to align and join the homologues and so stabilize them at a fixed distance apart.

The biogenesis of the central element has been, indeed still is, a contentious issue. One suggestion is that it is produced locally from the two lateral elements during the pairing process. This envisages that the transverse filaments form as extensions of the axial cores and interdigitate locally to form the central element. This view is supported by Comings & Okada (1971b) and by Moens (1973). An alternative proposal is that the precursor material for the central element is synthesized and assembled within the nucleolus and then transferred to the developing SC (von Wettstein, 1977). In the yeast *Saccharomyces cerevisiae* it has been possible to analyse SC formation by using mutations in which normal meiosis occurs but at a relatively slow rate. With this approach it has been shown that, before the onset of pre-meiotic S, many yeast cells contain a dense body which is not present in vegetative cells. Concurrent with pre-meiotic DNA synthesis, these bodies can be shown to contain structures reminiscent of the central elements of the SC. They have therefore been described as *presynaptons*. Zickler (1973) also notes the presence of synaptic structures in the nucleolus of the fungus *Podospora*. Additionally, there is evidence from the plant *Fritillaria* that components of the central element are synthesized in the nucleolus and then transported to, and laid down between, the pairing homologues. A variant of this form of assembly has been described in the nematode *Ascaris* (Bogdanov, 1977). Here there is a temporary accumulation of central region material in the form of cytoplasmically-located leptotene polycomplexes. These disappear as the SC is assembled.

Two studies claim that SCs are present in pre-meiotic cells. McQuade & Bassett (1977) reported them at pre-meiotic interphase in hexaploid wheat. Day & Grell (1976) made an equivalent claim for female *Drosophila melanogaster*. Both claims are generally assumed to be in error. The staging procedure used by McQuade & Bassett has been shown to be unreliable (Bennett, Toledo & Stern, 1979) while, in female *Drosophila*, Carpenter (1981) finds bulk DNA synthesis to precede SC formation and not to overlap it. Zickler (1973) reported rare SC-like structures in mitotic ascogenous hyphae of the ascomycete *Ascobolus* and this situation remains to be resolved. In *Podospora anserina*, which forms only one crossover per bivalent arm, no SCs are present at meiosis. A similar state of affairs exists in the fission yeast, *Schizosaccharomyces pombe*, and the fungus *Aspergillus nidulans* (Egel-Mitani, Olson & Egel, 1982), two additional ascomycetes which show normal pairing despite the absence of the SC and where, again, most chromosome arms have only one crossover. On this basis the authors conclude that SCs are not essential for crossing-over but might in some way be responsible for crossover interference which is lacking in these fungi. By contrast, monochiasmate short bivalents in higher eukaryotes show regular SC formation. Whether the absence of an SC in these fungi is the result of a technical artifact, stems from a fundamentally different chromosome ultrastructure, or simply relates to the fact that short SCs form only very locally, and so have escaped detection, has yet to be clarified.

In typical chiasmate meiosis the SC is eliminated after crossing-over has occurred, though short stretches are sometimes retained at putative chiasma sites. Solari (1970), for example, has described a convergence of lateral elements at diplotene in EM sections. In such regions, either there was a piece of retained central element present at early diplotene or, more commonly, the two axes came near to one another with the space between them becoming filled with chromatin fibrils. Solari believes that the former condition gradually transforms into the latter and argues that these regions where the axes converge represent chiasmata. Similar presumed chiasma-related remnants have been observed in water-spread mouse spermatocytes (Moses, 1977b). In *Neottiella*, too, there are short, structurally modified stretches of SC which are retained until late diplotene. Thereafter, they are eliminated and each 'chiasma' then consists of a continuous chromatin 'bridge'. Von Wettstein, Rasmussen & Holm (1984) therefore argue that an exchange of DNA occurs within the region of the SC at which a chiasma forms. When isolated rat spermatocyte SCs are treated with DNase II a fraction of chromatin is protected from digestion through its association with the chromosome cores (see Chapter 2A.1). If, as Moens & Pearlman (1988) have assumed, this SC-DNA involves core binding sequences, and if recombinant events occur only within the SC, then two rather different models can be envisaged to account for the selection of crossover sites. The one involves a preselection of such sites ahead of SC formation. This means that genes which lie at the SC in some cells of a given species may be in the chromatin loops of other meiocytes of the same species. The other model requires a more dynamic form of site selection in which gene loci which are in the loops at one time may be at the SC at a later time in the same cell. The occurrence of triple pairing involving two fully formed SCs (see Chapter 4C.5) argues against the SC-DNA representing the actual site of molecular pairing

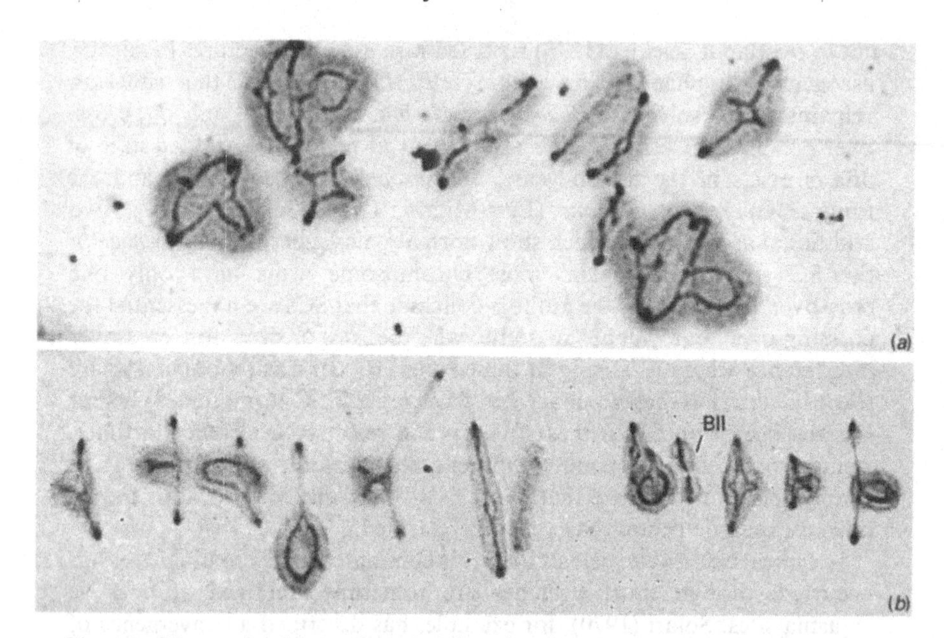

Fig. 4.8. Silver-stained metaphase-1 male meiocytes of two species of grasshopper (*a*) *Chorthippus juncundus* with 2n = 17♂, X0 and (*b*) *Arcyptera fusca* with 2n = 23♂, X0 plus two supernumerary (B) chromosomes. Both centromeres and chromatid cores are differentially stained. In autosomal bivalents the chromatin cores of sister chromatids are intimately associated except at chiasma points. In the bivalent formed by the two B-chromosomes, however, there is no actual contact between the cores of the two homologues although they are terminally associated (photographs kindly supplied by Dr Julio Rufas).

since there is no way for three DNA molecules to simultaneously undergo base pairing. From an analysis of the amount of DNA that can be sequestered in the SC, relative to the known recombination frequency, Lamb (1977) concluded that recombination must occur in regions of DNA that are not incorporated in the SC.

Moses (1977b) states quite categorically that there is no exchange of SC lateral elements at crossing-over. Wahrman (1981), on the other hand, has produced photographs of diplotene microspreads from the rodent *Apodemus mystacinus* which, he believes, demonstrate that the lateral elements of the SC are involved in chiasmata. This, if correct, would imply that the SC scaffold plays a role not only in the pairing process but also in the stabilization of the exchange event itself. The situation is complicated by the recent demonstration that a non-histone proteinaceous axis, which appears to represent the scaffold of mitotic chromosomes, can be identified in meiotic bivalents following silver staining and that this axis is most certainly involved in the structure of the chiasma (Fig. 4.8).

Shedding of the SC at the onset of diplotene always starts with the central region. However, with regard to the disappearance of the lateral components and the subsequent fate of the detached parts of the complex, different organisms appear to have evolved different strategies. These include:

(1) the shedding of both lateral and central elements as a structurally amorphous mass which is then degraded (*Neottiella*, *Lilium*, *Helix*);

(2) retention of the lateral component which returns to a position between the two sister chromatids (mouse, rat, man);

(3) the formation of multiple sheets of aggregated SCs (polycomplexes), frequently arranged into paracrystalline arrays which may be found free in the nucleoplasm (*Aedes*, *Ornithogalum*) or else trapped in the chromatin of the bivalents until first anaphase (*Chorthippus*).

Von Wettstein *et al.* (1984) suggest that, where SC material persists between sister chromatids until first anaphase, it may prevent chiasma terminalization. By contrast, in cases where the SC is completely eliminated, at or before diakinesis, chiasmata may be free to terminalize. However, the supporting evidence on which they base this suggestion is not particularly convincing. In *Bombyx mori*, coincident with the formation of the SC, the chromatin fibers undergo an orderly folding during zygotene. When this process is completed the chromatin loops extend radially and fairly evenly from the completely formed axial elements of the SC and are homogeneously distributed along the length of the homologues. A majority of these chromatin loops are transcriptively inactive but some do transcribe (Rattner, Goldsmith & Hamkalo, 1980). The SC is usually disrupted in regions where active loops are present. Some nucleolar RNA synthesis occurs during pachytene in the lily and, here too, these active RNA cistrons are excluded from the SC.

C.2 Interlocking

Traditionally, interlocking between meiotic associations has been considered to be a rare phenomenon since interlocked configurations are not often seen with a light microscope. There are, however, several known exceptions. For example, in the grasshopper *Chloealtis conspersa*, where pairing and SC formation is often incomplete in the long metacentrics, interlocking can sometimes be relatively common (Fig. 4.9). The frequency of interlocking is also normally low in bread wheat. However, as the dosage of the *Ph-1* gene in the long arm of chromosome 5B (5BL) increases from 2–6, by modifying the dose of 5BL, the frequency of interlocking increases (Fig. 4.10). Reducing the dose of 5BL also increases the frequency of interlocking (Yacobi, Mello-Sampayo & Feldman, 1982). No lines aneuploid for chromosomes other than those of group 5 exert any effect on interlock frequency but plants with 2–4 doses of 5BS (short

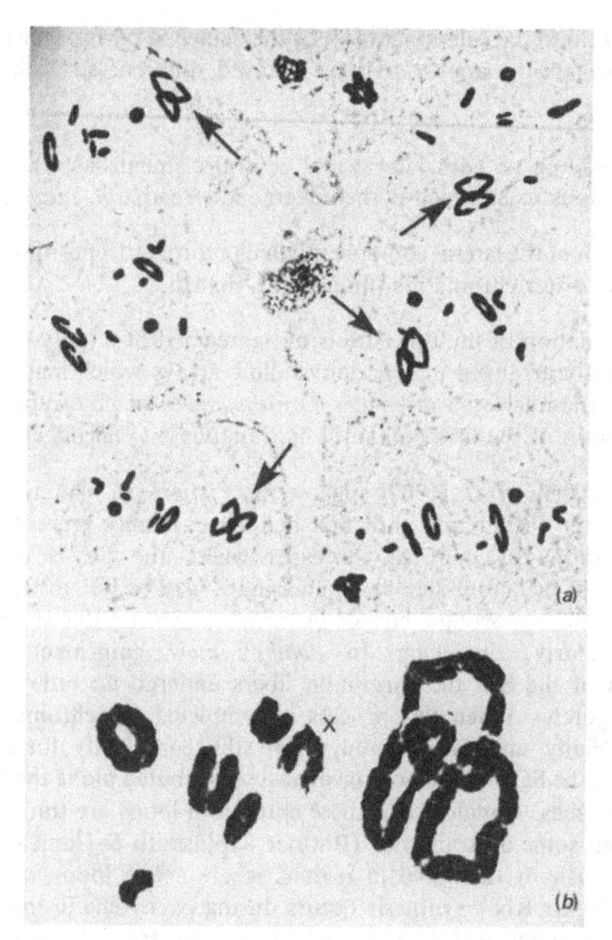

Fig. 4.9. Interlocked bivalents at metaphase-1 of male meiosis in the grasshopper *Chloealtis conspersa* (2n = 17♂, X0). In (*a*) four of the six cells illustrated have two interlocked metacentric bivalents (arrowheads). In (*b*) all three of the large metacentric bivalents are interlocked.

arm of chromosome 5B) have a lower frequency of interlocking than do those with 05BS at all levels of 5BL. Chromosomes 5A and 5D show comparable effects. That is, increases or decreases in the dose of *Ph-1* on 5BL increase the frequency of interlocking whereas 2–4 doses of 5BS, 5A and 5D all reduce interlock frequency. Deficiencies of 5A, on the other hand, give rise to an unusually high percentage of cells with interlocks.

Rather surprisingly when diploid meiocytes of male *Bombyx* were analysed by serial sectioning and three-dimensional EM reconstruction, SC interlocking was found to be a regular feature of zygotene pairing (Holm & Rasmussen, 1980). Two distinct forms of interlocking were evident (Fig. 4.11): (1) chromosome interlocks, where one lateral component of a partially paired bivalent was trapped between the two

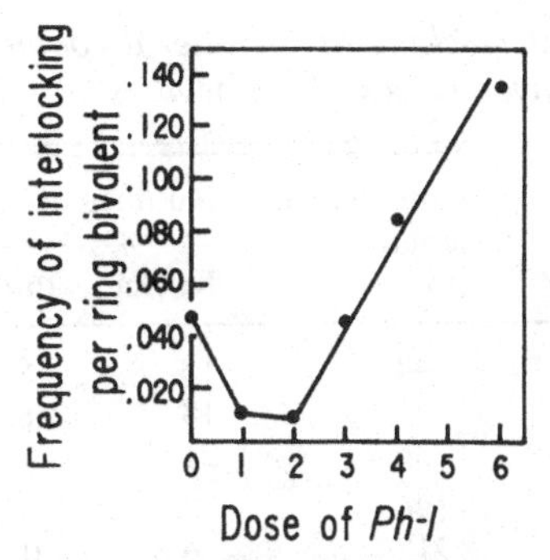

Fig. 4.10. The influence of different doses of the *Ph-1* gene of *Triticum aestivum* on the frequency of chromosome interlocking, i.e. on the trapping of a chromosome of one bivalent between the two homologues of another bivalent (after Yacobi *et al.*, 1982 and with the permission of Springer-Verlag).

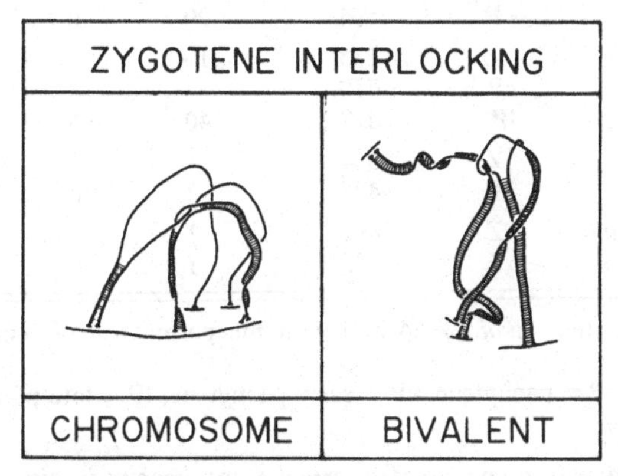

Fig. 4.11. The two categories of zygotene interlocking (after Rasmussen & Holm, 1980 and with the permission of *Hereditas*).

homologues of a second partially paired bivalent; and (2) bivalent interlocks, where both the lateral components of an unpaired interstitial segment were trapped in another bivalent.

Comparable interlocks have since been described in most organisms where complete reconstructions of zygotene nuclei have been carried out (Table 4.6). Indeed, Holm (1986) states that interlocking is a regular feature of zygotene pairing in all organisms (but see p. 151).

Chromosome and bivalent breaks, leading to the formation of major discontinuities of lateral components and SCs were also frequent at late zygotene in male *Bombyx* (Tables 4.6 and 4.7). Most, if not all, of these represent intermediates in the resolution of interlocks which implies that this process involves chromosome and bivalent breakage followed by repair. An analysis of the break–repair resolution process has, undoubtedly, been hampered in SC microspreads by the fact that chromosomes are represented only by their lateral components so that

Table 4.6. *Frequency of interlocking of the lateral components of the synaptonemal complex in five species of eukaryotes*

Species		Stage[a]	Complement length (m)	Number of		
				Nuclei	Interlocks	Breaks
Coprinus cinereus		Z	44	13	5	15
		eP	42	18	1	0
		lP	36	21	0	0
Bombyx mori	♀	Z	196	4	6	0
		P	212	6	0	0
	♂	Z	202	8	32	31
		eP	198	16	2	6
		lP	257	14	0	0
Homo sapiens	♀	Z	465	2	10	8
		P	504	20	0	0
	♂	Z	235	10	7	8
		eP	210	33	3	4
		lP	227	40	5	6
Secale cereale		Z	554	1	7	0
		P	485	3	0	0
Triticum aestivum		Z	1892	1	4	8
		P	1235	1	0	4

After von Wettstein *et al.*, 1984 and with the permission of Annual Reviews Incorporated.
[a] Z = zygotene, P = pachytene, eP = early pachytene, lP = late pachytene.

Table 4.7. *Mean frequencies of interlocks and breaks in eight spermatocytes of Bombyx mori*

Interlocks			Breaks		
Chromosome	Bivalent	Total	Chromosome	Bivalent	Total
1.6	2.4	4.0	1.6	2.1	3.8

After Rasmussen & Holm, 1982.

events within the chromatin itself cannot be assessed (Rasmussen, 1986). Certainly, the break–repair mechanism must be a very specific and efficient process as judged by the absence of rearranged chromosomes. Thus, in rye, three chromosome and four bivalent interlocks were found in one reconstructed zygotene nucleus but by pachytene these had been completely resolved so that seven bivalents were invariably present (Abirached-Darmency, Zickler & Cauderon, 1983).

The frequency of interlocking in the two human oocytes so far analysed was more than five times higher than in human spermatocytes (Bojko, 1983, 1985). Whether this depends simply on the fact that the SCs of oocytes are twice as long as those of spermatocytes is not known. In male *Bombyx* there was no correlation between the length of a bivalent and the frequency with which it interlocked with other chromosomes or bivalents (Holm & Rasmussen, 1980).

The breakage and resolution of interlocks is not related to crossing-over. Holm & Rasmussen (1980) speculate that it might be accomplished by type II DNA topoisomerase which is known to be able to accommodate the passage of one double helix through another by a transient double-strand break.

C.3 Synaptic adjustment

In mutant individuals heterozygous for structural chromosome rearrangements, or in species which are polymorphic for such rearrangements, zygotene synapsis is restricted to homologous regions (*homosynapsis*). When complete this leads to the formation of loop configurations in the case of heterozygous inversions, buckles in the case of duplication-deficiency heterozygotes and multiple configurations in the case of heterozygous translocations. Similarly, in polyploid individuals, with three or more sets of identical (*homologous*) or similar (*homoeologous*) chromosomes, multivalent configurations may arise given a sufficiently high cell chiasma frequency (see Chapter 2A.4).

Microspreading of SCs has, however, revealed an entirely unsuspected phenomenon, namely that a second phase of synapsis, which is indifferent to homology, may follow early pachytene. Moses *et al.* (1978), who first observed this heterosynapsis referred to it as *synaptic adjustment*. Such an adjustment results in duplication buckles and inversion loops being eliminated to produce straight, though non-homologously-paired, SCs by late pachytene.

Thus, in male mice heterozygous for the paracentric inversions In(1)IRk and In(2)5Rk, characteristic SC inversion loops are consistently formed at synapsis as observed by EM microspreads. These loops are equally consistently eliminated by late pachytene (Poorman *et al.*, 1981; Moses *et al.*, 1982). No asynapsis (*synaptic failure*) or antiparallel synapsis of the relatively inverted homologues was observed in such mice. Synaptic adjustment thus involves a desynapsis of the homologous SCs within the loop and a subsequent resynapsis of non-homologous regions leading to the complete obliteration of the inversion loop in all meiocytes (Fig. 4.12). A comparable desynapsis of homosynapsed regions occurs also in duplication heterozygotes but here it is accompanied by a differential shrinkage of the long axis to match the shorter one (Fig. 4.13).

In male hybrids (2n = 55) between *Lemur fulvus fulvus* (2n = 60) and *L.f. collaris* (2n = 51), 10 acrocentric chromosomes correspond to the

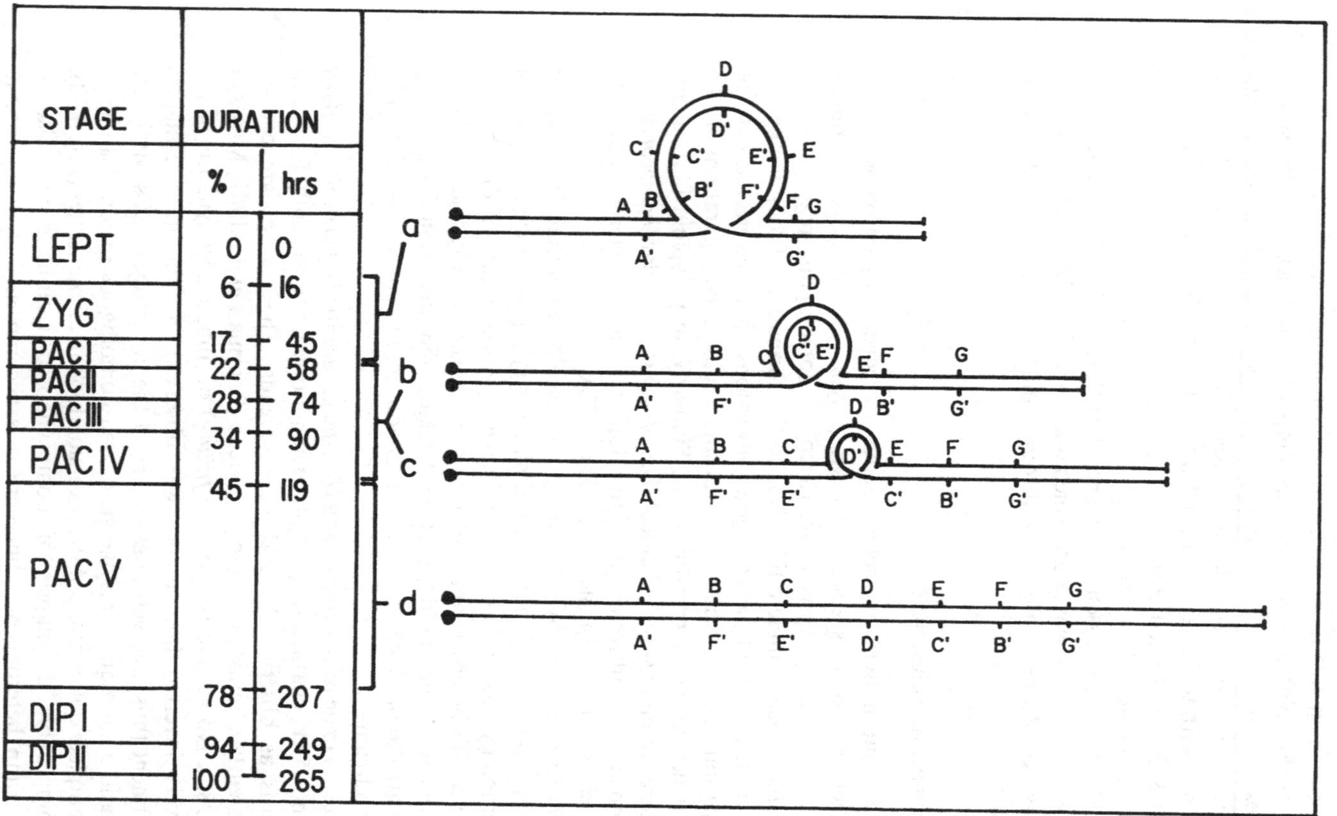

Fig. 4.12. Diagrammatic representation of the process of synaptic adjustment during prophase-1 leading to the abolition of a heterozygous paracentric inversion loop in *Mus musculus* with consequent heterosynapsis. The stages of prophase, and their duration, have been calculated in terms of their frequency per 1000 meiocytes assuming a duration of 265 h for the period leptotene to diplotene (after Moses & Poorman, 1984 and with the permission of Allen and Unwin).

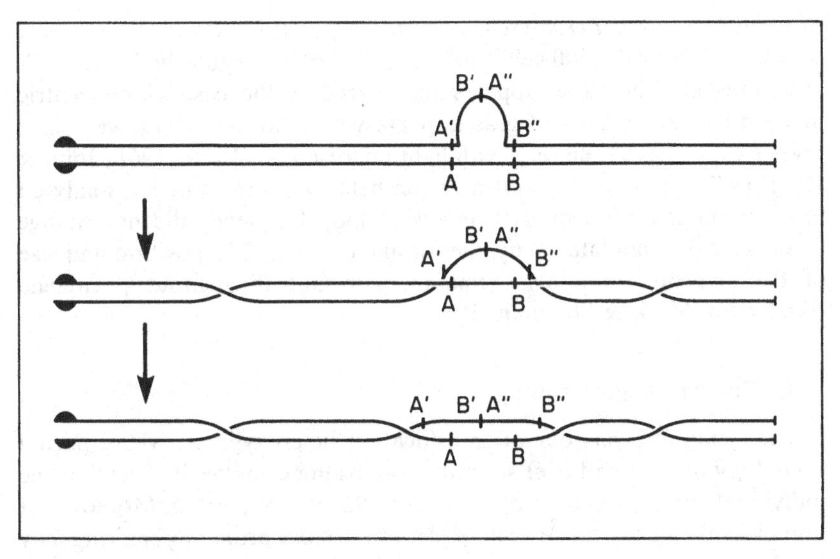

Fig. 4.13. Synaptic adjustment of the synaptonemal complex of a heterozygous tandem duplication. Homosynapsis at zygotene leads to an unpaired loop containing portions of the duplicate region. Shortening of the long axis of the homologue carrying the duplication then leads to the formation of relational twists between the homologous lateral elements of the complex. Initially these twists are asymmetrical but as equalization of the two homologues is completed they become symmetrical (after Moses & Poorman, 1981 and with the permission of Springer-Verlag).

individual arms of five metacentrics in terms of G-band pattern. Consequently, five chain of three multiples form at meiosis. Similarly, in *L.f. collaris* itself there is one such chain of three. In each of these multiples the short arms of the two acrocentrics involved are non-homologous and so do not pair at zygotene. Despite this they are fully paired at late pachytene as a result of synaptic adjustment (Moses, Karatsis & Hamilton, 1979).

Synaptic adjustment implies some form of topological instability of the SC as a result of the heterozygous rearrangement. In individuals heterozygous for a reciprocal translocation, on the other hand, while there may be some synaptic shift involving a small amount of heterosynapsis, the pairing configuration is sufficiently stable to preclude complete desynapsis and resynapsis. Indeed, synaptic adjustment is not a universal feature of all inversions. Some of the more complex inversions in the mouse, for example, remain homosynapsed (Chandley, 1982) while in other cases, including *Peromyscus maniculatus* (Greenbaum & Reed, 1984) and *P. sitkensis* (Hale, 1986), heterozygosity for pericentric inversion leads directly to antiparallel, i.e. non-homologous, synapsis. In pachytene meiocytes of two naturally occurring pericentric inversion heterozygotes of the deer mouse, *Peromyscus maniculatus*, possible reverse loops were seen in < 1% of the 500 cells analysed. Otherwise, regularly formed SCs were present. The same appears to be true in the case of pericentric inversion heterozygotes of grasshoppers where no, or at best very few, reverse loops have been seen with light microscopy (White, 1961; John & King, 1977). In four different inversion heterozygotes of maize, analysed in silver-stained microspreads, inversion loop frequency did not change from early through late pachytene for any of them. The position and size of the inversion loop also remained constant throughout pachytene (Anderson, Slack & Sherman, 1988).

C.4 Non-homologous pairing

The synaptic behaviour of translocation heterozygotes, where partial homology may extend over several chromosomes, makes it clear that the individual chromosome is not necessarily the unit of pairing. Moreover, if homologous regions are absent, or else not in close proximity, pairing may involve non-homologous regions or non-homologous chromosomes. Such non-homologous pairing is especially pronounced in one of two situations (Bojko, 1983): (1) where the possibility of homologous association is either non-existent (haploids) or else modified (triploids and structural heterozygotes); and (2) where homologous pairing is mechanically impeded by interlocks or chromosome breaks or else where homologous pairing extends into adjacent non-homologous regions.

McClintock (1933) was the first to demonstrate not only that non-homologous associations may develop in unbalanced chromosome complements but that, on occasion, they may even supersede true homology. Such non-homologously paired regions do not, however, usually persist. Non-homologous pairing is most frequently encountered in monoploid PMCs of haploid plants (2n = x) where each chromosome is represented singly. Here it leads to foldback pairing in which a portion of the univalent pairs, and forms a SC, with itself. This behaviour has been reported in a variety of haploid plants. Ting (1966), who described such foldback pairing at pachytene in haploid maize, initially attributed it to the presence of duplications and assumed it to involve strict homology. Subsequently, however, he showed it to be transient and non-homologous, although associated with SC formation (Ting, 1973). Evidently, therefore, not only may SC production not lead to crossing-over but, additionally, it may be a product of non-homologous, as well as homologous, pairing (Gillies, 1984). In the grasshopper *Chloealtis conspersa*, non-homologues sometimes pair terminally and these associations persist through to metaphase-1. Moens & Short (1983) showed that such associations involved the formation of short SCs between the terminal portions of one lateral element from each of two heterologous bivalents. They also concluded that these associations led to the formation of chiasmata but

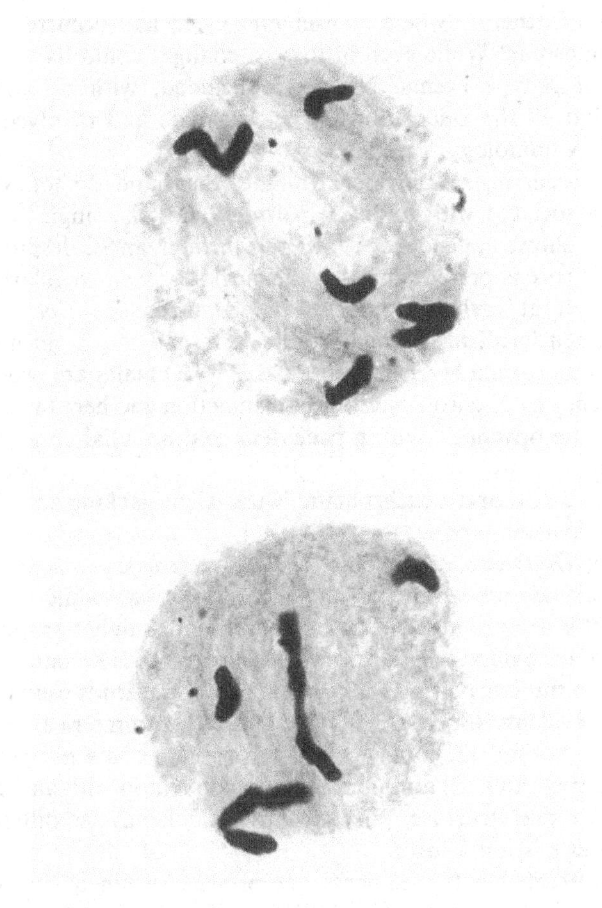

Fig. 4.14. Metaphase-1 in haploid meiocytes (2n = x = 7) of rye (*Secale cereale*). There is no congressed plate and a mixture of univalents, pseudobivalents, or less commonly pseudomultiples, are formed. In the two PMCs illustrated there are three univalents and two pseudobivalents per cell.

until recombination nodules which actually lead to crossing-over (see Chapter 4D.1) can be demonstrated to be present in these SCs this remains no more than a speculation on their part.

Associations of two, or more, chromosomes are also commonly observed at meiosis in haploids (Fig. 4.14). Neijzing (1982) studied a haploid rye plant in which C-banding allowed him to define a group of four chromosomes, with a C-band at each telomere, and a second group of three, where only one telomere band was present per chromosome. Using these C-bands as markers he was able to identify anaphase-1 bridge and fragment configurations which confirmed the occurrence of undoubted exchanges between two of the seven univalents in 19 out of 135

PMCs and a further 12 where an exchange event had occurred within a single chromosome. While both of these exchanges could have arisen by either U- or X-type events, Neijzing concluded, with no convincing evidence, that all the associations were chiasmate and involved regions sharing DNA homology.

Naturally occurring sex univalents, on the other hand, do not exhibit SC formation associated with foldback pairing. Thus the single X of male orthopterans shows no tendency to self pair through an SC despite the fact that an axial core is present and that the chromosome is regularly folded back on itself at early prophase-1. In the nematode *Caenorhabditis elegans*, the self-fertilizing hermaphrodite form has two X-chromosomes which regularly form a bivalent with a full SC. XO males are produced at a low frequency by X-chromosome non-disjunction and here the univalent X remains heteropycnotic during pachytene and no axial core forms. In a triple-X hermaphrodite strain, also produced by non-disjunction, an X bivalent with an SC and a euchromatic X univalent, lacking an axial core, develop (Goldstein, 1984).

In human (XO) sex monoploids, which are females, oocytes do not progress beyond pre-leptotene. In mice, however, while germ cell development is delayed in XO female foetuses, and a higher proportion of degenerating pachytene cells is observed compared to XX controls, in the cells which do survive, pairing configurations of three kinds were observed in the X univalents (Speed, 1986): (1) partial, symmetrical or asymmetrical, self pairing; (2) complete self pairing either as a hairpin or less frequently a ring; and, (3) non-homologous association with an autosome to form a tri-radial structure. Why sex univalents behave so differently in different species is not known.

Where heterochromatic segments are present within a complement, associations often occur between both homologous and between non-homologous chromosomes (Shaw, 1970; Drets & Stoll, 1974). Such associations usually lapse early in meiosis. In some cases, however, pairing of this kind persists to first metaphase. This is so in the grasshopper *Cryptobothrus chrysophorus* (John & King, 1985a), where it involves otherwise homologous chromosomes, and *Heteropternis obscurella* (John & King, 1982), where it involves non-homologues and gives rise to pseudomultiple configurations (see Fig. 2.27). In the case of *Cryptobothrus* one can be confident that the associations do not involve SC formation since the chromosomes concerned need not be in pairing contact for the associations to be effective (see Fig. 2.26).

C.5 Polyploids and hybrids

Conventional light microscopic studies on meiotic pairing in triploids have repeatedly demonstrated that while only two homologues pair at any

Table 4.8. *Number of partner switches in pachytene trivalents of Allium sphaerocephalon and pachytene quadrivalents of Allium vineale*

Number of	Number of switches									Mean
	0	1	2	3	4	5	6	7	8	
Trivalents[a]	1	9	28	28	19	13	8	2	3	3.4 per trivalent
Quadrivalents[b]	17	33	21	10	3	1	0	0	0	1.4 per quadrivalent

[a] After Loidl & Jones, 1986.
[b] After Loidl, 1986.

one site, pairing may switch between any two of the three homologues resulting in the development of a potential trivalent. An equivalent situation occurs in higher polyploids too. Thus, in tetraploids switch pairing between all four homologues gives rise to quadrivalents. EM studies confirm that, most usually, an SC is formed between two axes in any one region though one or more switches of pairing partners may occur giving rise to pachytene multivalents of varying size.

Occasionally, as in triploid chickens, triple pairing configurations, involving two SCs, complete with central elements, lying side-by-side, have been observed in as many as five chromosome groups (Comings & Okada, 1971a). Specific pairing leading to SC formation involving more than two homologues at the same region has also been observed in tri- and tetraploid oocytes of *Bombyx mori* (Rasmussen, 1987). These observations imply that there must be at least two sites for specific recognition at a given region in each chromosome of this species and that both of these synaptic initiation sites can sometimes function simultaneously. Additionally, triploid *Coprinus cinereus* and tetraploid yeast show regular triple pairing (Rasmussen *et al.*, 1981). Such cases, however, appear to be exceptional, though why this should be so is not known.

Triple pairing in non-polyploid organisms, while rare, has been reported in human oocytes which are trisomic for chromosome-21 (Wallace & Hulten, 1983) and in an aneuploid bull spermatocyte trisomic for chromosome-16 (Dollin & Murray, 1984). In the latter case the three lateral elements were completely associated for 85% of their lengths.

Using a surface-spreading technique, Loidl & Jones (1986) showed that in triploid *Allium sphaerocephalon* only two of the three axial elements of each set of three homologues were synapsed at any one site. The third homologue, however, was also intimately aligned with the paired partners and up to eight exchanges of pairing partners occurred in any one trivalent (Table 4.8). The unsynapsed axis was associated with the two paired axes at up to 50 intercalary sites per trivalent. Loidl and Jones therefore proposed that multiple potential pairing sites exist, only some of which

Table 4.9. *Number of trivalents present in 40 metaphase-1 PMCs of Allium sphaerocephalon*

Number of trivalents	0	1	2	3	4	5	6	7	8
Times observed	—	—	—	—	1	2	12	14	11

After Loidl & Jones, 1986.

actually lead to SC formation. The unused sites may then serve for the association of the unpaired homologue with its paired partners in triploids.

In *A. sphaerocephalon* initial SC formation was maintained until chiasmata formed, resulting in a high frequency of trivalents at first metaphase (Table 4.9). By late pachytene, however, all intercalary associations had lapsed. No triple pairing was observed. Likewise non-homologous regions of homologous chromosomes were rarely paired at early pachytene. When unpaired axes were released from alignment in mid-pachytene they were sometimes subsequently involved in foldback pairing, both terminal and intercalary in character. Heterosynapsis was also observed between unpaired axes of different trivalents. None of these non-homologous forms of association persisted to metaphase-1.

Using the same technique, Loidl (1986) identified equivalent intercalary associations between homologous axes at zygotene and pachytene in tetraploid *Allium vineale*. Here, however, they were less abundant and more short-lived and were abolished once an SC had formed.

The allohexaploid *Triticum aestivum* (2n = 6x = 42), common bread wheat, has arisen by two successive hybridization events, each coupled with a chromosome doubling, involving three different genomes (A, B and D) which, while not completely homologous, share considerable genetic similarity and thus constitute homoeologues. Despite this partial homology the allohexaploid behaves like a normal diploid and 21 bivalents are invariably present at both diakinesis and metaphase-1. This behaviour is known to be under genetic control and to depend on a locus, *Ph-1*, present on the long arm of chromosome-5B (see Chapter 6C.3). Not unnaturally, it was initially assumed that this diploidization depended on a restriction of zygotene pairing to strict homologues. This has proved to be incorrect. In a study of zygotene pairing behaviour from serial reconstructions of EM sections of single nuclei, Hobolth (1981) was able to demonstrate multiple initiation sites, coupled with shifts of pairing partners between different combinations of chromosomes, during SC formation at zygotene. This led initially to multivalent formation involving both quadrivalents and hexavalents. Although homoeologues could not be identified, it was assumed, quite reasonably, that the multivalent associations resulted from shifts of pairing partners between homoeologous chromosomes. Additionally, while pairing between homologues occurred preferentially at telomere regions, pairing between presumed homoeologues was, with few exceptions, initiated interstitially.

By pachytene all multivalent associations had been transformed into homologously paired bivalents. Hobolth concluded that diploidization was in fact achieved not by a restriction of zygotene pairing to structural homologues but by the conversion of zygotene multivalent associations, involving both homologues and homoeologues, into metaphase-1 bivalents involving only homologues. Precisely how this correction is regulated has yet to be determined but it clearly involves chiasma formation being restricted to homologues. The diploid behaviour of allohexaploid wheat was thus assumed to depend on two events:

(1) The initial production of multivalent associations involving SC formation between both homologues and homoeologues.

(2) A correction process whereby multivalent associations are resolved into homologous bivalents and crossing-over is suppressed between homoeologues. This implies that, despite the initial production of multivalent SCs, there must be a high stringency of effective pairing between homologues, i.e. pairing leading to crossing-over.

Also present at zygotene were numerous interlocks, of both chromosome and bivalent categories. In conjunction with this, both chromosome and bivalent breaks were again commonly observed in regions of interlocking, implying that their resolution involved transient breakage followed by rejoining.

On the basis of his observations, Hobolth (1981) postulated that the *Ph-1* gene controls the time of crossing-over relative to the correction of zygotene pairing. That is, *Ph-1* shifts the time of crossing-over to mid- or late pachytene thus enabling the correction mechanism to release interlocked bivalents. This effect, he argued, was dose-dependent so that while two doses postpone crossing-over to early pachytene, 4–6 doses delay it still further to mid–late pachytene. Holm & Rasmussen (1984) also concluded that, in normal wheat, crossing-over is delayed until the correction of multivalents into bivalents is completed and that the *Ph-1* gene on 5BL controls the time of crossing-over in relation to this correction of pairing. If this gene is absent (0 5BL) crossing-over may occur while multivalents are still present, while six doses of *Ph-1* delay crossing-over until conditions optimal for its occurrence have passed, so leading to partial desynapsis.

Yacobi *et al.* (1982) point out that, under this argument, plants with six doses of *Ph-1* might be expected to show no interlocking, which is not the case. Rather, as we have already noted, when the dosage of *Ph-1* is increased from two to six so too is the frequency of interlocking. They argue, therefore, that the correction mechanism that leads to the resolution of interlocks is constant in magnitude but that the extent of zygotene pairing is influenced by pre-meiotic events with the *Ph-1* gene controlling both somatic and pre-meiotic chromosome arrangements within the nucleus. There are numerous claims, of varying credibility, that homologous, homoeologous and even non-homologous chromosomes are arranged non-randomly in the interphase nucleus. The somatic pairing of homologues in dipterans and the association, through nucleolar fusion, of nucleolus organizing regions, whether located on homologous or non-homologous chromosomes, are the most convincing of these claims. Avivi & Feldman (1980), in reviewing the factors that influence the arrangement of chromosomes in somatic and pre-meiotic interphase nuclei of plants, argue that this arrangement determines the pattern of meiotic pairing. In the specific case of *Triticum aestivum* they claim that the *Ph-1* gene controls the location of both chromosomes and genomes in the pre-meiotic nucleus through its effects on spindle organization. They propose that two doses of the gene induces a partially divergent spindle leading to the partial separation of homoeologous chromosomes but with no effect on homologous sets. In the absence of *Ph-1* the spindle is said to be *convergent* with the result that all chromosome sets, whether homologous or homoeologous, are brought to the same polar region. Finally, in plants with six doses of *Ph-1* the spindle is split or *divergent*, resulting in the separation of all chromosome sets, both homologous and homoeologous. With the additional assumption that these pre-meiotic patterns are maintained into the first meiotic prophase, they claim that, with two doses of *Ph-1*, the exclusive pairing of homologues, coupled with the suppression of homoeologous associations, results in minimal interlocking. In plants deficient for *Ph-1*, both homologues and homoeologues are associated prior to zygotene so facilitating multivalent formation and interlocking. This, in their opinion, explains why interlocks of more than three bivalents were never observed in plants with a reduced *Ph-1* dosage. Extra doses of *Ph-1* result in partial pairing failure (*asynapsis*) but lead to increased interlocking in those associations that do form.

The matter has recently been resolved by Holm & Wang (1988) following an ultrastructural analysis of the effect of 5B on chromosome pairing and SC formation in *T. aestivum* using microspreads. They demonstrate that, contrary to the Feldman hypothesis, there is no evidence of any alignment of homologues at the leptotene–zygotene transition. Neither does the *Ph-1* gene affect the timing of crossing-over, as proposed by Hobolth. Rather, disomic inheritance in this allohexaploid is ensured by three factors: (1) a high stringency of chromosome pairing at zygotene resulting predominantly, though not exclusively, in the production of homologously paired bivalents; (2) a synaptic correction mechanism by means of which any multivalent associations that do form are converted into bivalents during zygotene and pachytene; and (3) the suppression of crossing-over between paired homoeologous segments. Even trihaploids of wheat are virtually achiasmatic, with only a low number of chiasmata between homoeologues of the A, B and D genomes at metaphase-1, despite extensive SC formation (Wang, 1988).

Wischmann (1986) found that, in plants carrying three isochromosomes of 5BL, pairing was arrested at early zygotene resulting in an average of only 25% pairing. Even so, there were a large number of associations of more than two, as well as SC formation between chromosomes which were not homologues. There was also a retention of chromosome and bivalent interlocks up to first metaphase, as well as pairing and crossing-over between presumed homoeologues. By contrast, in lines trisomic for 5DL there was no asynapsis and mostly bivalents were present. Wischmann concluded that the resolution of interlocks by breakage and rejoin requires complete pairing and that the incomplete synapsis produced by the presence of extra copies of the *Ph-1* gene therefore leads directly to a high interlock frequency.

In experimentally produced female polyploid forms of *Bombyx mori*, Rasmussen & Holm (1979) have shown that a correction mechanism, comparable to that operative in allohexaploid wheat, leads to consistent bivalent formation in tetraploid oocytes and to the consistent formation of bivalents and univalents in triploid oocytes.

Tetraploid females are fertile and show almost regular disomic segregation. When crossed to diploid males they produce viable, though sterile, triploid offspring. Ultrastructural analysis of tetraploid oocytes indicates that zygotene pairing is strictly homologous and leads to the production of both bivalents and quadrivalents, together with very much smaller numbers of trivalents and univalents. At pachytene, however, there is an almost total conversion of quadrivalents into bivalents (Table 4.10).

In the triploid, most of the 84 chromosomes were initially aligned in sets of three, two homologues being paired by an SC with the third lying parallel to the bivalent, though a short distance from it. In a number of such cases, exchanges of pairing partners led to trivalent formation. By mid-pachytene, however, only bivalents and univalents were present together with a number of non-homologous associations, including univalents with foldback pairing, associations of two chromosomes of unequal length and non-homologous configurations of three or four chromosomes (Rasmussen, 1977b). Consequently, triploid *Bombyx* females regularly produce unbalanced gametes containing one complete haploid genome together with a varying fraction of a second genome.

In experimentally produced male tetraploids, as in 4x females, multivalents again form during the early pairing phase. However, only

Table 4.10. *Chromosome configurations present in triploid and tetraploid meiocytes of Bombyx mori*

Stage[a]	Number of nuclei	Mean number				
		Univalents	Bivalents	Trivalents	Quadrivalents	Non-homologous associations
Triploid Oocytes						
Z–P	2	22.0	22.0	6.0	—	0
eP	2	23.5	26.0	1.0	—	2.5
m–lP	7	18.6	26.7	0.0	—	6.4
Tetraploid Oocytes						
eP	7	2.0	36.7	0.7	8.4	4.1
m–lP	11	1.4	51.9	0.6	0.9	3.4
Spermatocytes						
m–lZ		1.0	25.1	0.7	13.3	
P		0.0	37.0	0.0	8.7	
M1		0.0	42.15	0.0	6.7	

After Rasmussen & Holm, 1982; Rasmussen, 1987.

[a] Z–P = zygotene–pachytene transition, eP = early pachytene, m–lP = mid–late pachytene, m–lZ = mid-late zygotene, P = pachytene, M1 = first metaphase.

about 50% of these are converted into bivalents at the zygotene–pachytene transition. This striking, sex-dependent, difference in multivalent behaviour can be explained by the fact that, whereas the female is achiasmate, the occurrence of chiasma formation at male pachytene prevents the conversion of all multivalent associations into bivalents (Rasmussen, 1987).

An equivalent situation occurs in experimentally produced hybrids between *Lolium temulentum* and *L. perenne*. These two species share the same diploid number (2n = 2x = 14) but the latter has 50% more DNA (6.23 pg versus 4.16 pg), mostly in the form of repetitive DNA sequences, so that its chromosomes are very much longer. Despite this, pairing and chiasma formation between the two homoeologous sets is pretty regular in a majority of hybrid PMCs with up to 11–12 chiasmata per PMC. Consequently, up to seven asymmetrical bivalents are formed and, since the extra DNA in *L. temulentum* is distributed throughout all members of the complement, the smaller bivalents are markedly more asymmetrical.

From an analysis of EM sections, Jenkins (1985a) reports that SC formation was complete in about 50% of the F_1 bivalents. In the remainder the SCs incorporate a single interstitial loop. These loops presumably correspond to the pachytene loops observed in this hybrid by light microscopy. He also noted that a specific bivalent showed a SC loop in some nuclei but not in others. Moreover, except for bivalent 7, where all loops are close to the centromere, loops were not at the same site, or of the same size, in the same bivalent.

The capacity to form complete SCs depends, in this case, on an adjustment of the length differences between homoeologues. In part these differences are accommodated before synapsis, but final adjustment occurs during synapsis itself. This implies that SCs are flexible structures which can absorb large differences in length without affecting the integrity of the SC itself. Moreover, at early zygotene, pairing was not confined to homoeologues so that multiple configurations, including associations of three and higher, were formed between non-homologues in a proportion of nuclei. These associations, however, did not persist so that, here too, a correction mechanism must occur, presumably by breakage and rejoining. However, no interlocking was observed in this case. Despite the fact that SCs must have been partly non-homologously aligned in many cases, and that there was also asynapsis in a number of others, chiasma formation was successfully achieved in a large number of bivalents. Thus large differences in repetitive DNA content have little influence on the capacity of homoeologues to pair, to produce SCs or to form chiasmata.

The grass *Festuca drymeja* has some 50% more DNA than *F. scariosa* with which it shares the same diploid number (2n = 14). This DNA difference is distributed throughout the entire genome, though not equally so. While the chromosomes of *F. drymeja* are consistently larger than those of *F. scariosa* the size differential is most pronounced in the four smaller pairs. In the F_1 hybrid between these two species the three largest bivalents, where DNA differences are least, do not form loops at pachytene and show a complete SC with perfect matching of centromere regions. In these bivalents the lateral elements of the SC are evidently able to accommodate the length discrepancies which exist between homoeologues without disrupting the integrity of the complex. This reflects the fact that the additional DNA in *F. drymeja* is randomly dispersed along the length of each of the three largest chromosome pairs. In the four smaller bivalents, on the other hand, buckles and loops are regularly present at pachytene in F_1 hybrids and SC formation is abnormal and incomplete. In particular, there are differences in the length of the lateral elements in the short arms, which parallel an obvious difference in the content and concentration of DNA in these arms. Here, then, the extra chromatin does contribute to the length of the lateral elements. Moreover, at first metaphase, chiasmata are confined to the three larger bivalents (Jenkins & Rees, 1983; Rees, 1984).

The variation in loop position within the same bivalent observed by Jenkins in the *Lolium* hybrid implies that non-homologous stretches of axial cores must have been juxtaposed in the SC, as was sometimes indicated by the non-correspondence of centromere alignment. Thus, centromeres paired with non-centric regions in some 67% of the bivalents which displayed loops. Finally, unlike the X univalent of male orthopterans, in which the axial core does not self pair, that of the B-chromosome of *L. perenne*, when present as a univalent, does. Here the axial core folds back on itself to form two short stretches of SC representing some 4% of its length.

In tetraploids produced from the same interspecific hybrid by colchicine treatment, 0–5 multivalents form at metaphase-1. In the presence of B-chromosomes, however, pairing between the homoeologous chromosomes shared by *L. perenne* and *L. temulentum* is drastically reduced. In the diploid hybrid this leads to extensive univalent formation while in the tetraploid it produces bivalents composed of strictly homologous chromosomes (Evans & Taylor, 1976). That is, the tetraploid becomes diploidized (see also Chapter 6C.3).

At zygotene, the situation in the induced tetraploid, as revealed by EM reconstruction, was very different from the metaphase-1 light microscope picture. In the one multivalent-forming hybrid cell that was reconstructed by Jenkins (1986) there was 6.II + 1.IV + 1.X. While associations of four were expected, the association of 10 was not and there must again be some form of correction process which limits multivalents to IVs by first metaphase. Each of the six bivalents had a complete SC, while three of the chromosomes in the quadrivalent showed foldback synapsis in one arm and five short interstitial regions exhibited non-homologous synapsis with several other chromosomes.

Three early prophase-1 nuclei reconstructed from the bivalent-forming, B-containing, 4x hybrid had the following constitution:

Stage	Constitution
1. Zygotene	$7.II+1.I+1.III+2.V+1.IV_B$
2. Pachytene	$10.II+3.I+1.V+1.IV_B$
	$10.II+2.IV+1.IV_B$

and all the univalents showed extensive foldback pairing.

Triploid hybrids resulting from crosses between a 2n = 14+1B *L. perenne* and a 2n = 14 *L. temulentum* were also analysed (Jenkins, 1985b). Such triploids included two sets of *L. perenne*, two B-chromosomes and one set of *L. temulentum*. Here, as expected, metaphase-1 pairing was restricted to homologues, giving rise to 7.II+7.I+B.II. A reconstructed pachytene nucleus confirmed that, as at first metaphase, there were seven *L. perenne* bivalents and no association between the homoeologous *temulentum* and *perenne* chromosomes. However, there was extensive non-homologous association, with SC formation, both within and between the seven *temulentum* chromosomes. At early zygotene, on the other hand, six *temulentum* bivalents, a B bivalent and a trivalent were present. The transformation of multivalent pairing associations into bivalents in the presence of B-chromosomes in both tetraploid and diploid *Lolium* hybrids is reminiscent of the influence of the *Ph-1* locus in allohexaploid wheat.

In summary, in both natural and experimentally produced hybrids, multivalents formed at zygotene between homoeologues are unstable and are resolved during pachytene by desynapsis and resynapsis to give homosynapsed SCs. A similar phase of synaptic adjustment occurs also in Robertsonian hybrids. Using F_1s between the 2n = 34 and the 2n = 54 forms of the mole *Ellobius talpinus*, Bogdanov *et al.* (1986) analysed SC formation in a hybrid male heterozygous for 10 Robertsonian translocations with a potential for forming a maximum of 10 multiples of three. In microspreads of both zygotene and pachytene male meiocytes, they identified multichromosomal chains, with the equivalent of from 2–5 fusion multiples, as well as the expected individual multiples of three (Fig. 4.15). By late pachytene, however, only associations of three, or else bivalents plus univalents, were present.

An equivalent situation has been reported in a single tetraploid male of *Bombyx mori* in which synapsis was aberrant (Rasmussen, 1987). In this case, pairing and SC formation were arrested at about the 50% stage so that no completely paired meiocytes were observed. Moreover, except for a small number of regularly paired bivalents, pairing was highly irregular, leading to the production of associations ranging in size from three to complex chains involving maximally 65 chromosomes. Double, and even triple, SCs, combining three to four lateral elements in the same region, were also present and all but one of these included a set of telomeres.

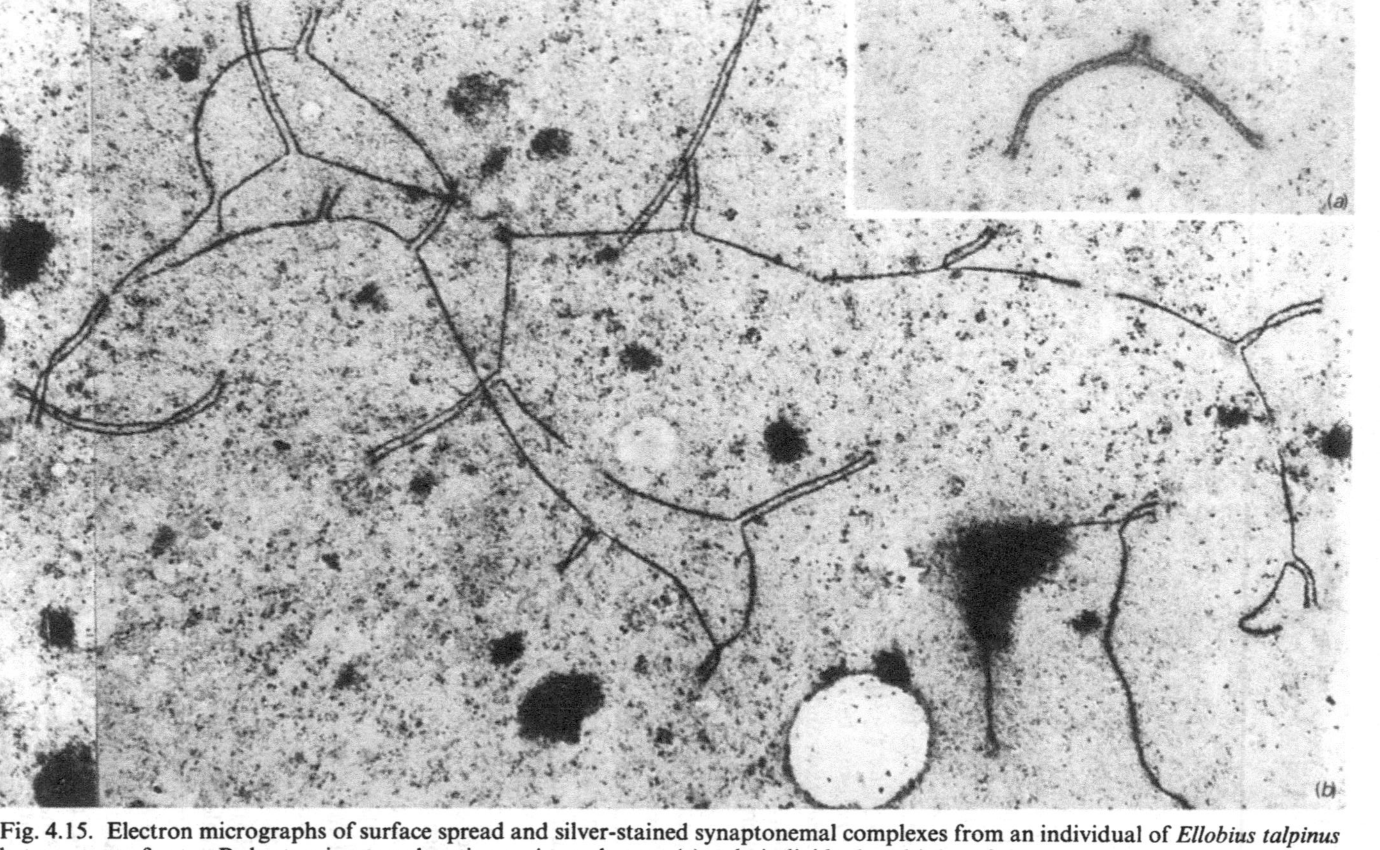

Fig. 4.15. Electron micrographs of surface spread and silver-stained synaptonemal complexes from an individual of *Ellobius talpinus* heterozygous for ten Robertsonian translocations. At pachytene (*a*) only individual multiples of three are present, each consisting of one metacentric and two acrocentrics. At late zygotene (*b*), however, from two to eight of the lateral elements of the short arms of the acrocentrics involved in these multiples of three are frequently joined into extended multiple chains (photographs kindly supplied by Dr Yu. F. Bogdanov).

Most of the SCs formed between non-homologous chromosomes. This applied not only to associations of more than four but additionally to the vast majority of associations of from two to four chromosomes and resulted in large differences in the length of the lateral elements involved. Thus, in this aberrant individual there was almost a total failure of specific homologue recognition prior to SC formation. Rasmussen suggested, therefore, that while in normal meiosis random contacts between recognition sites will lead to SC formation only if those sites are homologous, the non-specific and random SC formation in this one case resulted from a defect in the regulation of the timing of SC formation.

These several cases have led to the conclusion that, in both structural and numerical hybrids, synapsis and SC formation may involve two phases. In the case of inversion and duplication heterozygotes, zygotene pairing takes place between strictly homologous segments (*homosynapsis*) but a second phase of pairing during pachytene may extend to non-homologous regions (*heterosynapsis*). Consequently, differences in the length, or else in the form, of the SC may be adjusted so that the lateral elements of the two homologues in a given heterozygous bivalent attain the same length and pair straight regardless of homology. In polyploids, on the other hand, early appearing multivalent SC configurations may be corrected to produce bivalents while, in fusion heterozygotes, multiples in excess of the expected size revert to correct size.

C.6 Sex chromosome behaviour

Differentiated sex chromosomes must initially have originated from a pair of complete homologues. This involved the production of differential segments in which crossing-over was abolished. It has long been a convention to assume that, in the initial stages of their evolution, sex chromosomes would have shared common pairing segments in which chiasma formation ensured the segregation of the partially homologous partners (Darlington, 1939, 1958). It is also commonly assumed that pairing and differential segments still exist in the differentiated sex chromosomes of many extant eukaryotes so that while genes on the differential segment of the X can crossover in the female, those on the differential segment of the Y never crossover. A precise analysis of sex chromosome behaviour in the heterogametic sex is, however, rarely possible with the light microscope because the sex chromosomes behave differentially at meiosis. The situation has been most thoroughly analysed in male meiosis of mammals. Here, even the identification of the sex chromosomes at leptotene is often difficult by light microscopy because their heteropycnosis is not fully developed at that time (Reitalu, 1970). Two oval heteropycnotic masses, either independent or associated, can be distinguished by zygotene. Towards the end of zygotene these two bodies

Table 4.11. *The meiotic behaviour of XY chromosomes in male mammals*

Behaviour pattern	Species
Chiasmate XY bivalent formed after SC formation with parasynapsis	*Cricetulus griseus* *Cricetulus migratorius* *Cricetus cricetus* *Erethizon dorsatum* *Peromyscus sitkensis*[a] *Suncus murinus*
Terminally associated XY bivalent formed after SC formation with parasynapsis	*Chinchilla laniger* *Homo sapiens* *Mesocricetus auratus* *Mus dunni* *Mus musculus* *Rattus norvegicus*[b]
Terminally associated XY bivalent formed by telosynapsis	*Bandicota bengalensis* *Nesoka indica*
No synapsis, univalent sex chromosomes	*Baiomys musculus* *Psammomys obesus* All known marsupials[c]

After Raman & Nanda, 1986.
[a] After Hale & Greenbaum, 1986.
[b] After Joseph & Chandley, 1984.
[c] After Sharp, 1982.

fuse into a single, more condensed, mass of chromatin. Painter initially described this as a 'chromatin nucleolus'. Sachs (1953) subsequently introduced the term 'sex vesicle' in place of Painter's initial description since the structure that forms is certainly not analogous to a true nucleolus. While this body has an amorphous appearance with the light microscope, the electron microscope indicates that it is a mass of chromatin fibers packed differently from the other regions of the nucleus but with no definite boundary or membrane. For this reason, Solari & Ashley (1977) have pointed out that the term vesicle is both misleading and inappropriate. Even so the term continues to be used. In what follows we will refer to this distinctive structure as the *sex chromatin body*. The sex chromosomes resume a more regular morphology when the cell reaches mid- or late diplotene.

Within the sex chromatin body it is difficult to analyse the nature of the association between the X- and Y-chromosomes by light microscopy and most of what we know about this process is based on ultrastructural and microspreading studies. Prior to the adoption of EM analysis, most statements about the mode of association of the X and the Y were based on the appearance of the sex pair at diakinesis or metaphase-1; that is, on their appearance after the sex chromatin body has disappeared. At this time, the most common form is a simple, end-to-end association of the X and Y. A majority of the species where a visible chiasma occurs in the condensed sex bivalent possess large X- and Y-chromosomes. In theory, where a chiasma originates near the end of a bivalent, this chiasma may terminalize as the bivalent opens out during diplotene so leading secondarily to a strictly terminal union (see Chapter A.2.2). That this can and does occur in sex bivalents is suggested by the behaviour of the XY-bivalents of Chinese and European hamsters where the X and Y are sometimes terminally associated at metaphase-1 but at other times show a clear chiasma (Fredga & Santesson, 1964; Vistorin Camberl & Rosenkranz, 1977). The problem is most acute in forms where only terminal unions are present at first metaphase and where one has to assume either consistent terminalization, pseudo-terminalization or else the occurrence of a non-chiasmate association.

From ultrastructural studies it is clear that two patterns of behaviour can be distinguished (Table 4.11): (1) cases where the X and Y synapse to form a length of SC whose extent varies according to the species; and, (2) cases where no synaptic pairing occurs and no SC forms. Here there is either variable end-to-end association or no association at all.

C.6.1. Pairing with SC formation

In a majority of placental mammals, when the X- and Y-axes first become discernible at leptotene they lie some distance apart with their ends attached to the nuclear membrane. The attached ends then migrate in the plane of the nuclear envelope until homologous axes lie within 200 nm of each other. At this point, a SC forms along a limited length of the sex bivalent. This partial pairing marks the presumed region of homology between the X and Y, i.e. the presumed pairing segments. Synapsis is, however, delayed relative to the autosomes and begins at late zygotene or early pachytene with maximum pairing at pachytene.

The extent of pairing, that is the length of SC formed, and the subsequent behaviour of that SC, varies in different species (Moses, 1977b). Moreover, in many placentals, the unpaired portions of the X and Y undergo progressive thickening and apparent doubling after the partial SC has formed. In a small number of placentals, chiasmata are evident between the sex pair. This applies to the North American porcupine *Erethizon dorsatum* (Benirschke, 1968) and several rodents including the harvest mouse, *Micromys minutus* (Schmid, Solleder & Haaf, 1984), *Gerbillus pyramidum* (Wahrman *et al.*, 1983) and *Peryomyscus sitkensis* (Hale & Greenbaum, 1986). At least some of these cases can be explained on the grounds that the large sex chromosomes have originated through X.A translocations. In most other cases, although an SC forms over a limited area it then undergoes desynapsis and reduces to a terminal association. In the mouse, for example, most of the distal part of the Y initially pairs with the distal portion of the X. At late pachytene, following desynapsis, either only a very short pairing region remains or else the two chromosomes are attached end-to-end (Tres, 1977).

Likewise, in *Rattus norvegicus*, pairing and SC formation is markedly delayed relative to the autosomes and a precocious thickening of the X- and Y-axes may contribute to this. Here the SC involves the whole of the Y-chromosome and, during its formation, the lateral element of the Y thins considerably. So too does the paired portion of the X-axis, though to a lesser extent. By late pachytene and X- and Y-chromosomes have begun to desynapse and by diplotene they remain only as end-to-end associations (Joseph & Chandley, 1984). There is a comparable behaviour in man and the golden hamster where pairing is again reduced totally to an association of ends. In these cases there are only two possible interpretations of sex bivalent behaviour:

(1) Partial synapsis is followed by the occurrence of a single crossover which forms subterminally and so is either difficult to resolve (pseudoterminalization) or else must terminalize completely. On the assumption that synapsis reflects genetic homology, the presence of a SC between portions of the X- and the Y-chromosomes has traditionally been taken as evidence for their partial homology as predicted by classical cytogenetic theory, which assumes that sex chromosomes consist of both pairing and differential segments.

(2) Desynapsis is followed by a non-chiasmate association between the telomeric segments of the X- and Y-chromosomes. This implies either that the SC that forms is more representative of historical homology than of current genetic similarity and, like the SC of some achiasmate meioses, does not result in chiasma formation, or else that XY synapsis in most mammals may simply reflect a special form of non-homologous pairing which does not allow for crossing-over.

Most workers have followed Solari (1974) in assuming that the end-to-end connection represents a chiasma between near-terminal regions of the sex chromosomes which has thus either terminalized or else shows pseudoterminalization. Support for this interpretation has been inferred from the behaviour of *Sxr* mice. The Y-linked mutation *Sxr* causes genetic females to develop as phenotypic males with testes. Carrier males (X/Y*Sxr*) transmit *Sxr* equally to both XX and XY offspring. X/Y*Sxr* males have a high incidence of XY pairing failure which points to some alteration of the terminal pairing segments of the X- and Y-chromosomes. Such mice have a single abnormal Y which possesses an additional dark-staining distal G-band. During meiosis this band can be transferred to the X so that X/X*Sxr* sons of X/Y*Sxr* males have two kinds of X-chromosomes. One is normal while the other carries an additional G-band at its distal end. Evans & Burtenshaw (1982), who described this behaviour, argued that the transfer of the terminal band of X/Y*Sxr* mice to their X/X*Sxr* sons resulted from near-terminal crossing-over between the X and the Y*Sxr* despite the fact that, as already mentioned, a feature of meiosis in X/Y*Sxr* males is a high frequency (70–90%) of X,Y univalents compared to the 5–10% univalence of sex chromosomes in X/Y non*Sxr* sibs. Matsuda *et al.* (1982) report that F_1 hybrids in mice show varying degrees of X–Y dissociation ranging from 3 to 70% depending on the strain and the hybrid. In all cases the X- and Y-chromosomes are associated end-to-end at pachytene and start to dissociate at diplotene. To what extent this reflects differences in XY homology between the different strains remains to be clarified.

Holm & Rasmussen (1983a, b) showed that, in man, some 75% of the pachytene XY-bivalents contain either 1 or 2 recombination nodules (see Chapter 4D.1). This, coupled with the fact that SC segments are retained in both diplotene and diakinesis nuclei, convinced them that crossing-over occurs between human X- and Y-chromosomes. DNA sequence homology has since been demonstrated in the pairing region of human X- and Y-chromosomes (Buckle *et al.*, 1985; Cooke, Brown & Rappold, 1985; Simmler *et al.*, 1985) and the sequences concerned exhibit polymorphisms whose inheritance confirms the existence of recombination between the X and the Y at a site located near the telomeres.

Ashley (1984a, 1985, 1987), on the other hand, takes the view that XY-pairing in placental mammals may represent a special kind of non-homologous association and is achiasmatic. This runs contrary to the belief that XY-pairing is based on structural DNA homology. In Ashley's view, synapsis has been retained to ensure regular segregation of the sex bivalent in the absence of crossing-over. He makes the following points:

(1) In man, synapsis of the X and Y, as determined by the extent of SC formation, occurs between the short arms of both chromosomes and involves, on average, some 30% of the Y and 12% of the X when measured from the distal ends of those arms. In the mouse, where synapsis occurs between the distal ends of the acrocentric X and Y, up to 90% of the Y and 30–35% of the X synapse (Ashley, 1984a). In neither case, however, is there much evidence for homologous loci within the presumptive pairing segments. Most of the DNA sequences common to both X and Y do not lie in regions that synapse. Added to this, sequences which are unique to either the X or the Y may, nevertheless, lie within the region of synapsis as defined by the presence of an SC. Polani (1982) explained these incongruities by assuming the existence of homologous but 'silent' loci in the pairing

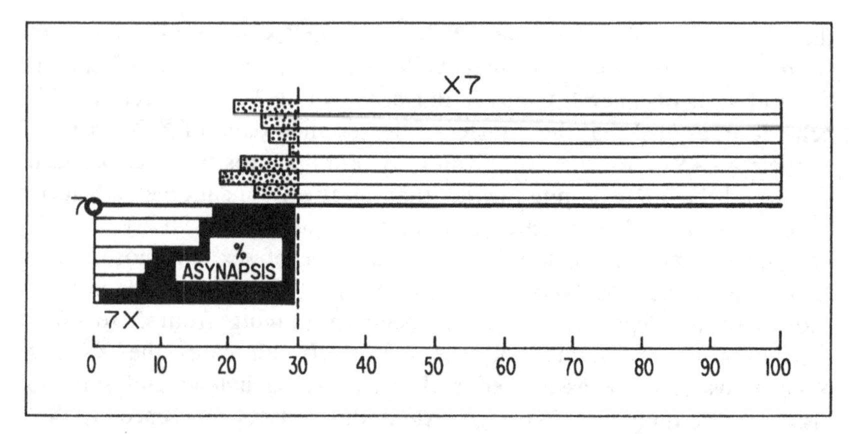

Fig. 4.16. Desynapsis of the 7X (solid area) and non-homologous synapsis of the X7 (stippled area) in seven R6 interchange multiples involving autosome 7 and the X-chromosome of *Mus musculus*. The assumed breakpoint on 7 is indicated by the vertical dashed line (after Ashley, 1984b and with the permission of Allen and Unwin).

segment of the Y but there is no hard data to support such an assumption. Indeed, 90 % of the sequences recovered to-date that map to Yp are specific in the sense that they do not map to the distal end of Xp (Rouyer *et al.*, 1986). An alternative explanation, and the one preferred by Ashley, is that much of the observed synapsis between the X and the Y must be of a non-homologous nature.

(2) In two X-7 translocations of mice the X and Y have been demonstrated to synapse non-homologously with an autosome. This synapsis involves not only those regions of the X and Y that normally form an SC with one another but, additionally, the differential portions of the sex chromosomes too (Fig. 4.16).

(3) If synapsis between the sex chromosomes is indeed based on homology then in XYY male mice the two Y-chromosomes might be expected to be at least as effective in forming a bivalent as does the X and one of the Ys. Yet XY bivalents form four times more frequently than do YY bivalents although a ratio of 2XY:1YY is expected if only bivalents form and they do so randomly. Of course, the formation of an XY bivalent need not preclude pairing between the proximal ends of the two Y-chromosomes which would lead to the formation of a sex trivalent. A trivalent was in fact present in 16 % of the spermatocytes analysed but in all of them the second Y was associated by its telomeric end to the junction of the X and Y. Added to this, the frequency of such trivalents decreased from early diakinesis to first metaphase resulting in the production of three univalents, which is not expected if the trivalent involved even one chiasma.

(4) XX males occur in humans with a frequency of approximately 1:20000 and a few of these are non-mosaic. By probing such males with Y-specific DNA fragments, Bishop *et al.* (1985) found that 66% of these non-mosaic XX males contained varying numbers of such fragments. Ashley interprets this to mean that non-homologous exchanges can and do occur between the X and the Y.

(5) In man, homology, and hence pairing, is assumed to be confined to the short, p, arms of the X and Y. However, Chandley *et al.* (1984) provide evidence from microspreads that up to three-quarters of the Y may be involved in SC formation. The actual involvement ranges from 11 to 71% with a mean of 34%. Since the Yp represents 23–27% of the total length of the Y, the synaptic region must extend into the long, q, arm of the Y which implies that such pairing is non-homologous. Additionally, double pairing, involving both the SC and a secondary association between the terminal regions of the X and Yq was observed in 4% of the cells. Since both these terminal regions contain tandemly repeated sequences, this may simply represent an association between regions of highly repeated DNA.

Ashley concludes that while some homology within the pairing region cannot be ruled out, much of the observed synapsis of the X and Y may well involve segments that are differential and non-homologous.

There is evidence for non-homologous sex chromosome pairing involving SC formation in other animals too. In chicken oocytes, which are ZW in constitution, unlike the situation in placental mammals, there is no sex chromatin body. At pachytene an unequal SC is present involving practically the whole of the shorter W-axis. Microspreads indicate that the longer lateral element of the Z-axis is twisted around the shorter W-axis which remains straight. Since the W in this species is heavily loaded with C-positive heterochromatin (Carlenius *et al.*, 1981), this may well indicate a high degree of non-homologous pairing between the ZW pair. Moreover, the W-kinetochore, which is medial in location, has no symmetrical counterpart in the Z which suggests that truly homologous regions, if they exist at all, must be restricted to the tip of the long arm of the sub-metacentric W in which both Z- and W-axes lie parallel and straight (Solari, 1977). Using microspreading, Rahn & Solari (1986) found a single recombination nodule (RN) very close to the paired telomeres of the ZW bivalent of the chicken. If this RN is indeed indicative of crossing-over (see Chapter 4D.1) then chiasma formation in this case is strictly localized and near-terminal.

Non-homologous pairing also plays a role in the association of the W- and Z-chromosomes in the female moth *Ephestia kuehniella*. Here the W can often be identified by its heteropycnotic remnants which remain

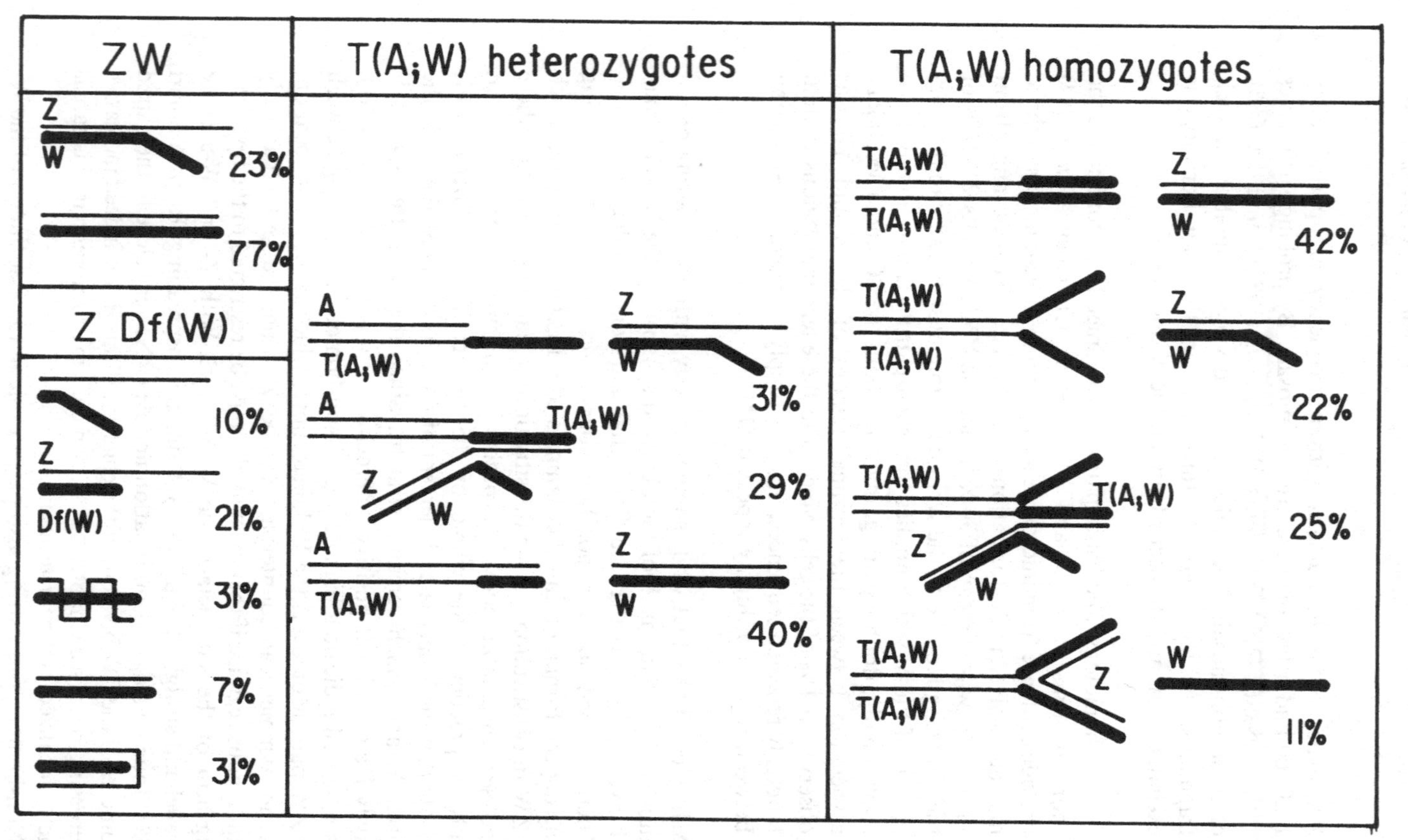

Fig. 4.17. The pattern and extent of synaptonemal complex formation between the Z- and W-chromosomes of the moth *Ephestia kuehniella* in normal (ZW) females, in females carrying a W deficiency [Z Df(W)] and in autosomal-W-chromosome [T(A;W)] translocation heterozygotes and homozygotes (after Weith & Traut, 1986 and with the permission of Springer-Verlag).

Type	4 ends	3 ends	2:2 heterologous	2 heterologous	2:2 autologous	2 autologous	0
Obs.	58	4	5	5	0	0	0

Fig. 4.18. The type and frequency of end-to-end associations between the axial elements of the X- and Y-chromosomes of *Psammomys obesus* in 79 late pachytene nuclei (after Ashley & Moses, 1980 and with the permission of Springer-Verlag).

associated with the lateral element as densely-staining material. In the wild type the W is 76% of the length of the Z and meiocytes show partial synapsis of the unequal axes. This is followed by non-homologous synapsis leading to adjustment, correction and subsequent complete pairing of these unequal chromosomes. In a deficiency, where the W has a pre-adjustment length of only 40% of the Z, five different kinds of SC pairing were observed (Fig. 4.17) including complete synapsis in 69% of the cells and complete equality of axes in 70%. Here adjustment involved both a shortening of the Z-axis as well as a lengthening of the W-axis. Finally, in translocations involving the W-chromosome and an autosome both heterozygotes and homozygotes again showed complete synapsis which confirms the occurrence of non-homologous SC formation since it now occurs between sex and autosome segments (Weith & Traut, 1986).

C.6.2 Absence of the SC

In the sand rat *Psamomys obesus*, the X- and Y-chromosomes associate during pachytene without SC formation. Heteropycnotic bodies, corresponding to the X and Y, are evident prior to pachytene. These may lie apart or together. By early pachytene there is a well defined sex chromatin body and microspreading indicates that all possible combinations of the four different ends of the X and Y are associated within this body with no preferred pattern (Ashley & Moses, 1980). In this case there is a clear tendency for the formation of non-specific end associations (Fig. 4.18). By first metaphase in some, but not all, meiocytes, there was an end-to-end connection between the sex chromosomes. This, presumably, represents a remnant of the association seen at pachytene and no chiasma forms. Despite this, segregation is normal in all cases. There are only two plausible explanations for such behaviour. First, that there are no homologous pairing segments, which have either been deleted or rendered non-functional. Second, since both the X and the Y are totally C-band positive, that the presence of heterochromatin may have led to the rejection of the normal synaptic mechanism. Since SCs are known to form in autosomal heterochromatic regions the first of these explanations appears more probable.

Three other rodents show a comparable behaviour. No SC forms in the southern pygmy mouse, *Baiomys musculus*, and, while the X and Y lie close to one another at diplotene, they are never in contact at diakinesis and by metaphase-1 are often widely separated (Pathak, Elder & Maxwell, 1980). Here, again, the acrocentric Y is totally C-band positive, as is the short arm of the sub-metacentric X. In *Mus dunni* and *M. agrestis*, in both of which the XY pair is similar in size and heterochromatin content to that of *P. obesus*, the X and Y are also associated terminally. In light microscope preparations of *M. dunni* the euchromatic long arm of the X has a short terminal segment that associates with the centric end of the totally C-band positive Y (Pathak & Hsu, 1976). In *M. agrestis* no form of pairing occurs between the sex chromosomes. Here the sex chromatin body lasts from mid-pachytene to the onset of diplotene and the sex chromosomes are unpaired at metaphase-1 (Zenzes & Wolf, 1971) yet segregate regularly.

Marsupials deviate markedly from the more general pattern of placentals and follow the behaviour shown by *Psammomys*. That is, no SC forms between the axes of the X and Y although a sex chromatin body is formed. Sharp (1982), who analysed the behaviour of the sex chromosomes in 22 species of Australian marsupials, demonstrated that, while axial elements of the X and Y associate at pachytene in all but one case, no SC is present in any of the species. In a majority, the X- and Y-chromosomes are separate at early pachytene. They then approach one another and their ends associate to form ring-like structures. In *Trichosurus vulpecta* there is no association at all between the X- and Y-axes. Here, neither pairing, as defined by SC formation, nor end-to-end association, is required for regular disjunction. Presumably, their initial association in a common sex chromatin body is sufficient to ensure their subsequent disjunction. Likewise, in the Viriginia opossum, *Didelphys virginiana*, only rarely are the X and Y associated at diplotene–diakinesis and then only loosely. No SC is formed between them (Pathak *et al.*, 1980). In cases like these, where no SC forms, there can be no question but that the terminal XY association is non-chiasmate.

X-chromosomes measuring approximately 5% of the female haploid set are common in placental mammals and have been assumed by Ohno (1967) to represent the original or basic X of placentals. End-to-end associations of the X and Y are found not only in species with a 5% X (man, mouse, rat, golden hamster) but also in some species with large Xs and small Ys. This includes *Chinchilla laniger* (Solari & Rahn, 1985) where the X is virtually duplicate in size. Here, the end-to-end association at first metaphase involves the telomeres of the short arm of the Y and the longer arm of the X. Microspreads indicate that, initially, both the X and the Y have a double axis with one of the X-axes close to the Y-axis. This is followed by a co-alignment of axes. When first formed, the SC covers 85–90% of the length of the Y-axis but it then undergoes desynapsis. While this desynapsis is in progress, the non-pairing ends may engage in an additional anomalous SC in some cells. Desynapsis reduces the SC to < 10% of the Y-axis. By late pachytene, both the X- and the Y-chromosomes are thinned, elongated and totally desynapsed. This is accompanied by a convergence of the four terminii. The delay in pairing of the X- and Y-axes in this species, the asymmetry of the axes, the subsequent return to symmetry and the formation of a secondary SC between the other terminii, suggest to Solari and Rahn that this is a case involving a relatively non-specific SC during the heterosynaptic phase.

The precise nature of the physical attachment between the ends of the X- and Y-chromosomes in cases where there is no synapsis remains to be established. In male marsupials the ends of the sex chromosomes appear to be joined to a modified site on the nuclear membrane but whether this depends on some specialized form of DNA-protein interaction is not known. Equally, the question of whether chiasma formation occurs in other cases with desynapsis and only terminal metaphase-1 associations remains to be fully resolved. Ashley (1984a) suggests that in many mammals there is an association between specific ends of the X- and Y-chromosomes that is independent of homology and that the precocious desynapsis that takes place has in fact evolved to minimize the chance of genetic exchange occurring between such non-homologously synapsed regions.

D CHIASMA FORMATION

D.1 Recombination nodules

Chromosome synapsis, as we have seen, is dependent on the production of a proteinaceous axis along each of the homologues. These axes subsequently adhere to one another in pairs to form the lateral elements of the definitive SC. While this SC provides the structural basis for recombination events, the actual process of recombination and chiasma formation is now believed to be associated with transient, small, dense bodies termed *recombination nodules* (RNs). These are associated with the central region of the SC and become evident during zygotene and pachytene (Fig. 4.19).

Strongly-staining, 100 nm wide bodies located in the central space of the SC were first described by Gillies (1972) and subsequently by

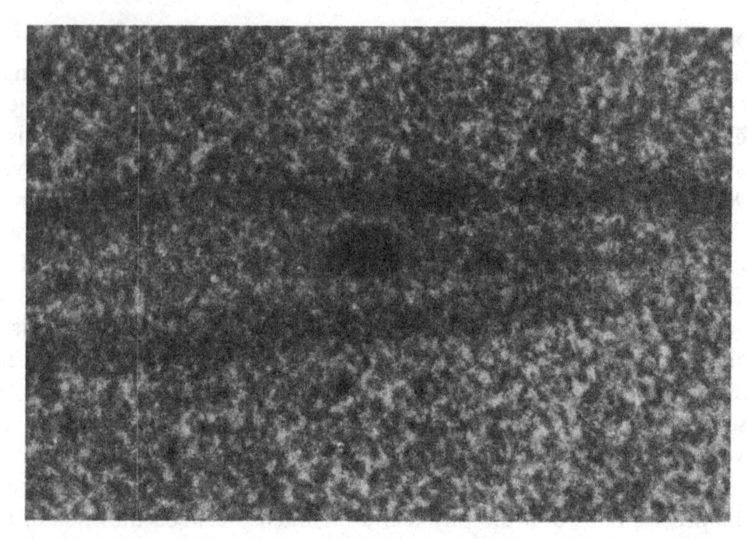

Fig. 4.19. A longitudinal electron microscope section of the synaptonemal complex in the fungus *Sordaria macrospora* demonstrating the presence of a recombination nodule in association with the central element of the complex (photograph kindly supplied by Dr Denise Zickler).

Carpenter (1975b) who proposed the term recombination nodule. These bodies are believed to represent multi-enzyme sites that either define, or else determine, precondition functions that are responsible for exchange between the DNA of non-sister chromatids (Carpenter, 1984b). The evidence in support of this belief is largely indirect and is based on five kinds of observations:

(1) In a variety of organisms, the number and distribution of RNs within bivalents correlates with either that of chiasma, or else with crossover, frequency (Table 4.12). This is particularly clear in species where the chiasmata are strictly localized. In *Allium fistulosum*, as we saw earlier (Chapter A.2.1), either one or two chiasmata occur proximal to the centromere of the metacentrics which characterize this species. Rarely, chiasmata may be interstitial or distal. In agreement with this, Albini & Jones (1984) report that either one or two proximal RNs are present in surface-spread PMCs. Similarly, Bernelot-Moens & Moens (1986) find that in *Choealtis conspersa*, where the three large metacentrics consistently form ring bivalents in which fewer than 1 % of the chiasmata are interstitial, all 57 RNs observed in these bivalents were sited in the terminal 10 nm at each end of the SC. Indeed 80 % of them were within the terminal 4 nm (Fig. 4.20). In female *Drosophila melanogaster* too, the distribution of RNs shows good agreement with the distribution of crossover events both in respect of total number

Table 4.12. *Quantitative data on recombination nodule (RN) numbers in seven species*

Species	Number of bivalents	Per nucleus: Xa/CO frequency	Per nucleus: RN number	Reference
Drosophila melanogaster ♀	4[a]	5.6 CO	5 (maximum)	Carpenter, 1975b
Neurospora crassa	7	18 CO	19	Gillies, 1979
Secale cereale	7	15–16 Xta	18–24	Abirached-Darmency *et al.*, 1983
Sordaria macrospora	7	18 CO	17–21	Zickler, 1977
Triatoma infestans	10	10 Xta	10	Solari, 1979
Lycopersicon esculentum	12	21.3 CO	20.5	Stack & Anderson, 1986b
Homo sapiens	23	50.6 Xta	69–73	Holm & Rasmussen, 1983a, b

[a] In this species only three bivalents (X, 2 & 3) form chiasmata, bivalent 4 is regularly non-chiasmate.

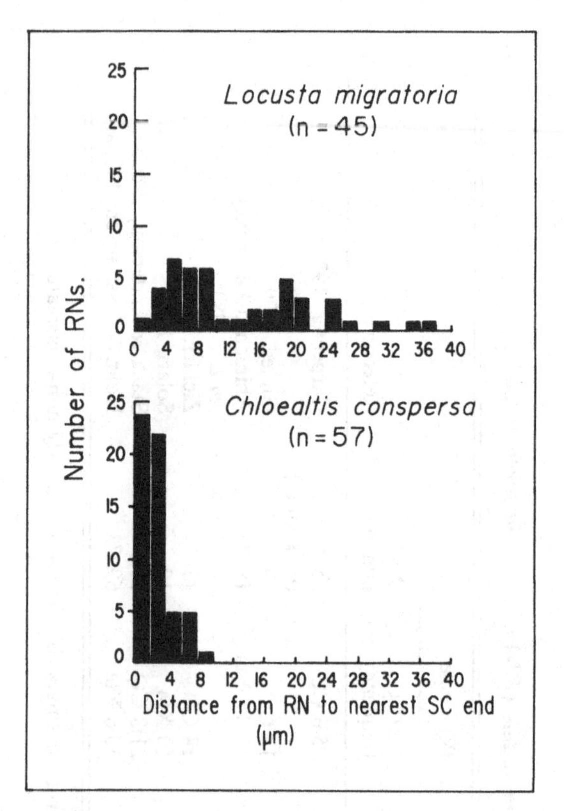

Fig. 4.20. Contrasting patterns of distribution of the recombination nodules (RNs) in bivalents of *Locusta migratoria* with random chiasma distribution and in the long metacentric bivalents of *Chloealtis conspersa* with distal chiasma localization (after Bernelot-Moens & Moens, 1986 and with the permission of Springer-Verlag).

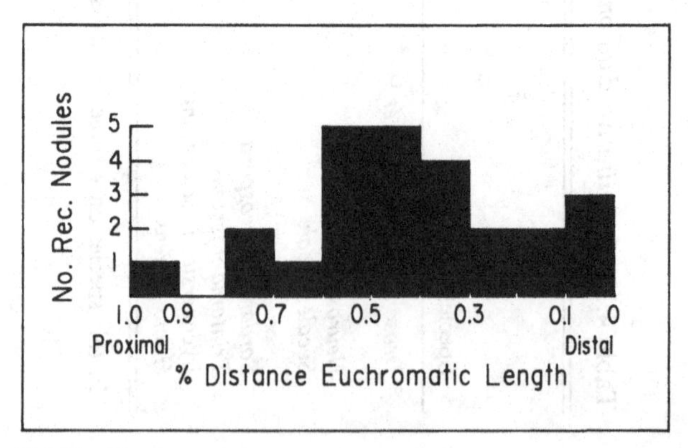

Fig. 4.21. The distribution of 25 recombination nodules (RNs) in the euchromatic arms of the major bivalents (X, 2 and 3) of *Drosophila melanogaster* females. In constructing this figure the euchromatic length of each bivalent arm has been normalized to 1.0 to simplify comparisons (after Carpenter, 1975b).

and location (see however pp. 172–3). Thus, the maximum number of five RNs per nucleus correlates well with the average expected number of exchanges. Moreover, most of the 25 RNs observed were located in middle–distal portions of the bivalent arms (Fig. 4.21), the precise location depending on the number of chiasmata per arm. Arms with one RN tended to have it located in the mid-region. Where two RNs are present one was proximal the other distal.

(2) The meiotic mutations *mei-41* and *mei-218* of *Drosophila melanogaster* (see Chapter 6A.2) lead to a parallel decrease in both recombination frequency and the number of RNs.

(3) Autoradiography demonstrates that DNA synthesis occurs in the vicinity of the RNs in female *D. melanogaster*. Thus Carpenter (1981) found a 2.6-fold excess of labelled thymidine grains over background in the region of the RNs.

(4) The SCs in female *D. melanogaster* have a different appearance in the pericentromeric heterochromatic regions (Carpenter & Baker, 1974). Here they are less thick, less distinct and supposedly less rigid than in euchromatin (Carpenter, 1975a). These heterochromatic regions do not crossover and they also lack RNs. The change from thick to thin SCs is gradual and so may also account for the lack of crossing-over in the proximal euchromatin. However, the small fourth chromosome, which has a euchromatic arm, also has a thick SC of euchromatic appearance, yet there is no crossing-over and no RNs in this chromosome at female meiosis.

(5) Finally, in *Bombyx mori* where crossing-over is limited to the male so are RNs, although SCs are present in both sexes (von Wettstein, 1984).

Unfortunately the data are not always so convincing. Abirached-Darmency *et al.* (1983) discuss the distribution of recombination nodules in rye. They point out that the number of nodules (18–24) correlates well with the number of chiasmata seen with the light microscope (12–21 with a mean of 15–16) and 10 of the 14 bivalents analysed in two pachytene nuclei had at least one distal nodule. However, as their Fig. 7 indicates, three of the seven bivalents in one nucleus have no recombination nodules anywhere near the distal ends. The authors themselves do not comment on this discrepancy.

Ashley (1987) draws attention to the fact that, while there is a good correspondence between the distribution and frequency of RNs and chiasmata in male human autosomes, this does not hold for the sex bivalent. Here, only 42% of the XY pairs analysed at late zygotene and

57% of those at early pachytene have RNs. Moreover, of the nine early pachytene spermatocytes analysed by Solari (1980), in three the RN was located in a segment which extended almost to the centromere. These would certainly not be expected to result in chiasma formation.

Precise details concerning the structure and function of RNs are even less consistent. Carpenter (1975b) initially suggested that RNs were located at crossover sites and represented structures essential for exchange. RNs do not, however, necessarily result in crossing-over since they have also been reported on SCs involved in non-homologous synapsis (Hobolth, 1981). In at least one case, the achiasmate oocytes of *Bombyx mori*, RNs are temporarily associated with the central region of the SC at early zygotene but are shed before pachytene (Rasmussen & Holm unpublished, referred to in von Wettstein *et al.*, 1984, p. 380). Moreover, there is a mutation in *Drosophila melanogaster* that reduces recombination but does not reduce the number of RNs (Carpenter, 1979b). Thus it is probably more correct to say that RNs, like SCs, are associated with and are a necessary, but not a sufficient, prerequisite for crossing-over.

Very little is known about the mechanism, or mechanisms, regulating the distribution of RNs within or between bivalents. Despite the two-fold difference in length between SCs of human oocytes and spermatocytes the number of RNs is the same in both sexes (Bojko, 1983). Evidently, in this case at least, the length of the SC is not a limiting factor in the binding of RNs to it. Carpenter favoured the idea that the information for the precise placement of the RN might be embodied in the nodule itself. An alternative possibility is that the nodules are initially distributed at random. Using a hypotonic bursting technique for spreading SCs, Stack & Anderson (1986a, b) found that at late zygotene in the tomato, *Lycopersicon esculentum*, the pericentromeric heterochromatic regions are still unpaired. These regions show late synapsis and do not form RNs. In the completed euchromatic SCs there are about 360 RNs which are distributed almost anywhere. There is, however, a 16-fold reduction in the frequency of RNs per unit length of SC by late pachytene. A complete set of 12 SCs has an average of 20.5 RNs at this stage, which compares well with the average of 21.3 crossovers per diploid set. Also, like chiasmata, RNs now tend to be located distally. Comparative data indicate that RNs are about twice as frequent in SCs in spread preparations compared to sectioned and reconstructed nuclei, though both show a comparable reduction in frequency between zygotene and late pachytene (Table 4.13). Stack & Anderson (1986b) also report on unpublished data from *Solanum tuberosum*, *Allium cepa*, *Tradescantia edwardsi* and *Psilotum nudum* (a 'primitive' seedless vascular plant), all of which show a high frequency of RNs per unit length of SC at zygotene but comparatively low frequencies at pachytene. Animals and fungi exhibit an equivalent decrease in RN

Table 4.13. *Recombination nodule (RN) frequency per μm of synaptonemal complex (SC) in Lycopersicon esculentum*

	RNs/μm SC	
Stage	Hypotonic bursting	3-dimensional reconstruction
Zygotene	1.41	0.74
Early pachytene	0.35	0.18
Late pachytene	0.085	0.05

After Stack & Anderson, 1986b.

Table 4.14. *Numbers of recombination nodules (RNs) and chromatin nodules (CNs) in spermatocytes of Bombyx mori*

	Number of			
Stage	Nuclei	RNs	CNs	Total nodule number
Late zygotene	8	91 ± 14	0	103 ± 21[a]
Early pachytene	16	58 ± 15	2 ± 3	61 ± 14
Mid-pachytene	8	37 ± 9	18 ± 10	55 ± 8
Late pachytene	6	26 ± 7	44 ± 11	70 ± 7
Pachytene–diplotene	3	16 ± 6	59 ± 8	75 ± 15

After Holm & Rasmussen, 1980.
[a] Number expected at completion of pairing from total SC length of 202 ± 17 μm.

number though the magnitude appears to be less striking (1.5 to 2-fold) and involves fewer RNs.

RNs have been described as varying in shape and size both within and between species. The between species variation is generally assumed to reflect the same plasticity in morphology that characterizes the SC itself. The variation within species, on the other hand, has been regarded as functionally significant. In diploid male *Bombyx mori* for example, RNs vary in shape and size, ranging from spheres to ellipsoids or dumbells. At mid-pachytene, some of the less dense RNs increase in size through the addition of material resembling chromatin. Holm & Rasmussen (1980) refer to these as chromatin nodules (CNs). These CNs increase in size and number during late pachytene while RNs become less frequent. Even so, the total number falls from zygotene to pachytene (Table 4.14). At the pachytene–diplotene transition most nodules are of the CN type which thus appear to evolve from RNs. By diplotene the number of CNs is similar to the number of chiasmata (Table 4.15). Nodules appear to be

Table 4.15. *The number of recombination nodules (RNs), chromatin nodules (CNs) and chiasmata (Xta) at late meiotic prophase-1 in spermatocytes of Bombyx mori*

Stage	Number of		
	Nuclei	RNs	CNs & Xta
Early diplotene	4	9	60±7
Mid-diplotene	6	2	62±7
Late diplotene	2	0	46
Diplotene–diakinesis	4	0	55±4
Early diakinesis	1	0	58
Mid-diakinesis	3	0	62±9
Late diakinesis	1		28 bivalents
Metaphase-1	5		28 bivalents

After Holm & Rasmussen, 1980.

evenly distributed along bivalents and are positioned at random with only a minor tendency for neighbouring nodules to be farther apart than expected on a random basis. The vast majority of bivalents have 1–4 RNs and 1–4 chiasmata. Only rarely do individual bivalents show more than four nodules or four chiasmata.

In human spermatocytes the total number of spherical nodules decreases from a mean of 68 at late zygotene to a mean of 30 at early diplotene. Additionally, an increasing number of spherical nodules transform into bars during pachytene. Holm & Rasmussen (1983a, b), who made these observations, concluded that, while a nodule represents a potential crossover site, the occurrence of crossing-over depends on a change from nodules to bars. In human oocytes, too, the initially spherical RNs undergo modifications in both ultrastructure and distribution after binding to the SC (Bojko, 1983). In particular, their fibrillar connections with the lateral elements of the SC become more prominent so that eventually each RN is converted into a fusiform bar which lies across the SC. These changes coincide with a reduction in the number of RNs resulting in the modification of an initially random distribution into one which is non-random and matches the distribution of chiasmata. Similar distributional changes have been noted in *Coprinus* and *Bombyx* (Holm *et al.*, 1981; Rasmussen & Holm, 1984).

Carpenter (1975a, 1979a) also found RNs of two different types – spherical and ellipsoidal – in oocytes of *Drosophila melanogaster*. These differ not only in shape and size but also in the time of their appearance. The smaller ellipsoidal type (40 d) occurs earlier in pachytene, are much more ephemeral and do not exhibit interference since they are randomly distributed in the euchromatin. The larger spherical category (50–175 d), found at mid-pachytene, are non-randomly distributed in the euchromatin and show interference. The maximum number of these spherical RNs in a wild type female is six, which agrees with the average crossover frequency of 5.6. The maximum number of ellipsoidal RNs is nine. From these, and other, facts (see Chapter 6A.2), Carpenter (1984b) has concluded that only spherical RNs are involved in crossing-over. The ellipsoidal RNs she believes to function in non-reciprocal conversion events (but see also p. 199).

In other cases, however, including the grasshopper *Chloealtis conspersa*, *Locusta migratoria*, the plant *Allium fistulosum* and the rat, *Rattus norvegicus*, only one type of RN is present and RNs may be either elliptical or rectangular according to the species involved. Clearly, these inconsistencies need to be resolved.

D.2 The molecular biology of meiosis

The process of crossing-over is evidently remarkably precise. Thus no mutations arise at the site of crossing-over. Neither is there any evidence for any local gain or loss of DNA at crossover sites. Thus whatever molecular events accompany the process there is no alteration in either the amount or structure of the DNA involved.

D.2.1 Molecular models

Discussions relating to molecular models of recombination are still largely speculative despite the fact that DNA recombination is amongst the most distinctive and common activity in the cell's biochemical repertoire. Molecular models were initially developed in the early 1960s in an attempt to explain the results which had accumulated from the analysis of recombination events in lower fungi and which are amenable to either tetrad or octad analysis. In these organisms it is possible to recover all the products of an individual meiosis. This applies both to the ordered eight-spored asci of *Ascobolus* and *Neurospora* and the sectored clones of four-spored *Saccharomyces* and *Schizosaccharomyces*.

Pre-meiotic DNA replication leads to a meiocyte which contains four DNA duplexes per potential bivalent. In the case of the ordered eight-spored ascus, meiosis, which generates a linear tetrad, is followed by a post-meiotic mitosis in each of the four meiotic products. This provides a unique opportunity to determine the genetic composition of each of the eight single DNA strands represented in the four duplexes produced by pre-meiotic DNA replication. This applies especially to cases where the marker genes employed affect the cellular phenotype of the meiotic products in terms of the shape of the ascospores or their pigmentation.

In *Ascobolus* and *Neurospora* a heterozygous gene marker (+/−) is

Table 4.16. *Meiotic gene conversion at 30 heterozygous sites in Saccharomyces cerevisiae*

Gene	Segreg-ations	6+:2*m*	2+:6*m*	5+:3*m*	3+:5*m*	ab4+:4*m*	Events (%)
pet1	4924	4	27	0	0	0	0.63
trp1	20826	62	47	12	8	0	0.63
mat1	23125	93	104	0	8	0	0.85
ura3	2315	14	10	4	1	0	0.95
ade6	1589	4	9	3	3	0	1.20
his5-2	2315	14	10	4	1	0	1.26
tyr1	8391	59	44	5	1	0	1.30
CUP1	18016	123	124	6	3	0	1.42
gal2	2416	36	17	0	0	0	1.78
leu2-1	3203	41	25	0	0	0	2.06
trp5-48	2315	23	31	1	0	0	2.38
met1	1589	18	17	18	6	0	3.72
met10	892	17	16	1	0	0	3.82
ura1	15014	331	275	11	15	1	4.23
ilv3	9487	230	239	9	6	0	5.10
lys1-1	2315	51	78	3	7	0	5.76
SUP6	892	22	33	0	0	0	6.16
thr1	21220	691	594	14	22	0	6.23
his4-4	12533	411	303	39	41	1	6.36
ade8-10	1118	48	30	0	0	0	7.41
met13	10454	459	474	1	0	0	9.07
ade8-18	14480	351	239	375	294	10	9.72
cdc14	892	48	39	1	1	0	9.99
ade7	1028	47	27	11	8	0	10.7
his2	2481	215	231	2	0	0	18.2
arg4-4	1188	5	13	0	1	0	1.6
arg4-3	2405	28	23	5	3	0	2.45
arg4-19	5352	87	69	2	6	0	3.04
arg4-16	14490	508	302	95	289	4	8.29
arg4-17	13476	485	541	38	20	0	8.12

After Fogel *et al.*, 1979 and with the permission of Cold Spring Harbor Laboratory.

expected to segregate 4+/4− following meiotic recombination and this is certainly true for distant linked markers, which regularly show a 4:4 segregation. In many fungi, however, it has long been apparent that a heterozygous marker may show aberrant segregation patterns. These are known collectively as gene conversions. In the four-spored asci of yeast the frequency of aberrant segregation varies from < 1% to upwards of 20% depending on the gene locus concerned. In *Sordaria* the frequency is much less, usually between 0.1% and 1.0%, though because of the presence of eight spores in the ascus, a greater range of aberrations can be identified (Fincham, 1983).

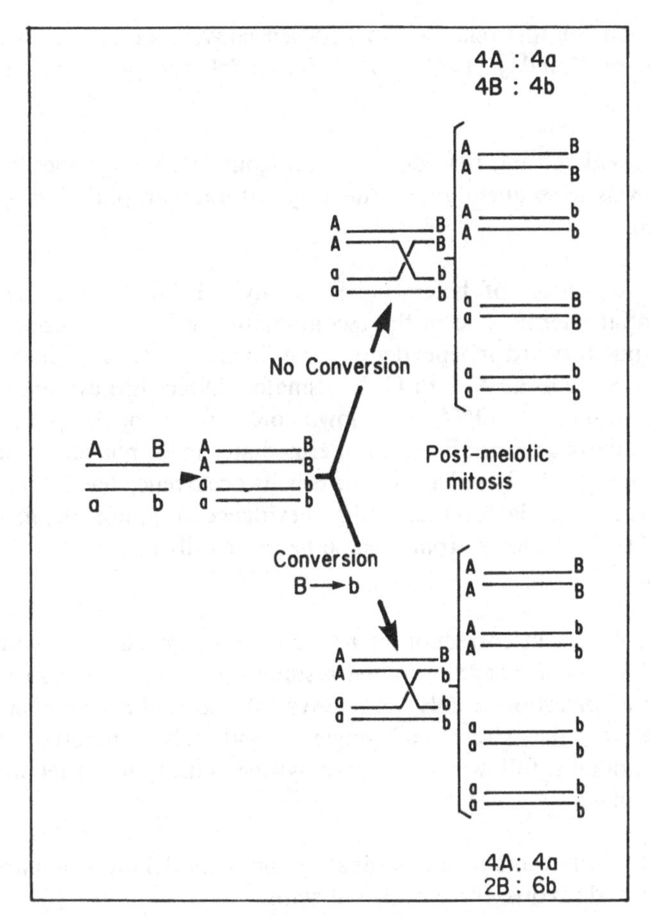

Fig. 4.22. The ratios of segregant classes in the presence and absence of gene conversion in fungi with ordered tetrads.

In yeast, the most common form of conversion results in a 6+/2− or else a 2+/6− pattern of segregation (Table 4.16). This is a consequence of a non-reciprocal transfer of genetic information from one DNA duplex to another (Fig. 4.22). A second form of aberrant segregation arises when the two strands of a single DNA duplex produced by meiosis carry different genetic information. Here a spore mother cell is produced that divides to give two genetically different daughters, an event termed *post-meiotic segregation*. When one pair of non-identical sister spores is present within an ascus a 5:3 segregation is produced. If, however, two pairs of non-identical sister spores are present there will be an aberrant 4:4 segregation.

Extensive tetrad analysis in fungi indicates that, when a site shows conversion, flanking markers undergo crossing-over at frequencies of up to 50%. The first widely accepted model to provide a molecular explanation for this relationship between conversion and crossing-over was that of Holliday (1964, 1974). His model incorporated three basic features:

(1) The breakage and reunion of homologous DNA molecules, an event that was subsequently confirmed by autoradiographic data (Taylor, 1965).

(2) The formation of heteroduplex or hybrid DNA (hDNA) as an essential intermediate in the recombination process, a proposal that was put forward independently by Whitehouse (1963). Such hybrid DNA contains an A–C or G–T mismatch. Direct EM evidence for the production of hDNA is known only from prokaryotes, more particularly as recombination intermediates in T_4 phages (Brooker & Lehman, 1971). In eukaryote meiosis its occurrence has been inferred from genetical data, though indirect evidence to support the formation of hDNA is known from chinese hamster cells in culture (Moore & Holliday, 1976).

(3) The subsequent correction or repair of mismatched bases within the hDNA region, resulting in the resumption of normal base pairing. Such a correction entails the removal of one of the two mismatched bases, or the stretch of single strand DNA involved in the heteroduplex, followed by a repair synthesis using the sister strand as a template.

The physical events of the original Holliday model are summarized in Fig. 4.23 and involve three principal steps:

(1) Initiation by enzymatic nicking of single strand breaks of the same polarity and at homologous sites resulting in co-parallel breakage.

(2) Formation of a heteroduplex joint by an exchange of non-sister strands, with the exchange partners being ligated into new positions. This generates a region of symmetrical hDNA on each of the two interacting non-sister strands and gives rise to a crossed strand, commonly referred to as a *Holliday structure* or *half chiasma*.

(3) Resolution of the resulting Holliday junction. In Figure 4.23 this is displayed in a chi-form to better illustrate the two possible modes of resolution. The strands involved in a Holliday junction can then swivel or isomerize by the cutting of either the originally crossed strands or the non-crossed strands. A vertical cut yields chromatids which are non-crossover for flanking markers. A horizontal cut leads to flanking marker recombination. When a marker site falls within the

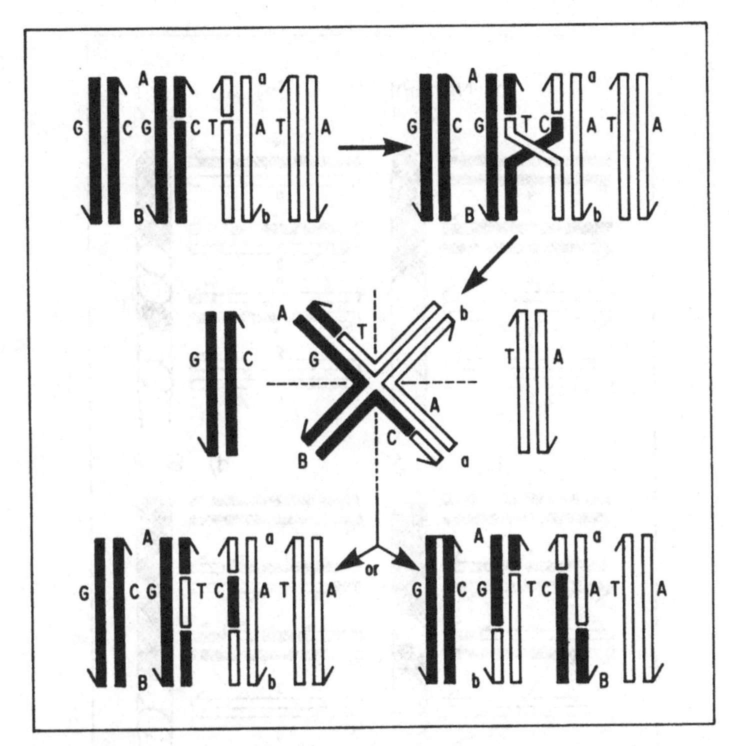

Fig. 4.23. The Holliday model for meiotic recombination. The Holliday structure resulting from the ligation of nicked chains of identical polarity is displayed here in chi form to illustrate the structural equivalence of the two modes of its resolution (after Fogel *et al.*, 1981 and with the permission of Cold Spring Harbor Laboratory).

region of hDNA, mismatches are created and aberrant segregation can occur. If the mismatches on both DNA duplexes are corrected in the same direction a 6:2 segregation results. If only the mismatch in one duplex is corrected a 5:3 segregation occurs while if neither mismatch is corrected an aberrant 4:4 segregation follows (Fig. 4.24).

According to the Holliday model, therefore, gene conversion results from the correction of the mismatched region of hDNA. More specifically, gene conversion results when one of the strands of a heteroduplex region containing mismatched bases is corrected using the opposite strand as a template. Isomerization of the crossed strand structure at a Holliday junction, followed by nicking and ligation, leads to reciprocal recombination accompanying gene conversion. The extent of reciprocal recombination thus depends on the extent of isomerization. Meiotic intrachromosomal gene conversion also occurs frequently in yeast but is not associated with reciprocal exchange of flanking markers (Klein, 1984).

In the original Holliday model it was assumed that heteroduplex DNA

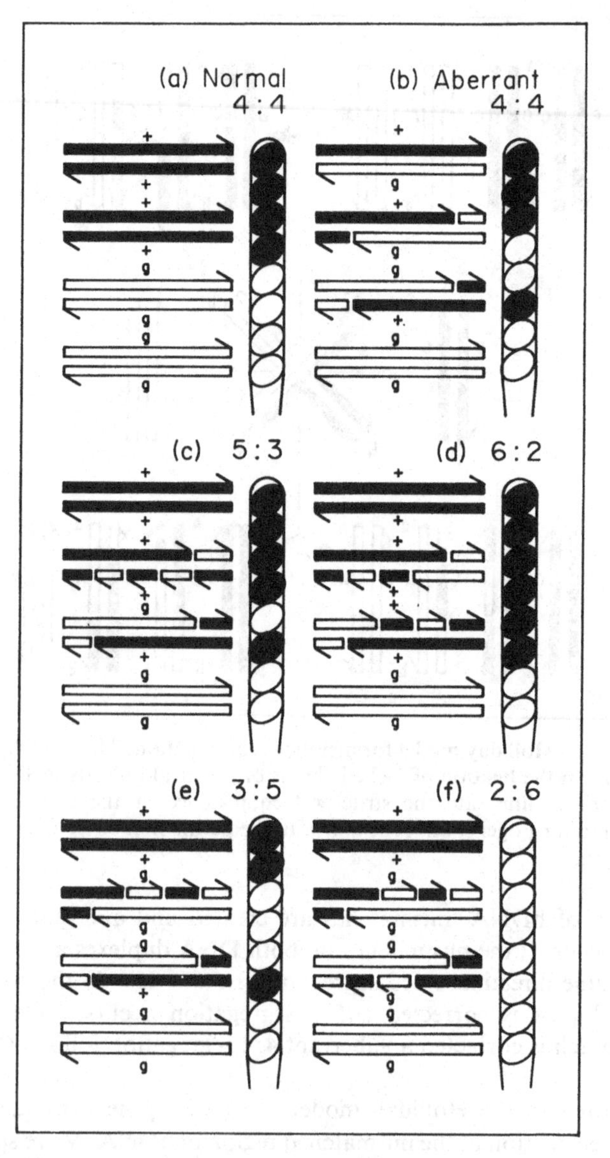

Fig. 4.24. Patterns of segregation for a grey (*g*) spore colour marker in the fungus *Sordaria fimicola*. (*a*) is normal behaviour. The formation of hybrid DNA accounts for post-meiotic segregation (*b*) while the subsequent correction of mispairing explains conversion. Thus a reciprocal exchange followed by mismatch repair in favour of wild type in one (*c*) or both (*d*) hybrid segments results in 5:3 and 6:2 segregation respectively. Mismatch repair in favour of the mutant in one (*e*) or both (*f*) hybrid segments leads respectively to 3:5 and 2:6 segregation (after Whitehouse, 1982 and with the permission of John Wiley & Sons).

formed equally in both chromatids. While some data certainly supports this assumption, studies on the *arg 4* locus of yeast and the *W17* locus of *Ascobolus* suggest that, for these systems, hDNA forms on only one chromatid. To accommodate this, Meselson & Radding (1975) proposed a model which allows for the production of a Holliday junction through a DNA hybridization event which is initially unilateral. In this model a single stranded segment is transferred from one chromatid to its homologue with the corresponding segment in the recipient being displaced and degraded. Coupled with this, the gap in the donor is then repaired by copying from its complementary strand. In this way, the recipient chromatid will have a region of hDNA, while the donor will have repaired its original homoduplex.

A number of other models have also been produced (reviewed in Catcheside, 1977 and Whitehouse, 1982) including, most recently, one whereby meiotic recombination may be initiated by double strand breaks (Szostak, Orr-Weaver & Rothstein, 1983). In all of these models the basic working concept is that a recombination event involves the formation of lengths of hDNA in one or more chromatids. Despite the plausibility of several of these models, none of them has been formally proven, nor is it known whether there is one, or more than one, recombination pathway within any given species (Carpenter, 1984b).

In the late 1950s two opposing hypothetical mechanisms were advanced to explain meiotic recombination. One was based on DNA strand breakage and reunion, the other involved strand copy choice during DNA replication (reviewed in John & Lewis, 1966). Although the copy choice model was eventually abandoned because of convincing evidence against it we now see it in a transformed, and acceptable, state. Copy choice replication is simply a local episode of repair synthesis which retains the ability to explain high negative interference over intragenic distances. Crossing-over at meiosis is, as we have seen, also characterized by positive interference. However, crossovers do not interfere with gene conversions. Neither do conversions interfere with one another (Holliday *et al.*, 1984). Molecular models provide no explanation for positive crossover interference and it appears unlikely that this phenomenon is directly related either to the inherent structure of DNA or to the molecular mechanisms responsible for crossing-over *per se*.

D.2.2 Meiotic metabolism

In terms of the models that have been developed for meiotic recombination it follows that the process should involve a sequence of biochemical events which includes:

(1) The breakage of DNA strands

(2) The formation of regions of hDNA

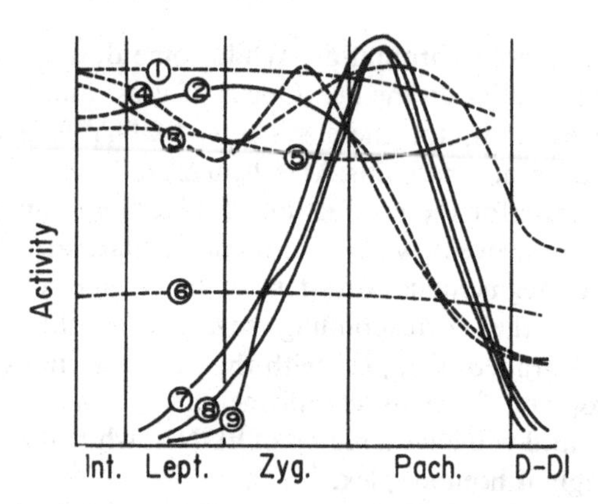

Fig. 4.25. Meiotic activity profiles of nine DNA-related enzymes in microsporocytes of *Lilium* including (1) DNA polymerase B, (2) Polynucleotide kinase, (3) Polynucleotide ligase, (4) Topoisomerase II, (5) Acid phosphatase, (6) Topoisomerase I, (7) R-protein, (8) U-protein and (9) Meiotic endonuclease. The stages referred to on the abscissa are Int = pre-meiotic interphase, Lept. = leptotene, Zyg. = zygotene, Pach. = pachytene and D–DI = diplotene–diakinesis (after Stern & Hotta, 1983).

(3) Localized degradation and synthesis of DNA, and

(4) Rejoining, or ligation, of free ends

These events require the participation of nucleases, polymerases, topoisomerases and ligases. Consequently, cells lacking any, or all, of these enzyme systems would be expected to be recombination defective (see Chapter 6A.2). Evidence in support of the occurrence of these biochemical events comes principally from a remarkable series of publications, spanning some 30 years, carried out by Herbert Stern and his colleagues on lily meiocytes. These studies have demonstrated that the zygotene–pachytene period is characterized by a distinctive phase of metabolism involving the synthesis of specific proteins and enzymes which are important for pairing and recombination, as well as the specific synthesis of small amounts of DNA at zygotene and pachytene. They conclude that these events are associated with the introduction of discontinuities in the DNA of the chromosome, and their subsequent repair, in such a way as to lead to recombination. While much of the evidence in support of their conclusions is indirect and circumstantial it is, nevertheless, compelling.

In lily anthers the meiocytes develop synchronously. Consequently, bud length can be used to predict and define the meiotic stage. Fig. 4.25 summarizes the patterns of DNA-related molecular activity in lily meiocytes during the interval between pre-meiotic S and the end of pachytene. Three major groups of molecules can be distinguished depending on their respective patterns of activity (Stern & Hotta, 1983). The first of these is active during pre-meiosis but then fluctuates to greater or lesser degrees during zygotene and pachytene. This group includes: DNA polymerase B (1); polynucleotide kinase (2); polynucleotide ligase (3); topiosomerase II (4); and, acid phosphatase (5). The second group is inactive during pre-leptotene but increases markedly in activity at the end of pachytene. This applies to the R (recombination) protein (7), the U (the DNA-unwinding) protein (8) and meiotic endonuclease (9). To the extent that members of this group have been characterized they appear to be specific to meiocytes and their transient presence during zygotene–pachytene points unmistakably to a role in the pairing–recombination events known to occur at this time. Precisely which chromosome sites regulate the transcription of these meiosis-specific proteins is not known. As far as the third group, which includes topoisomerase I (6), is concerned, its activity is maintained at a reasonably constant level throughout the entire period from pre-meiosis to diplotene.

Of the three meiosis-specific proteins, it is assumed that meiotic endonuclease is programmed to introduce nicks into the DNA at a time when pairing is completed. Indeed, the action of U-protein, which unwinds duplex DNA for a distance of up to *c*. 400 base pairs from an available 3′OH terminus, depends on the availability of these nicks as well as on the availability of ATP. Meiotic endonuclease itself forms 3′P terminii; consequently the phosphate so formed has to be removed before unwinding and repair synthesis can occur. The combined action of endonuclease, phosphatase and U-protein leads to the production of single stranded DNA at nicked sites and so can be expected to provide single strand stretches for heteroduplex formation. Reassociation of single stranded DNA is then catalysed by a combination of R-protein and ligase (Fig. 4.26). The R-protein is part of the lipoprotein complex of the nucleus and its activity is also dependent on phosphorylation. In the absence of homologous pairing, as in the case of the achiasmatic lily hybrid, cultivar Black Beauty, there is a marked decline in the activity of R-protein. Typical PMCs in this hybrid have a mean of 1.85 bivalents per cell, with a range of 0–6, and a mean cell chiasma frequency of 2.55, with a range of 0–8. Colchicine-induced tetraploid cells of this hybrid have a significantly higher mean number of SCs compared to the diploid controls (14 versus 5.7). The achiasmatic condition of the diploid hybrid is thus due to insufficient homology (Bennett *et al*., 1979). The achiasmate meiocytes do not undergo DNA nicking at pachytene. Neither do they show the high level of R-protein which characterizes chiasmatic lilies (Hotta *et al*., 1979). Induced tetraploidy, however, restores the level of R-protein to the normal diploid condition (Toledo, Bennett & Stern, 1979). The absence of nicking in the achiasmate diploid would thus appear to result from the unpaired condition of the homologues. This implies that homologues are

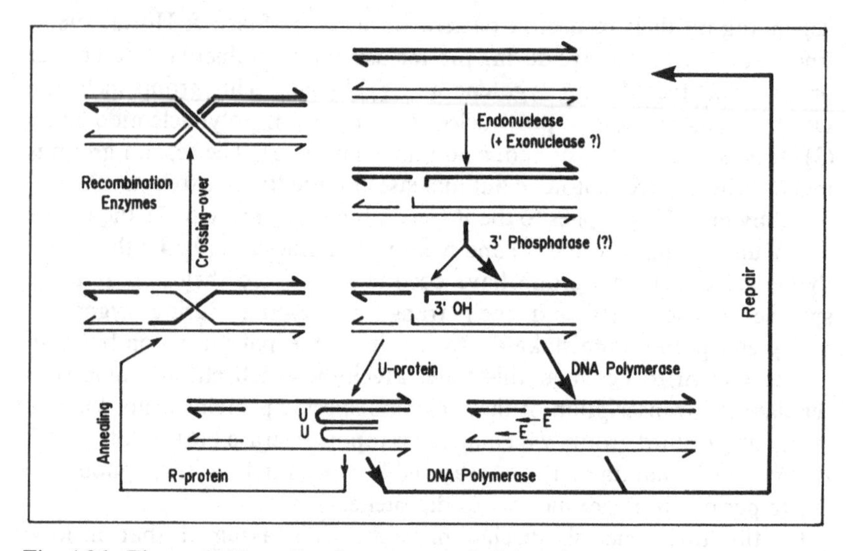

Fig. 4.26. The probable molecular events underlying the regulation of crossovers in eukaryote meiocytes (after Stern & Hotta, 1978 and with the permission of Annual Reviews Inc.).

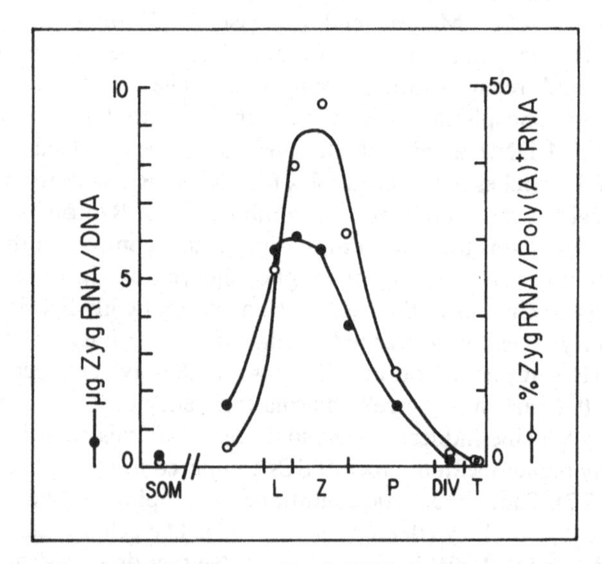

Fig. 4.27. Levels of PMC zygotene (Zyg.) RNA during meiosis in *Lilium*. The stages referred to on the abscissa are SOM = somatic, L = leptotene, Z = zygotene, P. = pachytene, DIV = remainder of meiosis and T = tetrads (after Hotta *et al.*, 1985c and with the permission of *Cell*).

not accessible to the activity of nicking enzymes since the developmental profiles of both endonuclease and U-protein activity are identical in chiasmate and achiasmate forms. Clearly, the cyclic activity of these meiotic-specific proteins is unaffected by the absence of homologous pairing. The data indicate that meiotic chromosome pairing can itself regulate metabolic activity in the pachytene nucleus, presumably by the chromatin in paired regions undergoing some form of conformational change.

One possibility suggested by earlier workers was that recombination in eukaryotes was initiated at pre-meiotic S, perhaps through a failure to complete DNA synthesis. This, in turn, would have led to the production of unjoined replicon ends which then became available for recombination. This can now be completely discounted. More recently, Chandley (1986) has proposed a model which couples pairing initiation and a commitment to recombination with early replication at the pre-meiotic S-phase. She suggests that early replicating (E) sites act also as sites for synaptic initiation and subsequently for crossing-over. This proposal carries the rider that full synapsis and SC formation are not required for crossing-over *per se*, though the completed SC could provide a stabilizing structure which ensures that E-sites remain in association until recombination is completed. This model, however, ignores the fact that, in addition to the bulk synthesis of DNA at pre-meiotic S, two distinctive groups of DNA sequences replicate during meiosis itself and in association with pairing and recombination, respectively (Hotta, Ito & Stern 1966; Hotta & Stern, 1971, 1976). Inhibition of these phases of meiotic DNA replication, which take place at zygotene and pachytene, respectively, leads to the production of chromosome breaks (Ito, Hotta & Stern, 1967).

The first class of sequences involved in meiotic replication is referred to as *Zyg-DNA* because its production is coordinated with chromosome pairing (Fig. 4.27). Its replication is semi-conservative in character and occurs in the vicinity of the SC (Kurata & Ito, 1978; Moses & Poorman, 1984). The DNA concerned represents only a very small fraction, *c.* 0.1–0.2%, of the total genome (Hotta *et al.*,1985c; Stern, 1986), is widely distributed in all chromosomes and consists of 2.5–10 kilobase segments (Stern & Hotta, 1985). Most of this Zyg-DNA has the characteristics of unique, or at least low copy number, DNA whereas much of the remainder of the genome is composed largely of repetitive sequence DNA. Although constituting only a very small fraction of the total genome the pattern of Zyg-DNA replication, while still semi-conservative, is also distinctive. Under the same conditions that govern its release at zygotene itself, unreplicated Zyg-DNA cannot normally be excised from chromatin by S1 nuclease treatment at any pre-zygotene stage. However, if pre-zygotene nuclei are treated with deoxycholate (DOC), which removes protein components of the nuclear membrane, incubation with S1

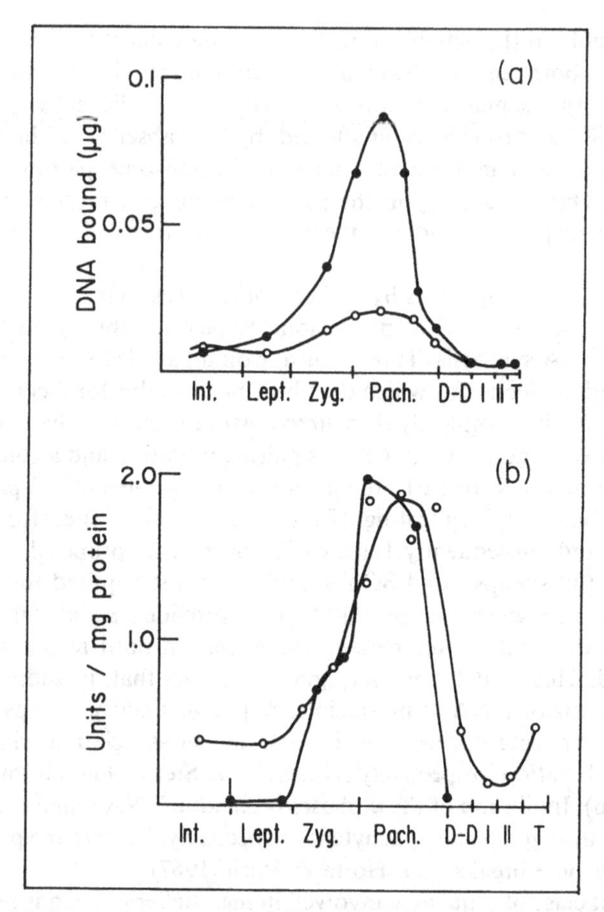

Fig. 4.28. (*a*) R-protein content and (*b*) endonuclease activity in PMCs of the triploid chiasmate lily cultivar 'Sonata' (closed circles) and the non-chiasmate diploid hybrid 'Black Beauty' (open circles). Note that while the peak concentration of R-protein in Black Beauty is only one-fifth that of Sonata, endonuclease activity is unaltered. The stages referred to on the abscissa include Int. = premeiotic interphase, Lept. = leptotene, Zyg. = zygotene, Pach. = pachytene, D–D = diplotene–diakinesis, I = remainder of first division, II = second meiotic division, T = tetrad stage (after Hotta *et al.*, 1979 and with the permission of Springer-Verlag).

nuclease will now release unreplicated Zyg-DNA as duplex fragments. Such treatment also makes it possible for these nuclei to replicate Zyg-DNA when incubated under appropriate conditions. The DOC-extractable factor that inhibits Zyg-DNA replication has been referred to by Hotta, Tabata & Stern (1984) as *leptotene-*, or *L-protein*. The ability of DOC-treated leptotene nuclei to replicate Zyg-DNA can be suppressed if they are exposed to L-protein or to an extract containing that protein.

The synthesis of L-protein must occur towards the end of pre-meiotic

S since DOC-extracts of pre-meiotic S nuclei have little L-protein. Somatic nuclei and nuclei from mid-zygotene and later stages of meiosis are also deficient in L-protein. The suppression of Zyg-DNA replication by L-protein thus coincides with the time at which there is an irreversible commitment of cells to enter meiosis. When bound *in vitro* to duplex Zyg-DNA, L-protein introduces a specific nick into one of the DNA strands at the site to which it binds. If this also occurs *in vivo* it could provide both the signal for initiating Zyg-DNA replication as well as producing single strands in homologous Zyg-DNA segments which would then permit the cross-hybridization of complementary DNA strands. Inhibition of Zyg-DNA replication leads to the arrest of chromosome pairing and of SC formation (Roth & Ito, 1967). Thus these Zyg-DNA sequences may provide sites for chromosome alignment and so play a key role in the synapsis of homologues. Suppression of Zyg-DNA synthesis coincides with an irreversible commitment of cells to engage in meiosis. Pulse labelling with tritiated thymidine during zygotene results in a concentration of label in the SC (Kurata & Ito, 1978). Stern & Hotta (1985) speculate that Zyg-DNA represents an axial component of each chromosome required for the alignment of homologues.

Zygotene replication is, however, incomplete since gaps in small flanking regions, which are also shielded from subsequent repair at pachytene synthesis, remain open until the end of first prophase (Hotta & Stern, 1976). This is so despite the fact that meiocytes are capable of repair replication at most stages of meiosis and especially during pachytene when there is active repair replication. Evidently, these gaps must in some way be protected from the activity of the pachytene repair mechanism.

Apart from its role as a structural component of homologue synapsis, a small proportion of the DNA sequences that replicate at zygotene are also transcribed at this time. Indeed, some 40% of the total poly(A)-RNA population of meiocytes consist of Zyg-DNA transcripts. By contrast, the achiasmatic lily hybrid has little Zyg-RNA (Hotta *et al.*, 1985), indicating a possible relationship between synapsis and the presence of such transcripts. Lily Zyg-RNA hybridizes with DNA from a wide variety of eukaryotes when tested under conditions of moderate stringency confirming its conservation over a wide phylogenetic spectrum. It is not known whether the sequences that are transcribed constitute a special subset of Zyg-DNA segments that function solely in transcription. Neither is it known whether the Zyg-DNA transcripts are subsequently translated.

The entry of cells into pachytene is marked by a switch from a semi-conservative mode of DNA synthesis. This switch is concerned with repair replication and involves the synthesis of a specific *pachytene*-, or *P-DNA* whose production relates to crossing-over and is concerned with the repair of sites of pachytene-DNA nicking (Hotta & Stern, 1974). Nicking at

Table 4.17. *Angiosperm DNA sequences which cross-hybridize with the P-DNA of the Asiatic hybrid cultivar Lilium variety Enchantment*

Species	1C-DNA (pg)	Number of repeats per P-DNA family
Lilium c.v. Enchantment	44.3	1000
Asparagus	1.9	375
Lilium longiflorum	34.0	320
Iris	3.6	230
Lilium henryi	30.5	180
Vicia faba	13.4	175
Lilium speciosum	29.7	140
Secale cereale	8.8	60
Zea mays	3.3	30
Triticum aestivum	5.8	25

After Friedman *et al.*, 1982 and with the permission of Springer-Verlag.

pachytene is dependent on the action of a meiosis-specific endonuclease as well as an exonuclease which extends the nicks into short single stranded gaps. These gaps occur at widely spaced intervals within the genome and are associated with the presence of a specific group of P-DNA regions. P-DNA consists of a number of distinct sequence families each consisting of virtually identical members (Bouchard & Stern, 1980). In *Lilium* c.v. Enchantment there are 600 such sequence families with an average of 1000 repeats per family and a sequence divergence within families of 1 % or less. Friedman, Bouchard & Stern (1982) find that P-DNA middle repetitive (mr) sequences from this lily cultivar are also highly conserved in other monocots as well as in the dicot *Vicia faba* (Table 4.17). Thus, sequence content shows < 6% divergence between *Lilium*, *Zea mays* and *Secale cereale*. Not only is there little divergence among the repeats within a P-DNA family but the families themselves have been conserved in the genome of even distantly related plant species. Additionally, the pattern of pachytene DNA metabolism in fractionated male mouse meiotic cells is virtually identical to that in lily microsporocytes (Hotta, Chandley & Stern, 1977).

Pachytene nicking is known to be regularly excluded from the highly repetitive (hr) satellite DNA sequences that make up some 10% of the mouse genome (Hotta & Stern, 1978b), as well as from equivalent sequences in the Arabian oryx (*Oryx leucoryx*) where they are four times more frequent (Stern & Hotta, 1980). This neatly accommodates the lack of chiasmata and crossing-over in heterochromatic regions. Most eukaryote genomes also include a substantial middle repetitive DNA sequence component. Since regions of this type also do not appear to recombine, it may be that only chromosome segments with extended homology become committed to exchange.

When total DNA is isolated from pachytene cells, the majority of P-DNA sequences can be shown to occur in regions composed of a modal length of 160 kb. In the lily such sequences are excluded from more than half of the genome. Moreover, the synthesis of P-DNA is localized within short end segments which appear to flank the P-DNA units. Hotta & Stern (1981) refer to these pachytene-DNA synthesis sites as *Psn-DNA*. Chromatin composed of Psn-DNA contains little or no histone but is associated with a non-histone Psn-protein as well as with Psn-RNA which is synthesized during zygotene–pachytene. Nicking, exonuclease activity and repair synthesis during pachytene occur only in regions of Psn-DNA (Hotta & Stern, 1984).

Synapsis is accompanied by distinctive changes in the chromatin which houses the Psn-DNA segments (Hotta *et al.*, 1985a). These changes regulate the accessibility of Psn-DNA regions to endonuclease and it is this that determines the site selective nicking that occurs at pachytene. How P-DNA metabolism relates to the formation and function of recombination nodules is unknown. Certainly the number of P-DNA sites exceeds the number of crossover sites by several orders of magnitude. As has already been mentioned, there is evidence that the synapsis of homologous regions in some way controls the sequence of events involved with pachytene nick-repair activity. Thus, if synapsis is reduced, either by hybridization or by treatment with colchicine, then the levels of endogenous pachytene nicking, repair synthesis, R-protein and Psn-RNA are all significantly reduced compared to those in cells with normal chiasma frequency (Stern & Hotta, 1983). In the absence of homologous pairing the changes that normally occur in Psn-DNA evidently do not occur. Consequently, there is an absence of nick-repair activity despite the normal levels of endonuclease (Fig. 4.28).

What is lacking is information on the factors that transform potential crossover sites into chiasmata (Stern & Hotta, 1984). DNA discontinuities and the availability of single strand gaps, while required for recombination do not ensure it. Moreover, synapsis can only accommodate pairing of the intimacy required for crossing-over in a small section of the total chromatid. While either DNA matching or protein matching may function in the alignment of homologues, DNA regions must be involved for the more precise form of pairing necessary for recombination. Given that the SC is a device to hold homologues in register, the base pairing required for crossing-over is achieved at only a small number of sites. Because of the DNA compaction evident at zygotene–pachytene it has been argued that potential crossover sites must in some sense be preselected before becoming entrapped in the SC. How such preselection is achieved, if indeed it occurs, is unknown.

Rec A mutants of *E. coli* were first isolated on the basis of their complete deficiency of recombination. This, in turn, led to the identification of a Rec A-protein, a small specific protease that regulates an inducible pathway of

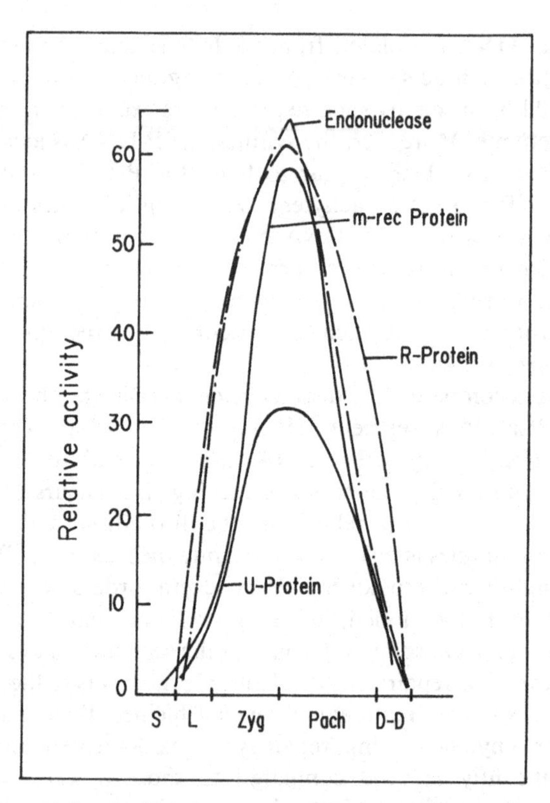

Fig. 4.29. The activity profile of m-rec protein relative to that of the three other principal recombination-related proteins of *Lilium*. The stages referred to on the abscissa are S = somatic, L = leptotene, Zyg. = zygotene, Pach. = pachytene and D–D = diplotene–diakinesis (after Stern & Hotta, 1987 and with the permission of Academic Press Inc.).

DNA repair. Using the energy of ATP, this protein also promotes both the *in vitro* pairing of single stranded, or partially single stranded, DNA with double stranded DNA, as well as the displacement of a strand from an original duplex and the assimilation of a new strand (Flory *et al.*, 1984). Interestingly, Hotta *et al.* (1985) have described the purification of two types of Rec A-like protein, one from somatic (s-rec) and the other from meiotic (m-rec) cells of both lily and mouse. They find that a major increase in m-rec protein coincides with entry into meiosis, with a peak of activity at early pachytene (Fig. 4.29). There is a possibility, therefore, that the R-protein functions only in effecting synapsis and has no immediate role in recombination. In support of this is the fact that U-protein, but not R-protein, has a stimulatory effect on recombination activity in meiotic extracts. Additionally, using plasmids to assay recombinogenic activity, these same authors have been able to show that

Table 4.18. *Factors affecting in vitro recombination frequency of prophase-1 cell extracts from lily (zygotene–pachytene microsporocytes), mouse (early + late prophase spermatocytes) and yeast (cells in sporulation medium for c. 6 h)*

Incubation medium	Relative frequency of recombinants (%)		
	Lily	Mouse	Yeast
Complete medium	100	100	100
+ Rec A (E. coli)[a]	140	250	160
+ m-rec (lily)[b]	140	250	120
+ U-protein (lily)[b]	175	290	190
+ m-rec (lily) + U-protein (lily)	225	300	190

After Hotta *et al.*, 1985b and with the permission of Springer-Verlag.
[a] 3 μg added to 10 μg extract.
[b] 5 μg added to 10 μg extract.

meiotic extracts of yeast, lily and mouse have the capacity to produce high frequencies of recombination (Table 4.18).

Among the proteins tested, lily U-protein had the most marked effect. For the lily, extracts were prepared from mixtures of zygotene–pachytene microsporocytes. In the case of the mouse, extracts came from mixtures of early and late prophase spermatocytes, while in yeast extracts were obtained from cells that had been in sporulating medium for approximately six hours. In the case of the lily, the rise in recombination frequency did not begin until cells entered zygotene. By using the achiasmatic cultivar Black Beauty it was possible to show that the recombinogenic activity profile was the same regardless of whether homologously synapsed chromosomes were present.

A protein similar to the Rec A-protein of *E. coli* has also been characterized in *Ustilago maydis* (Holliday *et al.*, 1984). The *Ustilago* Rec 1-protein promotes homologous pairing and strand exchange between a variety of DNA substrates (Kmiec & Holloman, 1984). This protein is absent in *Rec 1*-mutants of *Ustilago* which are defective, though not totally so, in both mitotic and meiotic recombination. *Ustilago* diploids homozygous for both *Rec 1*- and *Rec 2*-mutations have a much more severe deficiency and here meiosis is blocked.

Male mice that carry X-autosome translocations are sterile as a result of spermatogenic failure. So too are the tertiary trisomic derivatives of these mice. Some autosome–autosome translocations are also sterile. In all these male steriles, spermatogenesis is arrested at or before metaphase-1. From an analysis of DNA metabolism at first meiotic prophase in such

Table 4.19. *Germ line variant forms of rat histones*

Histone class	Germ line variant
H1	H1t, H1a
H2A	TH2A, H2A.X
H2B	TH2B
H3	TH3
H4	No variants

After Meistrich & Brock, 1987.

male sterile mice, Hotta *et al.* (1979) found that the truncation of male meiosis was accomplished by a disruption of pachytene metabolism. Specifically, they were able to show that:

(1) Endogenously-generated pachytene nicks in DNA were seven times more intense in steriles due to a persistence of nicking activity. Nicking also tended to be randomized in distribution.

(2) In normal males, pachytene repair synthesis does not occur in satellite regions but, in all three types of steriles, satellite DNA underwent pachytene nick repair.

(3) The repair capacity of spermatocytes from sterile males was identical with that of fertile forms so that the DNA repair system was obviously not implicated in the production of sterility.

The authors conclude that spermatocyte abortion in mice is associated with increased single strand nicking but not with a decreased capacity to repair DNA damage. Whether the persistence of nicking is responsible for the subsequent germ cell failure is not clear, though this is a view favoured by Hotta *et al.* themselves. It certainly points to a deregulation of endonuclease activity but whether this is a cause or an effect of the disturbance that leads to sterility remains to be clarified.

Transitions in chromosomal proteins are known to occur concurrently with changes in meiotic morphology and such transitions too may play some role in regulating meiotic metabolism. DNA replication at zygotene and pachytene, like DNA replication at pre-meiosis, is likely to involve localized changes in chromatin structure in specific groups of DNA sequences. So too must the transcription of meiosis-specific messages (Hotta *et al.*, 1985a). Meiosis-specific histones are present in a number of species (Sheridan & Stern, 1967; Bogdanov, Strokov & Reznickova, 1973). Thus rat testis contains six specific histones (Table

4.19) closely related to somatic histones 1–3, respectively (Meistrich & Brock, 1987). Their precise role has yet to be defined.

D.3 Systems of recombination

Recombination events at meiosis are, as we have seen, of two kinds – reciprocal gene crossing-over and non-reciprocal gene conversion. Crossing-over changes the linkage relationships of substantial chromosome segments. Gene conversion is much more restricted, leading to aberrant segregation at a single site or several close sites.

Conversions can only be defined and analysed when all individual meiotic products can be recovered. Although, strictly speaking, it is not possible to directly demonstrate conversion events in half tetrads, conversion provides the only plausible explanation for the data obtained from the analysis of half tetrads in female *Drosophila melanogaster*. On this basis it is possible to demonstrate three clear parallels between crossing-over and conversion in this species (Finnerty, 1976; Hilliker & Chovnick, 1981; Hilliker, Clark & Chovnick, 1988):

(1) Both occur only in female meiosis, which is chiasmate. Neither occur in the achiasmate male meiosis.

(2) Both also occur in mutant heterozygotes but not in homozygotes.

(3) The wild types generated by gene conversion are indistinguishable from those generated by crossing-over.

However, conversion in female *Drosophila* shows some differences from the process identified in fungi. While post-meiotic segregation and polarized exchange are important features of fungal conversion they appear to be less important to conversion in *Drosophila*, though post-meiotic segregants have been identified in a recombination deficient genotype (*mei-9*) by Carpenter (1982). Hurst, Fogel & Mortimer (1972) estimate that there are some 150–200 conversion events per meiosis in yeast with 75–100 associated reciprocal exchanges. On the other hand, Hilliker *et al.* (1988) report that in the rosy locus of *D. melanogaster*, all intragenic crossovers originate as conversions and that approximately 20% of these result in crossing-over. The rosy locus represents a single structural element with a physical size of 4000 nucleotides and the size of the average converted sequence is approximately 400 nucleotides. Since conversion is responsible for the non-reciprocal transfer of some 10 kb DNA in each female meiosis, there must be a minimum of 25 conversion events per meiosis.

In eukaryotes, however, recombination is not a phenomenon that is confined to meiosis. Rather, a variety of processes are known which fall under the rubric of 'recombination'. All of them, in some way, generate DNA molecules whose sequence specificity derives from more than one 'parental' source. In mitotic cells, for example, three different categories of chromatid exchange have been identified including *mitotic recombination, sister chromatid exchange* and *translocation*. All of these, like meiotic crossing-over, involve X-type physical exchanges between whole chromatids. It is worth paying some attention to these alternative systems of recombination since they confirm the unique nature of the meiotic process.

Mitotic, or more correctly *somatic*, recombination was discovered by Stern in *Drosophila melanogaster* as long ago as 1936. It differs from meiotic crossing-over in this same species in four respects:

(1) Unlike crossing-over, which is restricted to the female, it occurs in both sexes though probably at a significantly lower frequency in males compared to females

(2) It does not require SC formation

(3) Recombination at mitosis is relatively more frequent near the centromere, and

(4) Minutes, which increase mitotic recombination, do not affect meiotic recombination. Likewise, the mutation *c(3)G* which virtually abolishes meiotic crossing-over has no effect on mitotic recombination

Since Stern's early study, mitotic recombination has been detected in a variety of fungi. It is best known in yeast where it occurs with spontaneous frequencies some 100–1000-fold greater than those for spontaneous mutation and again occurs in the absence of SC formation (Olson & Zimmermann, 1978). As already mentioned (see Chapter 4A), yeast grows vegetatively in either a haploid or a diploid phase. There are two haploid mating types, a and α, which by hybridization give rise to diploid cells which divide mitotically – the so-called *vegetative diploids*. Diploid cells are larger than haploids but otherwise are very similar to them in organization. Vegetative diploid cells can be induced to enter meiosis by transferring them into a sporulating medium. Meiosis gives rise to an ascus containing four haploid ascospores which, on germination, either produce vegetative haploids or else form diploids by fusing with one another (see Fig. 4.1). Thus yeast can be maintained in either a haploid or a diploid state by asexual budding. Here, each mitotically dividing cell produces a small outgrowth which increases in size until it is about half

Table 4.20. *A comparison of the frequency of allelic recombination in mitosis and meiosis of Ustilago maydis*

Heteroallelic combination	Frequency of allelic recombination			
	Wild type		rec-1 strain	
	Mitosis	Meiosis	Mitosis	Meiosis
nar 1-1/nar 1-6	1.7×10^{-6}	1.1×10^{-3}	9.5×10^{-6}	2.1×10^{-3}
inos 1-4/inos 1-5	2.9×10^{-7}	1.9×10^{-2}	4.9×10^{-6}	1.9×10^{-3}

After Holliday, 1977 and with the permission of the Royal Society.

that of its parent. At this point it is cut off as a daughter cell. The distinctive life cycle of yeast means that mitotic and meiotic recombination can be compared in the same hybrid. Using this approach (Kunz & Haynes, 1981; Esposito & Wagstaff, 1981) it has been possible to show that:

(1) Spontaneous mitotic recombination can be as much as 1000-fold less than that of mieotic recombination. In other fungi the difference in frequency may range from 600- to 6000-fold (Table 4.20).

(2) Mitotic recombination is not associated with SC formation which implies that what is lacking in mitotic cells is the machinery of synapsis.

(3) In addition to absolute differences in frequency, the distribution of exchange events at mitosis is also different. As in the case of *Drosophila*, regions close to the centromere have a higher rate of mitotic recombination.

(4) The association of outside marker recombination with conversion at mitosis ranges from 2 to 41 % (average 16 %). The equivalent values for meiosis are 18–60 % (average 32 %). Additionally, conversion frequencies are very much lower at mitosis than at meiosis (Lamb, 1977).

(5) Spontaneous mitotic recombination in yeast involves unduplicated chromosomes, occurring either at G_1 or else in the unreplicated regions of chromosomes during the S-phase, though resolution of the exchange requires progression through G_2 (Fig. 4.30). Thus, unlike meiotic gene conversion, the equivalent mitotic process does not require replicative DNA synthesis. Stern's original interpretation of

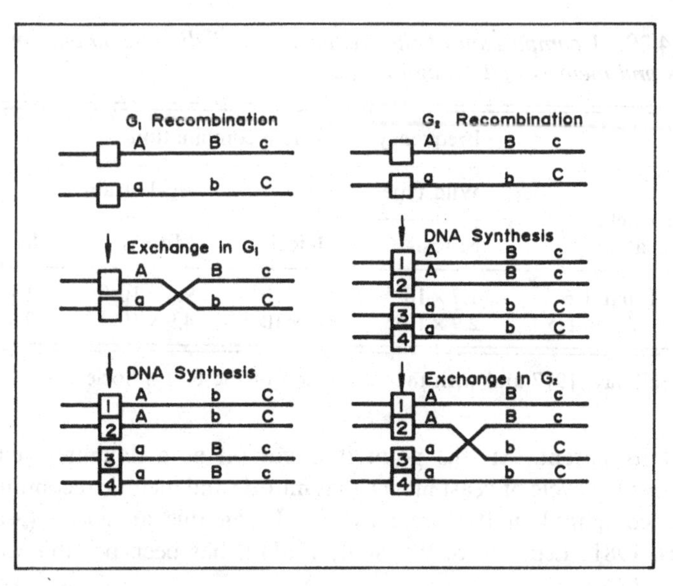

Fig. 4.30. Reciprocal mitotic exchange at G_1 and G_2 compared. Following exchange at G_1, between unreplicated homologous chromosomes, there are two possible modes of chromatid separation. Either 1 and 3 separate from 2 and 4 or else 1 and 4 separate from 2 and 3 in the ensuing mitosis. Neither of these lead to a sectored colony, or to twin spotting, for the recombination markers b and c. Recombination between homologous non-sister chromatids at G_2 leads to a sectored colony, or else to twin spotting, when chromatids 1 and 3 separate from 2 and 4 in the ensuing mitosis. The exchange event cannot be detected, however, when chromatids 1 and 4 separate from 2 and 3 (after Esposito & Wagstaff, 1981b and with the permission of Cold Spring Harbor Laboratory).

mitotic recombination was that it occurred between non-sister chromatids at a four strand stage. Initially, this interpretation was accepted for yeast too since reciprocal recombination between unreplicated homologues was not expected to generate reciprocal segregation of markers distal to the site of exchange. As Esposito (1978) and Esposito & Wagstaff (1981) indicate, this is possible following the production of a Holliday structure at a two strand stage. Since this is not cleaved, it persists and is resolved by DNA replication and not by endonucleocytic cleavage. Consequently, both mitotic and meiotic reciprocal exchanges can be explained by the production and resolution of Holliday structures. Becker (1976) points out that Stern's original interpretation of mitotic recombination does not even hold in *D. melanogaster* since here it is impossible to distinguish between the results of two and four strand exchange.

(6) The DNA heteroduplexes involved in mitotic recombination differ from their meiotic counterparts by virtue of the fact that they are often

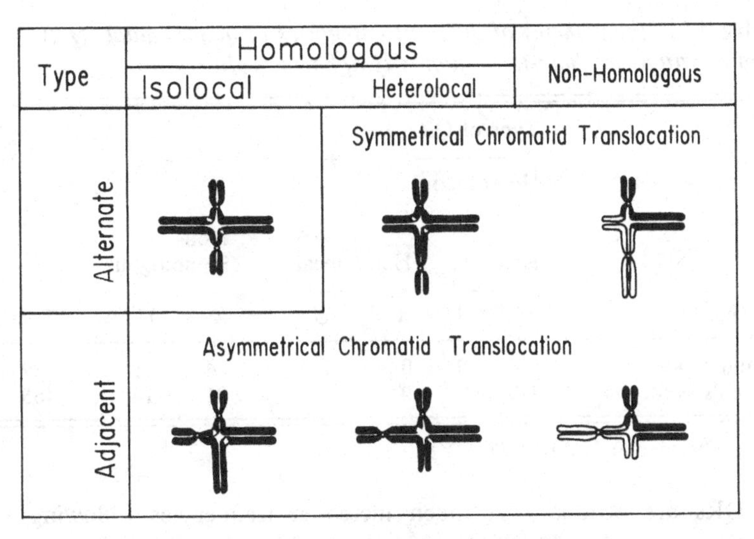

Fig. 4.31. A classification of mitotic quadriradial (QR) configurations. The homologous, isolocal, alternate category is sometimes referred to as a 'mitotic chiasma' (after Therman & Kuhn, 1976 and with the permission of Elsevier Scientific Publishing Company).

symmetric, whereas meiotic heteroduplexes are almost always asymmetric. Mitotic heteroduplexes are also more extensive.

Direct cytological evidence for spontaneous somatic exchange events has been provided in both normal human cells, where they occur at a frequency of 0.1–1.0/1000 cells, and especially in cases of Bloom's syndrome, where the frequency is as high as 5–150/1000 cells (Chaganti, Schonberg & German, 1974; Therman & Kuhn, 1976, 1981). These exchanges take the form of quadiradial (QR) configurations (Fig. 4.31). Several types of QRs can be distinguished depending on whether they are X- or U-type (referred to as alternate and adjacent, respectively, by mammalian workers), whether they involve homologous or non-homologous chromosomes and whether they are isolocal or heterolocal. Those which are isolocal, involve homologous chromosomes and are of the X-type, and hence have corresponding centromeres at alternate positions, are commonly referred to as *mitotic chiasmata* and have been assumed to provide direct evidence for a crossing-over process at mitosis equivalent to that which operates at meiosis. They certainly differ from conventional chromatid translocations in several respects:

(1) X-irradiation at G_2 does not increase their frequency though it does increase the incidence of other QR types.

Table 4.21. *Frequencies of different categories of quadriradial (QR) configurations in short-term human lymphocyte cultures*

	Type of QR						
	Homologous						
	Isolocal		Heterolocal		Non-homologous		
Category	X	U	X	U	X	U	Totals
Spontaneous	14	1	0	2	4	8	29
Bloom's syndrome	419	8	0	3	21	14	465

After Therman & Kuhn, 1976.

(2) QRs are produced at much increased frequencies following the treatment of cell cultures with the antibiotic mitomycin C. Such treatment, however, has the opposite effect to X-irradiation, promoting the formation of 'mitotic chiasmata' at the same sites that they occur spontaneously

Even so, while homologous, isolocal and alternate QRs may indeed lead to a recombination of heterozygous linked markers, the description of them as mitotic chiasmata is nothing but an interpretation based on a parallelism with meiotic chiasmata. As in the case of mitotic recombination, neither SC formation nor RN production is involved and the most noncommittal statement that one can make is that they represent isolocal mitotic exchange events which may result in mitotic recombination.

As already mentioned, a particularly high frequency of mitotic exchanges are present in Bloom's syndrome. This is a rare human autosomal recessive disorder characterized by retarded growth, a UV-sensitive eruption on the face, an immunopathology and a predisposition to cancer. There is also a marked chromosome instability in the lymphocytes and, to a lesser degree, in the dermal fibroblasts. Although a variety of QRs can be observed in patients with Bloom's syndrome one class predominates, namely that which leads to the production of so-called 'mitotic chiasmata' (Table 4.21). This class is from 100–500-times more frequent in diplochromosomes than between homologous chromosomes in diploid cells. Such diplochromosomes are present in the first mitosis after a nucleus has undergone *endoreduplication*. That is, in nuclei where two successive phases of DNA synthesis intervene between two mitoses. This gives rise to bundles of four homologous chromatids at the ensuing mitosis. Endoreduplication occurs sporadically in many plant and animal tissues but is especially common in human placental cultures obtained from spontaneous abortuses (Therman, Denniston & Sarto, 1978). In diploid cells the corresponding regions of homologous chromosomes must find each other by chance unless, as in *Drosophila*, there is somatic pairing of homologues or else the formation of heterochromatic chromocentres. In diplochromosomes that chance is obviously greatly enhanced. The incidence of homologous isolocal X-type exchanges in the diplochromosomes formed in patients with Bloom's syndrome is much higher than in diplochromosomes formed in normal humans. Since homologues are identically paired in both cases it is evident that pairing is not sufficient to explain this difference. Using cells from patients with Bloom's syndrome, Kenne & Ljungquist (1984) were able to isolate, and partly purify, a DNA recombinogenic protein analogous to the bacterial Rec A-protein. Such cells also show many more sister chromatid exchanges (SCEs) than do cells from normal individuals (Chaganti *et al.*, 1974).

Differential staining of the two sister chromatids in each homologue of a meiotic bivalent can be achieved through the incorporation of the thymidine analogue BrdU into the replicating DNA at the penultimate S-phase and then staining the bivalents with Giemsa and a fluorescent dye (see Chapter A2.2). Chromatids which contain BrdU in place of T in both strands of the double helix then stain more weakly than do chromatids with BrdU in only one DNA strand. This makes the identification of SCEs a simple matter.

A majority of SCEs identified in this manner are probably induced by the incorporated BrdU but a small number may occur spontaneously. Their frequency is increased both by mutagens and by carcinogens, which suggests that SCE formation may involve recombinational repair events similar to that which occur at crossing-over. In support of this, Moore & Holliday (1976) provide evidence for the formation of hybrid DNA during the production of mitomycin-induced SCEs in chinese hamster cells.

A number of distinct events are thus now known to involve the breakage and rejoining of DNA molecules. These include meiotic and mitotic recombination, the production of sister chromatid exchanges and chromosome abnormalities as well as DNA repair systems. It is of particular interest, therefore, to ask whether these several systems share similar molecular mechanisms or whether meiotic exchange is different in principle as well as in kind.

In the male grasshopper *Stethophyma grossum*, where chiasmata are proximally localized, all non-proximal exchanges identified by thymidine labelling patterns in meiocytes must result from SCEs (Fig. 4.32). Using this criterion, Jones (1971) reported that from 0–3 SCEs were present per anaphase-1 chromosome with a mean frequency of 0.35/chromosome, a value similar to that found also in mitotic spermatogonial cells of this species. The same holds true for *Locusta migratoria* (Tease &

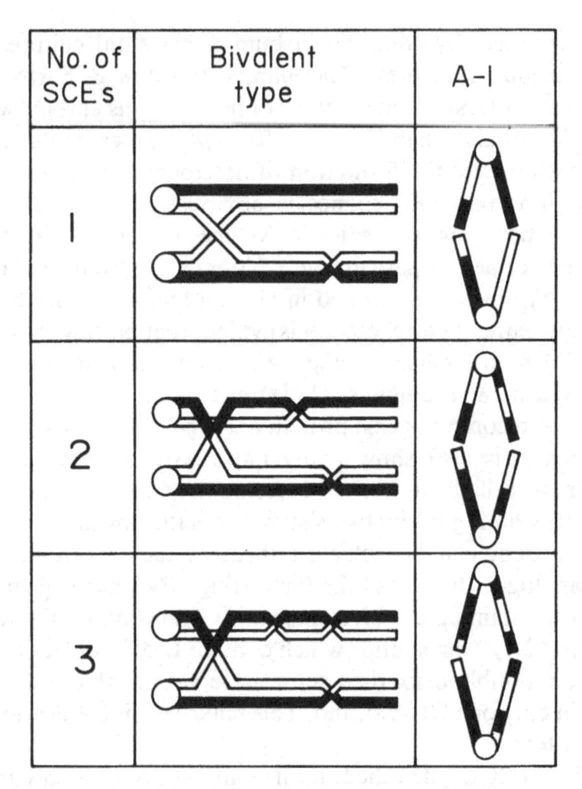

Fig. 4.32. The influence of sister chromatid exchanges (SCEs) on the anaphase-1 segregation of differentially labelled sister chromatids in male bivalents of *Stethophyma grossum* with a single proximal chiasma (after Jones, 1971 and with the permission of Springer-Verlag).

Jones, 1979) where there is a similarity between SCE frequencies, identified by BrdU labelling, in male meiocytes and female ovariole mitoses (Table 4.22). These two cases indicate that, despite the presence in meiocytes of the enzymatic machinery responsible for crossing-over, meiotic chromosomes do not form more SCEs than do mitotic chromosomes. Moreover, the contrasting behaviour between SCEs and crossovers in the spermatocytes of *L. migratoria*, both in respect of their frequency and their distribution, does not support the concept that the two processes share a common molecular mechanism.

In *Drosophila melanogaster* the meiotic mutations *mei-9* and *mei-41*, both of which decrease meiotic recombination and increase the frequency of spontaneous chromosome aberrations, do not show any increase in the frequency of SCEs (Table 4.23). Gatti, Pimpinelli & Baker (1980) and Gatti (1982) conclude, therefore, that chromatid interchanges have more in common with meiotic recombination than do SCEs.

Finally, gene conversion and reciprocal recombination (crossing-over)

Table 4.22. *Frequency of sister chromatid exchanges (SCEs) in bromodeoxyuridine (BrdU)-labelled cells of Locusta migratoria*

Chromosome class	Male mean Xa frequency	SCE frequency: Male meiocytes: Diplotene/ Diakinesis	Anaphase-1/ Metaphase-2	Female ovariole wall mitosis
Long	2.71	—	0.447	0.393
Medium	1.25	0.144	0.140	0.226
Short	1.00	—	—	—
X	0.00	0.371	0.200	—

After Tease & Jones, 1979. Note a majority of these SCEs were probably induced by the incorporated BrdU.

occur in both vegetative cells and in meiotic cells of the yeast *Saccharomyces cerevisiae*. Whereas these are interrelated at meiosis they are separable events at mitosis (Fogel *et al.*, 1983). Here gene conversion can occur at either G_1 or G_2 (Roman & Fabre, 1983). When it takes place at G_1 it is accompanied by recombination at G_2 with frequencies that indicate an interdependency of the two events. Whether this dissociation of conversion and recombination in vegetative cells means that they involve different mechanisms is not clear. Neither is it known whether conversion and recombination at mitosis is a result of the same mechanism that operates in meiotic recombination.

The assumption underlying the molecular models we considered earlier was that there is only a single pathway for meiotic recombination and that conversion is essentially a by-product of the process that governs chiasma formation. An alternative point of view has been expressed by Carpenter (1987) who proposes that conversion and crossing-over are distinct events and that conversion is required specifically for the homology recognition that subsequently leads to crossing-over. Since conversion requires the formation of a length of biparental (heteroduplex) DNA, it is ideally suited for monitoring extended DNA sequence homology. The essence of Carpenter's argument is that homology is not itself part of synaptic initiation. However, once two unpaired lateral elements come into close contact, there is an immediate check for homology via conversion events involving the early forming RNs. If this check confirms homology then extended SC formation takes place. If it fails to reveal homology the associated lateral elements fall apart and, in her opinion, the presence of non-homologous SCs at zygotene is a reflection of the large number of transient synaptic trials. In bacteria, Rec A-protein is known to facilitate the rapid search for DNA homology between duplex DNA molecules,

Table 4.23. *The influence of two repair and recombination defective mutations on exchange phenomena in Drosophila melanogaster*

Mutation	Effect on ♀ crossing-over		Genotype	Effect on neuroblast mitoses[a]		
	Frequency	Distribution		% SCEs	% chromatid deletions	% isochromatid deletions
			Control FM7/Y	10.91	0.14	0.29
mei-9	8 % wild type	Unaltered	mei-9^{b}/Y	7.44	7.93	0.99
			mei-9^{AT1}/Y	7.05	5.60	0.64
mei-41	50 % wild type	Altered	mei-41/Y	10.48	11.70	5.92
			mei-41^{195}/Y	10.87	13.62	8.36

After Gatti *et al.*, 1980; Gatti, 1982.
[a] Exposure of brain ganglia to BrdU (9 μg ml^{-1}) for two rounds of replication.

yielding a stable joint in cases where homology is detected. As has already been mentioned, eukaryotes are known to have Rec A-like proteins. Interactions between non-homologous regions cannot generate heteroduplex DNA and so do not normally lead to synapsis.

Carpenter also assumes that the search for homology is mediated by the numerous RNs formed during zygotene, or even earlier (Stack & Anderson, 1986b). In *D. melanogaster* where early RNs are not present until pachytene, there is no conventional zygotene since cells enter meiosis with homologues already paired somatically (see Chapter 4C.1). Implicit in Carpenter's proposal is the belief that exchange events leading to chiasma formation are mediated after completion of synapsis by the late-forming RNs, though she accepts that some of these may also generate conversions. Her argument leaves unexplained the occurrence of mitotic conversions which cannot be required for monitoring DNA homology in anticipation of synapsis and where neither SCs nor RNs are present.

CEN 3 DNA of yeast is capable both of repressing meiotic crossing-over and reducing the frequency of gene conversion in its vicinity (Lambie & Roeder, 1988). This ability is, however, reversed by a single base pair mutation, *cen 3**, located in the CDE III region, which stimulates adjacent crossing-over and gene conversion some 2.5-fold. It also stimulates mitotic recombination at least 10-fold above background level. Other known yeast mutations include some that increase, and others that decrease recombination (see Chapter 6B). Some influence conversion but not crossing-over, though the converse is not true. This is understandable if, as suggested by Kunz & Haynes (1981), the early steps of conversion and crossing-over involve the same, or a similar, set of enzymes whereas additional enzymes are required for the subsequent steps of isomerization and strand breakage involved in crossing-over.

5

Chromosome disjunction

I like uniformity, but when I cannot have it I go with nature.
Bruce Nicklas

A MOVEMENT AND ORIENTATION

The controlled distribution of chromosomes during meiosis, as in mitosis, depends on the orderly behaviour of chromosomes on the division spindles. This is achieved principally by two series of coupled movements which lead to the development of first metaphase orientation.

A.1 Prophase movements

Chromosome movements in the first prophase nucleus are dominated by an interaction between the chromosomes and the centrosome or, in cases where there is no centrosome, by an equivalent spindle-forming centre. At the onset of meiosis the centromeres are often gathered close to that centre, in Rabl orientation. This is a passive form of orientation, indicating that little movement has occurred since the preceding pre-meiotic telophase. *Bouquet formation*, which commonly replaces Rabl orientation, involves an active movement of chromosome ends in relation to the centrosome or its equivalent. The centrosome itself separates into two daughter centres during first prophase and these take up positions on opposite sides of the nucleus. It is this movement which, in effect, first establishes bipolarity within the cell.

In some species the bouquet forms during leptotene. Here all chromosome ends become associated with the side of the nucleus close to the undivided centrosome with the remainder adopting a looped arrangement within the body of the nucleus. This involves an interaction between chromosome ends and the centrosome, with all ends moving along the inner surface of the nuclear membrane relative to the extranuclear polarizing centre represented by the centrosome or its equivalent. In mantids, the leptotene bouquet lapses only to reform as a double bouquet during pachytene as a result of a double polarization which develops in association with the division of the centrosome

(Hughes-Schrader, 1943). In other cases the bivalents become polarized into two clusters during diakinesis, in a manner reminiscent of the bouquet arrangement. Here, however, the chromosomes already occupy peripheral positions close to, and attached at, the nuclear membrane. Then, in conjunction with the separation of the centrosome, some of the condensed bivalents follow its movement around the nuclear envelope. This has been especially well documented in male spiders (Revell, 1947 and see Chapter 5B.1.2).

Rickards (1975) studied the behaviour of the condensed chromosomes of the house cricket, *Achaeta domestica*, in relation to the nuclear membrane and just prior to its breakdown, using living spermatocytes and time lapse photography. He confirmed that chromosome movements at this time were focussed on the centrosomes and that they proceeded along paths delineated by the microtubules (MTs) which focussed on the centrioles contained within the centrosomes. Consequently, by the end of first prophase the chromosomes accumulate in two groups within the nuclear membrane and close to the two centres. This system shows four characteristics (Rickards, 1981):

(1) The movements are polarized radially in relation to the spindle centres but follow the curve of the nuclear envelope

(2) Most frequently the chromosome ends, but sometimes the centromeres, lead the movements

(3) Movements are bidirectional and occur away from, as well as toward, a given centre, and

(4) When spermatocytes are treated with the anti-microtubule drug colchicine, or its less active derivative colcemid, the asters disappear and chromosome movement ceases. Treated cells can, however, recover and resume normal movements.

The bulk of nuclear and chromosomal non-histone proteins in both mitotic and meiotic cells are contractile in character and include actin and myosin as their major components. These molecules have often been assumed to play some role either in chromosome condensation or else in spindle formation and functioning. Thus, a large amount of actin, representing some 15% of the total nuclear protein, is synthesized during, and then transported to the nucleus late in, G_2. Rickards suggested that the prophase chromosome movements he observed were based on actin–myosin interactions rather than on microtubules acting directly as force producers. He postulated an interaction between the MTs of the asters, the nuclear envelope and actin filaments located inside the nuclear envelope. Myosin, adsorbed into the chromosome ends and the centromeres, was then assumed to interact with the actin filaments. Chromosomes also undergo movements within the prophase-1 nuclei of male crane flies. Astral MTs are abundant in these cells during the later phases of diakinesis when chromosome movements occur but this movement is not related to the asters and is also insensitive to colcemid (La Fountain, 1985).

It is not without interest that, in certain dinoflagellates, mitotic anaphase movements take place under conditions which parallel those of the prophase movements in insect spermatocytes. That is, they take place within an intact nuclear membrane and without any direct connection with the extranuclear microtubules (Kubai, 1975).

A.2 Pro-metaphase movements

It is clear from micromanipulation studies that microtubules (MTs) play an integral part in the mechanism of chromosome movement on the division spindle. The reader will recall from Chapter 1C that in a majority of eukaryotes the spindle includes four categories of MTs: *polar*, which have one end located at a pole, with the other end free in the spindle; *kinetochore*, which have one end attached to a kinetochore and the other end attached to a pole; *interdigitated*, with one end attached to a pole and the other associated laterally with a microtubule attached to the opposite pole; and, *free*, which are relatively short and are completely unattached (Fuge, 1977; Pickett-Heaps *et al.*, 1986).

With the breakdown of the nuclear envelope the two asters, when present, invade the nuclear area, become individually associated with each of the two spindle poles and give rise to the two developing half spindles which form at this time (Fig. 5.1). The kinetochores then establish a mechanical link with the poles. This, as we saw earlier, (Chapter 1C) probably involves the capture of pre-existing MTs present in the two developing half spindles. However, the presence of short MTs, located at a variety of angles to the kinetochores themselves, suggests that direct nucleation by the kinetochore may also occur. Whatever their origin there is no doubt that the force that causes chromosomes to move on the spindle is transmitted, if not produced, by the kinetochore fiber. Although much of the evidence for this conclusion comes from anaphase movement it is generally believed that pro-metaphase movements are also caused by kinetochore fibers.

Most bivalents achieve bipolarity immediately on meeting the developing spindle. This is not surprising when one recalls that, as first pointed out by Östergren (1951), the kinetochores of partner half bivalents are constrained by their organization to face in opposite directions so that there is a natural tendency for sister centromere pairs, and the

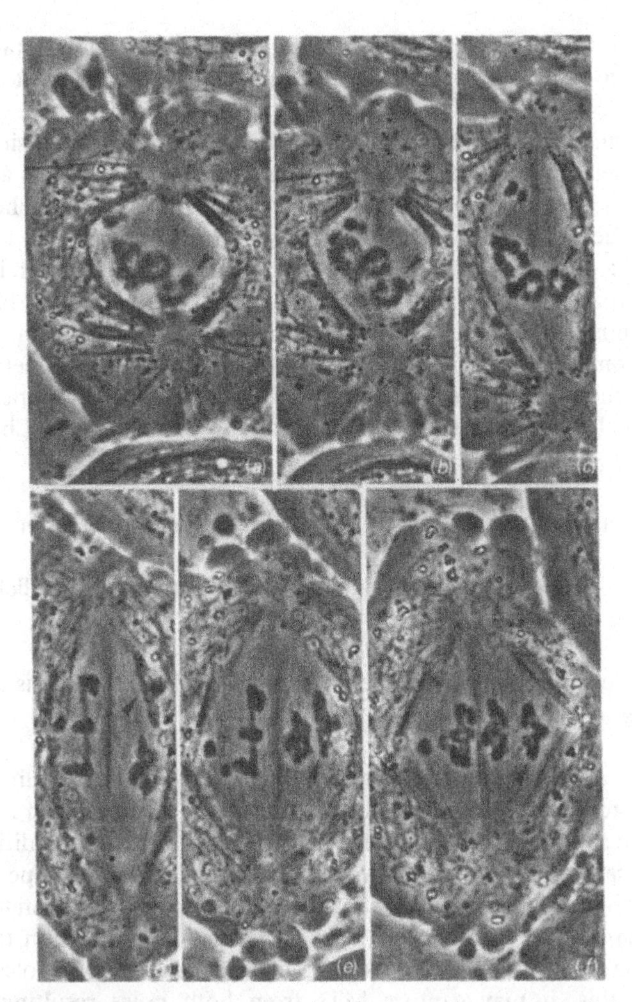

Fig. 5.1. Pro-metaphase-1 movements in a living male meiocyte of the crane fly *Pales ferruginea* (2n = 8, XY) as seen by phase contrast microscopy. The asters which define the spindle poles are especially clear in (*a*), (*b*) and (*c*). The arrow identifies the same bivalent throughout the sequence and so defines its congression behaviour. Initially it moves from the lower pole towards the equator, then to the upper pole and finally back to the equator. The two sex univalents lie close to one another and move together before congressing with the autosomes. The entire sequence covers a period of 2 h 10 min (photographs kindly supplied by Dr Roland Dietz).

kinetochores which form at them, to orient to the poles which they more nearly face. Hence, they normally orient to opposite poles at pro-metaphase-1 (Nicklas, 1977). More recently, this back-to-backness of homologous centromeres has been related to the existence of distinct pairing faces between homologues during the formation of the SC. In this sense, a bivalent may be pre-oriented, its incipient pro-metaphase-1 orientation being determined before it associates with the first division spindle.

Drosophila melanogaster may be something of an exception since here at male meiosis there are large protruding kinetochores which, at initial attachment, are rarely oriented to a single pole. Rather, a kinetochore that is clearly facing one pole can also have MTs directed towards the opposite pole. As a result, pro-metaphase-1 movement in this species is more complex than that of grasshoppers and crane flies where most bivalents achieve immediate bipolar orientation. In *D. melanogaster* over 50% of the autosomal bivalents are regularly maloriented during pro-metaphase-1. Consequently, bivalents are delayed in achieving bipolar orientation and several unique types of chromosome movement occur (Church & Lin, 1985). These include:

(1) Simultaneous reorientation of both half bivalent kinetochores

(2) Lateral bivalent movement in addition to movement parallel to the spindle axis, and

(3) Poleward movement of bivalents when neither kinetochore is actually facing a pole

Church & Lin relate these features to the unusual structure of the kinetochore which is exceptionally large and is not recessed in a cup of chromatin so that its entire surface is exposed to the spindle. Additionally, the three layers of the kinetochore assume a hemispherical shape at pro-metaphase-1 and this does not transform into a disc-like configuration until metaphase-1. Consequently, no matter in which direction the pro-metaphase kinetochore faces, its sides are exposed to both poles. As a result of this, it may capture MTs from both poles resulting in an unorthodox arrangement of MTs and, hence, also to an unorthodox pattern of behaviour. This presumably explains why bivalents in this species may occasionally reorient even at full metaphase-1.

In other organisms, too, one or more bivalents may initially be maloriented, with both centromere pairs directed to the same pole. Such monosyntelic orientation if uncorrected would, of course, lead to non-disjunction (Fig. 5.2). In most cases, however, monosyntelic orientation is corrected by a reorientation process. This is a complex event which involves a loss of kinetochore spindle fiber attachments to one half-bivalent followed ultimately by the re-establishment of new microtubule connections with the opposite pole. While the pro-metaphase-1 orientation adopted by a multiple chromosome configuration, like that of a bivalent, may be influenced by centromere placement at diakinesis, its

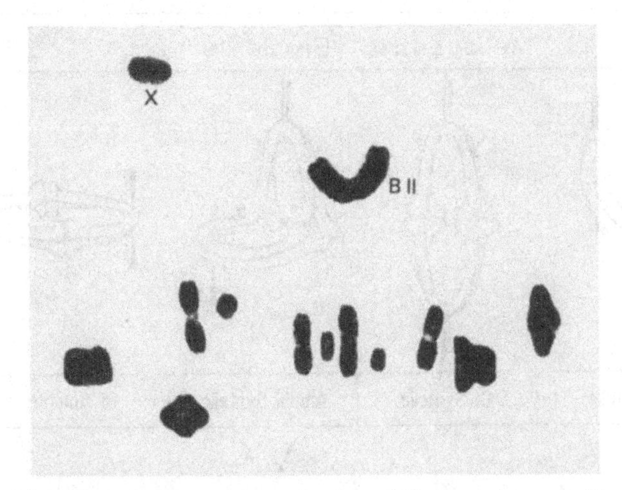

Fig. 5.2. Metaphase-1 in a male meiocyte of the grasshopper *Phaulacridium vittatum* carrying two telocentric supernumerary (B) chromosomes (2n = 2x + 2B = 11.II + X.I + B.II). The B-bivalent lies in monosyntelic orientation off the spindle equator and nearer to the pole to which both its centromere pairs are oriented. In this respect it parallels the behaviour of the X-univalent which is also oriented to the same pole.

final orientation is not necessarily achieved when it first meets the spindle. As will become apparent shortly, reorientation most certainly does occur in multiples too (see p. 217).

Reorientation may not lead to immediate bipolar orientation since a variety of modes of bivalent orientation are possible (Fig. 5.3). The probable source of bivalent stability is commonly assumed to depend on the tension which results from bipolar kinetochore connections (Nicklas & Koch, 1969). However, there has been some debate on whether such tension itself stabilizes the MT attachments of a kinetochore to a pole or whether tension simply maintains the position of a kinetochore so that it continues to face a particular pole and so precludes MT interactions with the opposite pole. It may, of course, serve both these functions.

When bivalents are experimentally detached from the spindle by micromanipulation they invariably reattach to it and re-enter the spindle (Nicklas, 1967). Detachment leads to a complete loss of prior spindle connections. By combining micromanipulation with electron microscopy Nicklas & Kubai (1985) have carried out a three-dimensional reconstruction of male meiocytes in the grasshopper *Melanoplus* to correlate microtubule arrangements with the chromosome movements which immediately precede fixation. This involved detaching both half bivalents of a given bivalent within the spindle and then displacing the bivalent as far as possible from the spindle and into the surrounding microtubule-free cytoplasm. It was then turned so that the sister centromere pair of one half

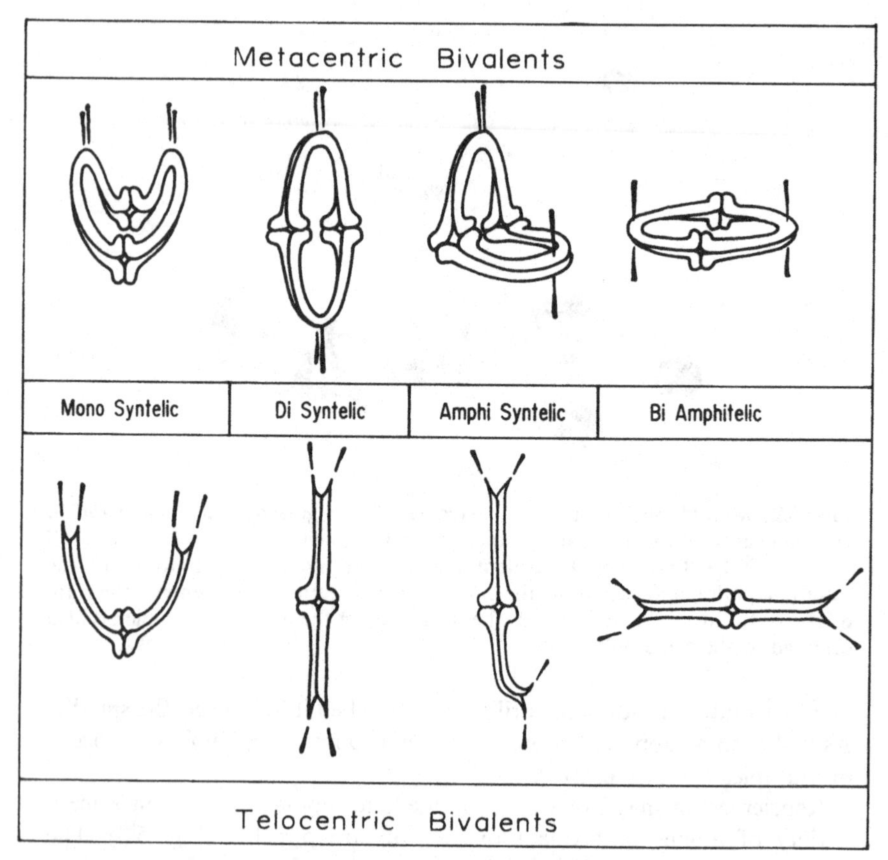

Fig. 5.3. Possible modes of bivalent orientation at first meiotic metaphase.

bivalent faced the spindle with the other half bivalent facing the cell membrane. When the micromanipulation needle was removed, the detached bivalent moved backwards towards the spindle and, at a selected point in this movement, the cell was fixed for electron microscopy. The pattern of reorientation so obtained is summarized in Fig. 5.4.

During this reorientation, microtubules extending to both poles reappeared at the kinetochore pair facing the spindle with an average of 11 kinetochore MTs compared to averages of 23 per half bivalent at prometaphase-1, 37 at metaphase-1 and 45 at anaphase-1. This was followed by movement of the bivalent to the equator until it reached the edge of the spindle. At this point the outer, unattached, kinetochore swung around, forming one or two long kinetochore MTs which were connected to one of the spindle poles. While these long microtubules intermingled with the numerous spindle MTs, only rarely did they extend all the way to a pole. As rotation continued the long MTs were augmented by others with an equal distribution. This experiment demonstrates that a single long

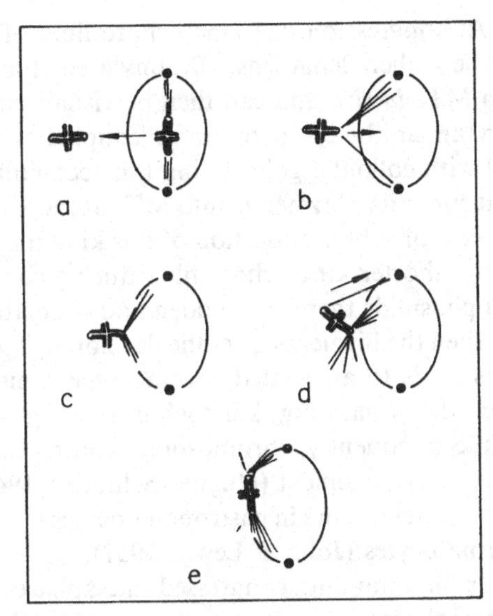

Fig. 5.4. Chromosome movement and kinetochore microtubule behaviour following detachment of a telocentric bivalent from the first division spindle by micromanipulation (after Nicklas, 1985 and with the permission of the Plenum Publishing Corporation).

kinetochore MT is initially all that is required for renewed chromosome movement after detachment. Additionally, it confirms the claim of Östergren (1951) that movement initiates reorientation since it is evident that it is rotation that leads to a shift in the MTs attached to the inner end of the displaced bivalent and so assures a return to bipolarity. Following reorientation of the bivalent, there is a transformation of the kinetochore MTs at the original proximal end. Thus, in this experimental system, reorientation involves both chromosome movement and microtubule instability.

Unlike chromosomes released for reattachment well outside the spindle boundaries, those released within the spindle re-establish kinetochore connections to the pole they face regardless of their location relative to the poles (Nicklas, 1967). Moreover, experimentally detached chromosomes will orient on spindles other than the one from which they are detached if transferred to a second spindle present in a fused cell combination. Indeed, it is possible to detach a bivalent and transfer it to a metaphase-2 spindle. Here, the bivalent reorients on the second division spindle in the same manner as on the original metaphase-1 spindle. The reverse transfer is also possible (Nicklas, 1977).

During pro-metaphase-1, chromosomes move to and fro in the spindle in a pole-to-pole direction. These movements become slower as pro-metaphase-1 progresses so that chromosomes gradually congress to the spindle equator. At congression one kinetochore fiber of a bivalent thus shortens while the other lengthens. Biotinylated tubulin is readily incorporated into MTs *in vivo* and can then be visualized in the electron microscope when an antibody to biotin is coupled with a secondary antibody labelled with colloidal gold. Using this technique, Mitchison *et al.* (1986) find that subunits polymerize into MTs at the kinetochore. They suggest that a process of subunit insertion of this kind may be involved in the extension of the shorter kinetochore fiber during congression.

In mantids and phasmids there is a sudden and dramatic separation of the kinetochores when the bivalents meet the developing spindle at the end of diakinesis. This leads to associated chromosome arms being forcibly stretched between the separating kinetochores, so generating a pre-metaphase stretch. Subsequently, chromosomes contract and congress to a typical metaphase-1 arrangement (Hughes-Schrader, 1947). An equivalent pre-metaphase stretch occurs in gastropod oocytes (Staiger, 1954) and in cockroach spermatocytes (John & Lewis, 1957).

A crucial factor in attaining congressed metaphase-1 bipolarity is tension on the kinetochores. This was first proposed by Östergren (1951), who argued that stable congression reflects isometric forces on the bivalent. More recently Nicklas (1988a, b) has suggested that tension forces associated with spindle development regulate its structure by altering MT length and stability. Tension forces on MTs might then explain chromosome orientation, reorientation and movement in terms of the stability of MT arrays, though the molecular mechanism involved in such a stabilization is not known. Under tension, for example, the subunits of a MT might be expected to be pulled further apart and to provide sites for additional subunits. This, Nicklas proposes, may account for the adjustment of MT length associated with pro-metaphase congression, which itself probably results from unequal poleward forces on each chromosome pair within a given bivalent. By contrast, at metaphase-1 when bivalents lie midway between the two poles these forces are in balance. On this view, the positioning of bivalents at metaphase-1 is determined by a balancing of traction forces on opposing kinetochores. That is, a bivalent in bipolar orientation remains at the equator because of the balanced traction applied to its kinetochores. There are six lines of evidence in support of this:

(1) The application of artificial tension to unstable forms of malorientation results in their stabilization (Fig. 5.5). When this tension is released, reorientation follows (Nicklas, 1974). Unipolar orientation of a bivalent is also easily induced in grasshopper spermatocytes by micromanipulation. When this tension is released reorientation follows within minutes (Nicklas & Koch, 1969).

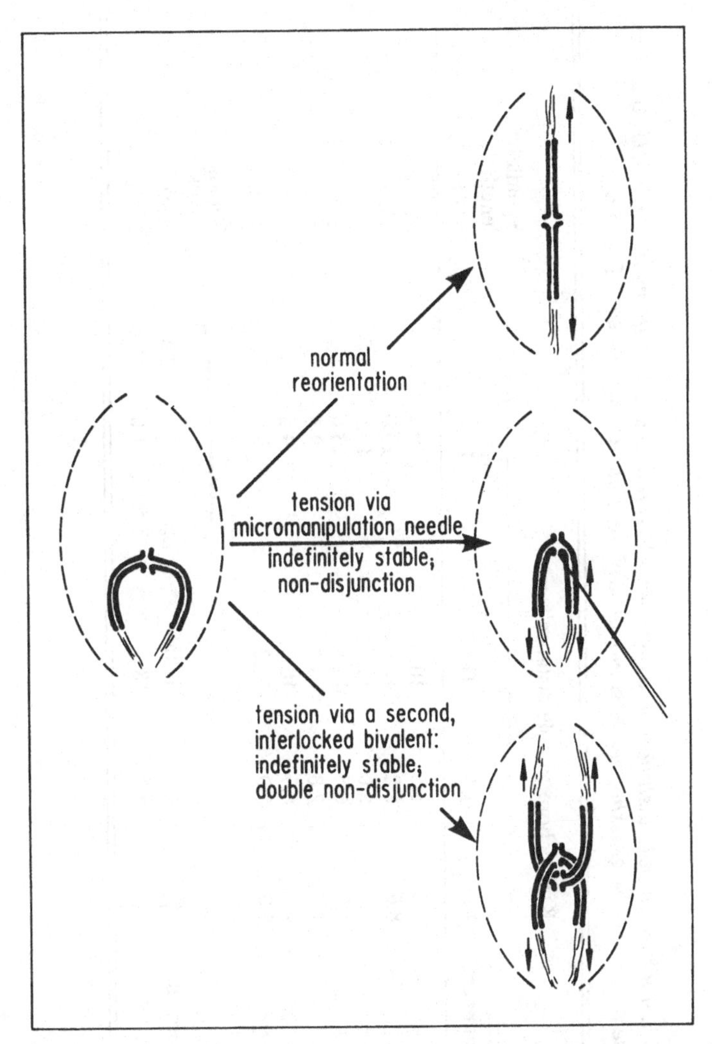

Fig. 5.5 The consequences of bivalent malorientation under normal and experimental conditions (after Nicklas, 1974 and with the permission of *Genetics*).

(2) Interlocked, maloriented unipolar bivalents produced following heat shock at first meiotic prophase, also achieve a stable orientation (Buss & Henderson, 1971). Such interlocked bivalents also show normal congression to a mid-equatorial position.

(3) Experimentally generated 'trivalents' in *Melanoplus differentialis* invariably assume an asymmetric orientation on the metaphase-1 spindle with two chromosomes oriented to one pole and the third

Table 5.1. *A comparison of kinetochore to pole distances in experimentally-produced associations of three (III) and four (IV) chromosomes in male meiocytes of the grasshopper Melanoplus differentialis (compare with Fig. 5.6)*

Orientation	Cell number	Kinetochore to pole distance (μm)						Spindle length (μm)
		L_1'	L_1''	Total	L_2			
2:1 III	1	6.0	4.5	10.5	10.5			27.5
	2	8.0	8.0	16.0	10.0			25.5
	3	5.5	5.0	11.5	11.5			25.0
	4	6.0	6.5	12.5	13.0			27.0
	5	7.5	2.0	9.5	9.5			25.0
	6	4.0	6.0	10.0	9.5			25.0
	7	6.0	5.0	11.0	9.0			24.0
	Cell number	L_1'	L_1''	Total	L_2'	L_2''	Total	Spindle length (μm)
2:2 IV	1	9.0	9.5	18.5	7.5	11.0	18.5	25.0

After Hays *et al.*, 1982.

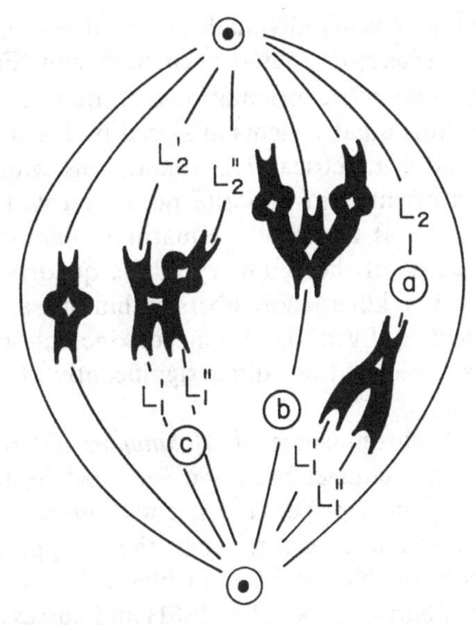

Fig. 5.6. The influence of centromere number on the position of telocentric chromosome configurations within the first division spindle.

directed to the opposite pole. Such 'trivalents' lie closer to the pole to which the two kinetochore pairs are attached (Fig. 5.6*a*) so that the sum of the lengths of the two kinetochore fibers oriented to the same pole approximates to the length of the single kinetochore fiber oriented to the opposite pole (Table 5.1). Likewise, in an interchange configuration in the same organism, analysed by Wise & Rickards (1977), where three chromosomes were oriented to one pole, with the fourth directed to the opposite pole, the multiple again lay closer to the pole to which the three chromosomes were oriented (Fig. 5.6*b*).

Such behaviour supports the two propositions of Östergren (1951) that: the position adopted by a configuration at first metaphase of meiosis reflects the summation of forces towards opposite poles; and the force at a given kinetochore is proportional to its distance from the pole. Expressed in more modern terms it could be argued that the force acting on a kinetochore at any given position in the spindle would be proportional to the number of MTs attached to it. Under these circumstances the tension generated would be expected to depend on the structure, and hence the mode of origin, of the configuration. For example, centric fusion might sometimes produce an enlarged centromere through the combination of two progenitor telo- or acrocentrics into a single metacentric (see Chapter 2A.4). Chain of three multiples involving one such a fusion dicentric would then be expected to lie co-oriented at the spindle equator and this is precisely what happens in individuals of the grasshopper *Oedaleonotus enigma* which are heterozygous for centric fusion (Fig. 2.35).

By contrast, symmetric orientations of quadrivalents are more stable. Thus of nine quadrivalents analysed by Hays, Wise & Salmon (1982), seven had symmetrical 2:2 orientations with equal numbers of kinetochores oriented to opposite poles and with the congressed quadrivalents lying at the spindle equator in line with the bivalents (Fig. 5.6*c*). Because of the geometry of the quadrivalent, the actual lengths of individual kinetochore fibers within the same spindle can be significantly unequal. Even so, the sums of kinetochore fiber lengths in opposite half spindles did not differ significantly (Table 5.1).

(4) The univalent X-chromosome of *Melanoplus differentialis* does not congress. Rather, it courses back and forth between the two spindle poles, averaging four trips per cell, and with up to eight such trips in all, with reorientation occurring when the X approaches a pole. It finally remains in the half spindle in which it happens to lie when all bivalents have congressed (Nicklas, 1961) and passes at first anaphase to the pole of that half spindle. The same is true of irradiation-induced autosomal univalents (Hays *et al*., 1982). This means that when a unidirectional force is applied to a chromosome it may fail to achieve a stable equilibrium position.

On the other hand, Ault (1984) reports that, with few exceptions, the X univalent in living meiocytes of *M. sanguinipes*, establishes a stable unipolar orientation during early pro-metaphase-1 and does not undergo reorientation. Indeed in this case the X did not move more than a few microns during the 260 min it was observed. It was possible, however, to induce reorientation – forcing the kinetochore to face a different pole – by micromanipulation. This case indicates that orientation stability need not involve tension-created forces from opposing kinetochore spindle fibers since the X is in permanent syntelic orientation. This contrasts with other univalent X-chromosomes, including that of *M. differentialis*, where, in the absence of tension, orientation is transitory and univalents reorient several times during pro-metaphase-1 as a consequence of unstable unipolar orientation. An equivalent behaviour was found to apply also to two autosomal univalents which occurred spontaneously in one cell. Subsequently, Ault (1986) argued that, in the case of *M. sanguinipes*, the constraint to reorientation resulted from MTs entangled in the X-chromatin since there were far fewer of these MTs in *M. differentialis* where the X is more condensed.

(5) A majority of male grasshoppers, like the species of *Melanoplus*, are also X0 in constitution and the X is either acro- or telocentric. Some species, however, have XY males where the X is metacentric having

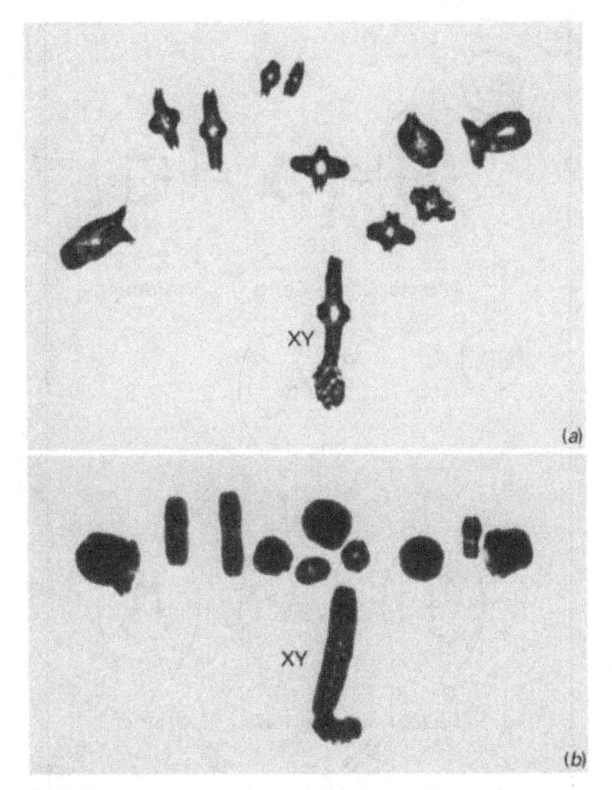

Fig. 5.7. Neo XY bivalents with a metacentric X-chromosome, produced by centric fusion, and a telocentric Y, frequently lie off the spindle equator at metaphase-1 of meiosis as demonstrated by the behaviour of bivalents of this type in males of (*a*) *Podisma pedestris* and (*b*) *Tolgadia infirma*.

originated by the fusion of the acro–telocentric X with an equivalent autosome. Moens (1979) found that metacentric autosomes in grasshoppers, which are fusion products between acro–telocentric autosomes, have approximately twice the number of MTs as do the acro–telocentrics themselves. If this holds for the X and Y of XY-males then one would predict that the tension on the X and Y components of the XY bivalent would be unequal because of the different number of MTs associated with the two chromosomes. Significantly, therefore, the XY bivalent of at least some species frequently lies off the spindle and nearer to the pole to which the metacentric X is directed (Fig. 5.7).

(6) Where multiple configurations are part of a normal cytogenetic system, either in a permanent or a polymorphic sense, they consistently show a high frequency of balanced or alternate orientation so that successive members of the multiple subsequently segregate to opposite

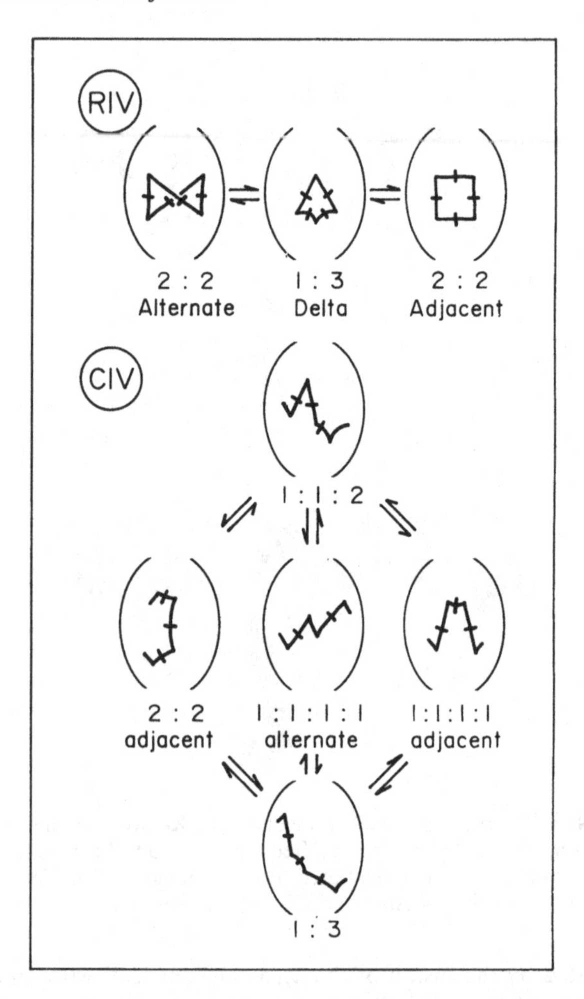

Fig. 5.8. Patterns of orientation and reorientation in ring and chain translocation multiples observed by Rohloff (1970) in living male meiocytes of the crane fly *Pales ferruginea*. In Rohloff's terminology the number of centromeres within the multiple which are oriented to one pole is separated by a colon from the number connected to the opposite pole in a sequential fashion, which in the case of chain multiples begins arbitrarily at one end of the chain.

poles. This contrasts with the behaviour of spontaneous multiples, as well as with that of many experimentally produced multiples (John, 1987). The principal modes of orientation adopted by multiples in permanent or polymorphic situations are those that have the highest stability so that the frequencies of different forms of multiple orientation are not those predicted by chance. Indeed, it has long been argued that no naturally-occurring multiple system could have arisen

Type	Mono-syntelic (U or J)		Reorientating Disyntelic						Amphi-syntelic		Bi-amphi-telic		Totals
Appearance	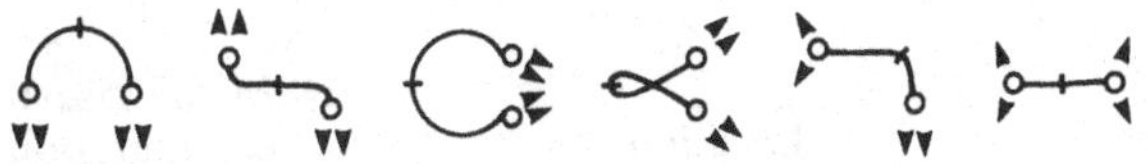												
Individual	A	B	A	B	A	B	A	B	A	B	A	B	
Equatorial	5	2	4	8	7	4	1	0	7	5	1	1	45
Intermediate	1	3	1	1	2	0	1	0	0	1	0	0	10
Polar	1	4	1	0	1	1	0	0	1	0	0	1	10
Totals	7	9	6	9	10	5	2	0	8	6	1	2	65
	16		15		15		2		14		3		

Fig. 5.9. The kinds, and locations within the first division spindle, of maloriented bivalents produced by heat treatment in two individuals (A and B) of the desert locust *Schistocerca gregaria* (after Buss & Henderson, 1971 and with the permission of Springer-Verlag).

unless the multiple or multiples involved were characterized by a high frequency of disjunction at their inception (Lawrence, 1963). Alternate orientation leads to a uniform tension on each kinetochore in a multiple and this also supports the tension hypothesis.

Even so, the tension argument cannot explain all observed situations. In a study of three experimentally-induced interchange heterozygotes in living meiocytes of *Pales ferruginea*, Rohloff (1970) showed that prometaphase-1 reorientation, in both ring and chain multiples, led to an increase in alternate configurations in all cases (Fig. 5.8). However, a substantial number of these reorientations clearly involved configurations, like the 2:2 adjacent rings and the 1:2:1 chains which, on the basis of the tension hypothesis might have been expected to be relatively stable. Similarly, biamphitelic orientation of a bivalent is also expected to produce a stable pattern of orientation, with equal tension on all kinetochores (see Fig. 5.9), yet is not observed in normal meiosis though it has sometimes been seen under experimental conditions (Buss & Henderson, 1971). In the three induced interchanges studied by Rohloff, amphitelic orientations were also observed in both rings and chains of four, as well as in chains of three, in up to 2% of the meiocytes. They were most frequent in the terminal centromeres of chain multiples.

Background genotype, operating through variation in chiasma frequency and hence also on chiasma position, is known to influence the orientation behaviour of multiple configurations in rye. Selection for high and low frequencies of disjunctional arrangement in rye multiples was

Table 5.2. *The occurrence of interchange polymorphism in male individuals of three South Australian populations of buthid scorpions all with 2n = 14*

Species	Location	Metaphase-1 configuration	Number of males
Lychas marmoreus	Kangaroo Island	7II	24
		⊙4+5II	9
		⊙6+4II	1
		⊙10+2II	2
Lychas variatus	Overland Corner	⊙12+1II	7
Isometroides vescus	Middleback Station	7II	1
		⊙4+5II	2

After Shanahan, 1986.

found to be positively correlated with chiasma frequency (Rees & Sun, 1965; Sun & Rees, 1967). Overall, there was a 26% difference in the frequency of alternate orientation between interchange multiples containing five versus four chiasmata, with the decrease in chiasma number favouring alternate orientation. Whether this difference influences the capacity of multiples to reorient is not known but it would be well worth examining.

In scorpions, achiasmate male meiosis occurs in species with either monocentric chromosomes (Scorpionidae) or holocentric chromosomes (Buthidae). Male buthids are also characterized by the occurrence of polymorphism for interchange heterozygosity (Table 5.2). Though achiasmate, complete SCs are formed and meiosis in male buthids is not inverted. Rather, a pronounced ditelocentric activity, with both ends upturned on the metaphase spindle, is evident at both meiosis and mitosis (Shanahan, 1986). The combination of an achiasmate meiosis and a holocentric organization in this case leads to the production of multiples that cannot be compared to those of monocentric systems since they do not open out. Despite this, anaphase-1 chromosomes move in a parallel fashion to the poles with either one or both ends leading, so that disjunction is regular.

To account for this, Shanahan (1986) proposes that there is a single pairing face for each lateral element of the SCs which are present in a complete form throughout the multiples, despite the occurrence of regions of pairing partner exchange. This, coupled with the ditelocentric activity of the chromosomes at meiosis means that, provided both ends of a given chromosome are oriented to the same pole, with adjacent chromosomes oriented to opposite poles, then a balanced segregation will result (Fig. 5.10). Equivalent pairing faces have been identified in chain of three

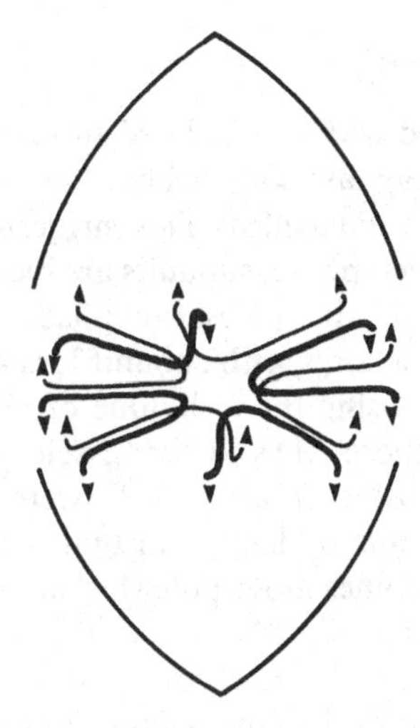

Fig. 5.10. Presumed metaphase-1 orientation behaviour of the holokinetic and achiasmatic multiple chromosome configuration found in male buthid scorpions (after Shanahan, 1986).

multiples of *Lemur* hybrids (Moses *et al.*, 1979) and in a balanced Robertsonian polymorphism in *Sigmodon fulviventer* (Elder & Pathak, 1980). In both cases, the centromeres of the two acrocentrics in the chain of three always lie in a *cis*-configuration relative to the centromere of the single metacentric; that is, on the same side of the pairing configuration. This has been taken as evidence in support of a single pairing face on the axis of the metacentric, which would predispose the three members of the chain to segregate regularly. This interpretation, if correct, provides additional evidence, of a novel kind, in support of the tension hypothesis.

Pro-metaphase movements thus contrast with those of prophase in three respects:

(1) Since it is the function of the kinetochores to mechanically link the chromosomes to the poles, it is the kinetochores, and not the chromosome ends, which regulate pro-metaphase movements

(2) The unidirectional response of each kinetochore pair to a single spindle pole contrasts with the bidirectional prophase movements, and

(3) The rate of movement during pro-metaphase-1 is much reduced in magnitude compared to the movements which chromosomes evince at prophase

A.3 Anaphase movements

When cells are injected with biotinylated tubulin followed by colloidal gold, essentially all metaphase kinetochores are labelled. At anaphase, however, some 60% are unlabelled. This suggested to Mitchison *et al.* (1986) that, whereas at metaphase, subunits are incorporated slowly at the kinetochores and lost to the poles, anaphase movement occurs by disassembly at the kinetochores with subunit loss at the poles playing, at best, only a minor role. Using the technique of photobleaching to mark segments of the kinetochore MTs in living cells just after the onset of mitotic anaphase, Gorbsky *et al.* (1987) were able to analyse the movement of such segments by hapten-mediated immunocytochemistry. They found that chromosomes move poleward along stationary MTs with the driving force arising directly from the kinetochore by active depolymerization at this site. Were chromosomes to be dragged passively by traction forces applied to the kinetochore microtubules they would be expected to move poleward during anaphase and to be depolymerized at the poles. Since this did not occur, Gorbsky *et al.* conclude that an interaction of the kinetochore with MTs must be responsible both for the attachment of the chromosomes to the spindle and for the movement of the chromosomes at anaphase.

Precisely what destroys the equilibrium governing the stable metaphase-1 co-orientation, and so triggers the onset of first anaphase, is not known. Forer (1974) has argued that actin-like filaments, present in the spindle system, produce the force for anaphase movement, based on an assumed interaction between actin and the kinetochore microtubules, though whether this also provides the trigger mechanism that initiates the separation of half bivalents remains unclear.

The anaphase-1 movements of the different half bivalents which pass to the same pole must, in a measure, be independent of one another since they move at different rates. Moreover, micromanipulation or ultraviolet (UV) irradiation may affect the movement of one half-bivalent without any influence on the others in the same polar group. Despite these facts, anaphase-1 movements are evidently coordinated in the sense that all movement is initiated at the same time. This contrasts with pro-metaphase movement where bivalents behave individually to a large extent. There is also some coordination between homologous half bivalents since irradiation of the kinetochore fiber of one half-bivalent not only stops its movement but, additionally, results in the cessation of the movement of the partner half bivalent to the opposite pole (Forer, 1966). Irradiation of the interzonal region has no effect unless carried out prior to irradiation of the kinetochore fiber of one of a pair of half bivalents when the coordination between the homologues is again uncoupled.

B SYSTEMS OF SEGREGATION

The segregation of homologous chromosomes which are physically linked to one another at the onset of pro-metaphase-1, whether with or without chiasmata, can, as we have seen, be accounted for in terms of a co-orientation of their centromere systems. Bivalent formation is thus normally an integral part of successful segregation. Moreover, it is a convention to assume that each bivalent orients independently of all other bivalents in the same spindle system and so assorts at random.

Any mechanism that guarantees a regular separation of homologous kinetochore pairs to alternate pole at first anaphase will, however, ensure segregation. While bivalent formation provides the most common means of achieving segregation it is not always necessary. Added to this, segregation is not always a random event. We turn now to such exceptions.

B.1 Non-chiasmate segregation

Unconnected, homologous, univalents result either when there is no synapsis (*asynapsis*) or else when synapsis is not maintained (*desynapsis*). In both of these situations, the univalents produced are expected to be distributed to the poles independently of one another. There are, however, a number of exceptions where mechanisms have evolved to ensure the non-random assortment of univalents. Here, chromosome orientation is in some sense independent of conventional meiotic synapsis. This is true both for certain autosomal univalents and, more especially, for sex chromosome univalents.

B.1.1 Autosomal univalents

In the flatworm *Mesostoma ehrenbergii ehrenbergii* there are five chromosome pairs, only three of which actually give rise to bivalents at male meiosis. The other two sets of homologues, one of which is metacentric (pair 2) and the other of which is acrocentric (pair 5), do not form SCs at pachytene (asynapsis) and so are consistently represented as four univalents. Yet anaphase-1 cells invariably show correct segregation, a behaviour which is readily confirmed from second division cells. This suggests that the species is, in reality, a stabilized hybrid which incorporates a system of non-chiasmate segregation.

A study of univalent behaviour in living male meiocytes of *M. ehrenbergii ehrenbergii* indicated that both kinds of univalent underwent a series of rapid pole-to-pole migrations during the early and middle phases of metaphase-1. Each migration was completed within 1–2 min, usually with only one univalent moving at any one time. This was

Table 5.3. *Non-chiasmate systems of sex chromosome segregation*

Category	Occurrence	Reference
Hereditary sex univalents		
(1) Preferential movement of syntelic sex univalents – distance pairing		
(a) At 1st division	Neuroptera	Hughes-Schrader, 1969; Nokkala, 1983
	Oedionychina (Coleoptera)	Smith & Virrki, 1978
(b) At 2nd division	*Nabis flavomarginatus* (Heteroptera)	Nokkala & Nokkala, 1984
(2) Coordinated movement of amphitelic sex univalents	Tipulid flies (Diptera)	Dietz, 1969
	Altica spp. (Coleoptera)	Smith & Virrki, 1978
(3) Post-reduction following persistent anaphase-1 bridges	*Hyperaspis* spp. (Coleoptera)	Smith & Virrki, 1978
	Phryne spp. & *Mycetobia* spp. (Diptera)	Wolf, 1950
(4) Persistent joint polarization to a single spindle pole	Numerous spiders (Arachnida)	Revell, 1947; Madison, 1982; Wise, 1983
Pseudo-bivalent or pseudo-multiple formation		
(1) Persistent non-homologous pairing		
(a) At 1st division	Dermaptera	Callan, 1941; Ortiz, 1969; Henderson, 1970a
	Drosophila melanogaster (Diptera)	Cooper, 1964
	Tsetse flies (Diptera)	Southern & Pell, 1973
(b) At 2nd division	Homoptera	Ueshima, 1979
(2) Development of a specialized segregation body	Polyphaga (Coleoptera)	John & Lewis, 1960; Smith & Virrki, 1978

followed by a 5–10 min interval in which no further univalent movement took place. During these movements, the univalents appeared syntelically-oriented and were attached to a single pole. Reorientation involved detachment and retachment of univalent kinetochores to the opposite pole. In conventional stained preparations all combinations of distribution of the four univalents could be observed including 3:1 and 4:0 arrangements. These, however, were invariably corrected to give a regular 2:2 separation at first anaphase (Oakley, 1983, 1985).

The univalent pairs showed a clear difference in reorientation efficiency with the acrocentrics achieving a 1:1 arrangement first. When the correct distribution of all four univalents had been attained, the cells often stabilized. In some cases, however, this did not lead directly to a stable state, in which case homologous univalents simply exchanged poles. Throughout the phase of univalent migration, the three bivalents remained stationary and correctly aligned between the two poles. In this example the cell must somehow be able to recognize when the correct balance of centromeres has been attained in the absence of any apparent tension effect operating between the homologous univalents. In female meiosis all five pairs of chromosomes form bivalents, though these are achiasmate (Oakley, 1982). The control of bivalent formation thus differs between the two 'sexes' in this hermaphrodite. Not only are no univalents formed but in post-pachytene oocytes there is no trace of the SC that was formed in the earlier stages. Coupled with this, homologues at first metaphase are associated only at localized distal regions. Thus, unlike chiasmate systems, where sister chromatids are splayed at anaphase-1, they remain closely associated until anaphase-2 in female *Mesostoma.* This is true also of the achiasmate systems present in the dipterans *Drosophila pseudoobscura, Phryne fenestralis* and the copepod *Tigriopus*. Maguire (1978, 1982) argued that sister chromatid association at metaphase-1 was a function of the SC but this is clearly not the case in these four achiasmate species.

B.1.2 Non-chiasmate sex chromosome systems

In animals a variety of non-chiasmate systems are known in which different mechanisms exist for the preferential distribution of sex chromosomes. These fall into two basic types, those involving hereditary sex univalents and those which depend on consistent pseudobivalent or pseudomultiple formation (Table 5.3).

Hereditary sex univalent systems

Distance pairing. An orientation that is independent of conventional meiotic pairing occurs in a number of species. When the chromosomes involved are separated by a considerable distance this has been referred to as 'distance pairing'. In other cases, the chromosomes concerned may actually come together for a brief period. This is referred to as 'touch and go' pairing (Schrader, 1940). These two situations are not always easily distinguished since some distance conjugation systems may themselves involve prior touch and go pairing. In neuropteran insects, as exemplified by *Hemorobius marginatus*, the X- and Y-chromosomes are positively heteropycnotic and only loosely associated at pachytene. By diakinesis some cells include a sex chromosome pseudobivalent whereas others are already characterized by two sex univalents. By pro-metaphase-1, however, only sex univalents are present and these initially congress as univalents. They then take up stabilized positions, one at each of the two poles, in late metaphase-1. Thus, no persistent pairing is involved in the segregation process (Nokkala, 1983).

When B-chromosomes are also present in *Hemorobius* they tend to be located near the sex chromosomes at pachytene but by diakinesis have separated from them. If two B-chromosomes are present they sometimes form a bivalent and sometimes remain as two univalents. These B-univalents may disjoin preferentially from one another, presumably by the same mechanism that ensures regular segregation of the sex univalents. However, the presence of a B-univalent may also lead to the non-disjunction of the sex univalents with the B-univalent segregating preferentially from one of the sex univalents. These modifications of the segregation process again do not require any form of permanent prophase pairing.

Given that a kinetochore is most likely to establish a connection with the pole to which it is nearest, or to which it faces, when the spindle forms, it follows that the behaviour of a chromosome prior to spindle formation is likely to influence its pattern of orientation.

In the neuropteran system, the only pairing that occurs is that observed at pachytene. Consequently, although this does not lead to chiasma formation or to the production of a bivalent, it may still facilitate the regular orientation of unpaired univalents if, after their desynapsis, they move away from one another in such a way that their kinetochores face towards different poles at the onset of pro-metaphase-1 and maintain this relationship until anaphase-1. This is the view adopted by Nokkala himself to explain the distance pairing of sex univalents at metaphase-1. He argues that a temporary association between a B-univalent and one of the sex univalents would be expected to disturb normal prophase movements. This, in turn, might result in both sex univalents orienting to the same pole, with the B-univalent segregating preferentially from that sex univalent with which it had temporarily associated at pachytene.

The heteropteran *Nabis flavomarginatus* shows a distinctive form of distance pairing between the X- and Y-chromosome since it occurs at second, not first, meiosis (Nokkala & Nokkala, 1984). In the earliest meiotic stage observed in this species the autosomes are in a diffuse stage whereas the two sex chromosomes are positively heteropycnotic, sometimes separate and sometimes associated. When the autosomal bivalents condense out of the diffuse stage they each consist of two parallel-aligned homologues and are clearly achiasmate. The chromosomes are also holocentric. They move parallel to the equator and do not show the telokinetic activity typical for holocentric hemipteran male meiosis. The X and Y are invariably separate at first metaphase and each divides equationally at first anaphase. There is an interkinesis, usually lacking in male hemipterans, and at second division the X- and Y-chromatids do not associate but show distance pairing and so segregate at second rather than at first division.

The coordinated movement of amphitelic sex univalent pairs. The telocentric X- and Y-sex chromosomes of tipulid flies are heteropycnotic at the outset of male meiosis and are associated in a sex chromocentre (John, 1957). This persists into diakinesis but, with the onset of pro-metaphase-1, the sex chromosome move apart. This may be facilitated by the pronounced pre-metaphase stretch that occurs at this time.

The sex univalents in the crane fly spermatocytes are initially syntelically-oriented (unipolar) at early prometaphase-1. During pro-metaphase, frequent changes of orientation occur as the result of their migration between the poles and subsequently they each adopt an amphitelic (bipolar) orientation and come to lie side-by-side on the metaphase-1 spindle (Fuge, 1985). The amphitelically-oriented sex univalents, however, show the same anaphase-1 behaviour as the syntelically-oriented autosomal half bivalents in that they too segregate to opposite poles though their segregation is delayed until that of the autosomes has been completed (Fig. 5.11). The sex chromosomes of the fleabeetle *Lysanthia ludoviciana* behave in the same way (Virrki, 1989a). The sister kinetochores of each of the amphitelically-oriented tipulid sex univalents are interconnected by a bridge of electron-dense material which stems from the kinetochores. This bridge is retained during first anaphase (Fuge, 1974). Its function is unclear. Fuge suggests that it provides a mechanical link which prevents the separation of sister chromatids. It labels with scleroderma 5051 serum up to metaphase-1 and so appears to consist of the same protein as is present at the kinetochores (Bastmeyer *et al.*, 1986).

Anaphase movement of the sex univalents usually starts with one of them moving poleward for a short distance before its partner also starts to move. Additionally, during the initial phase, both univalents may move in the same direction before regular segregation begins. From this point on, however, the two univalents move synchronously to opposite poles while still retaining their amphitelic orientation so that during the whole sequence the kinetochores occupy a lateral position on the chromatids.

The kinetochore of tipulids is not well defined and MTs simply insert

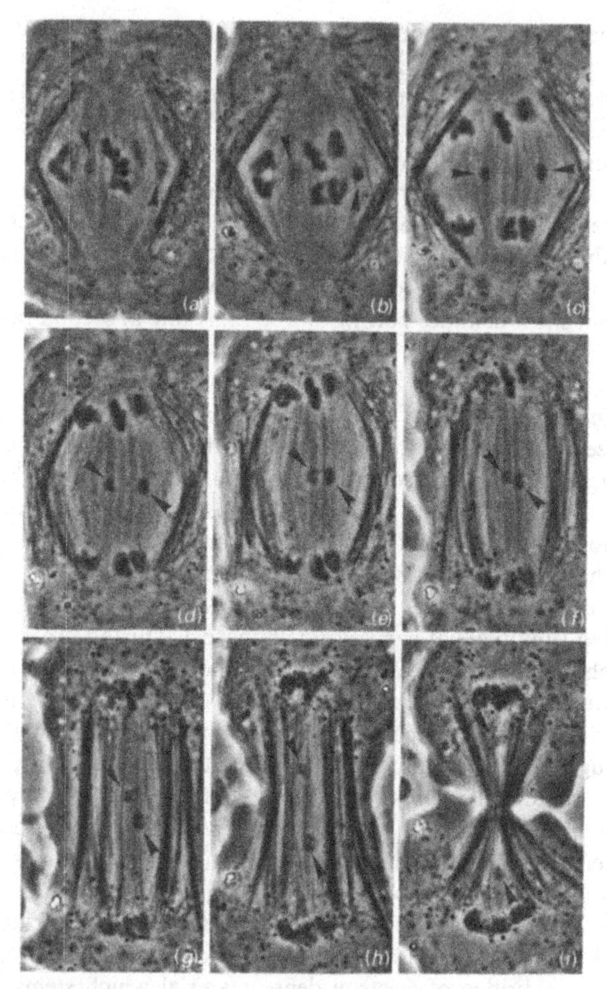

Fig. 5.11. Anaphase-1 behaviour of the X and Y sex univalents (arrowheads) in living meiocytes of the crane fly *Pales ferruginea* as seen by phase contrast microscopy. The entire sequence covers a period of 1 h 25 min (photographs kindly supplied by Dr Roland Dietz).

into a region that has less contrast than the conventional chromatin. At the poleward side, the segregating univalents are relatively flat with the interzonal side more pointed. This reflects a differential stretching of the amphitelically-oriented kinetochores (Fuge, 1972). During anaphase-1 movement a series of lamellae form in conjunction with the sex chromosomes. Their development is associated with a system of parallel MTs which originate from the masses of MTs that accumulate around the chromosome at first metaphase and then enter the lamellae. Fuge, Bastmeyer & Steffen (1985) suggest that these extra-kinetochore MTs may determine the pattern of sex univalent segregation. This suggestion is based on the fact that some of the free microtubules (fMTs) run obliquely to the bulk of the MTs and these Fuge terms *skew fMTs*. The degree of disorder of skew fMTs is some 10–15 % higher in the active kinetochore end compared to the passive kinetochore end of each sex univalent.

Tipulid sex univalents, as already mentioned, move polewards only after the three autosomal half bivalents have completed their movement. Coordination of the behaviour of the autosomes and the sex chromosomes is confirmed by two other series of observations:

(1) Half bivalents of *Nephrotoma suturalis* and *N. ferruginea* when detached by micromanipulation at first anaphase and then reattached syntelically to the original pole had no effect on sex chromosome segregation. When half bivalents were amphitelically-oriented at reattachment, however, the amphitelic autosomal half bivalent segregated against one sex univalent while the other remained at the equator (Forer & Koch, 1973). When reattached half bivalents were syntelically-oriented to the opposite pole, most commonly one sex chromosome moved normally to the spindle pole with the two homologous autosomal half bivalents while the other sex univalent did not move at all.

These experiments demonstrate the occurrence of interchromosomal coordination and indicate that the movements of the sex univalents are not unconditionally predetermined. Rather, the behaviour of the amphitelic sex univalents depend on the previous movements of the syntelic autosomes. Consequently, sex chromosome segregation can be modified by changes in the pattern of autosomal segregation. While the precise basis of this coordination is not known, spindle fiber interactions are presumably involved, as the next series of experiments indicate.

(2) When a UV microbeam was used to irradiate the kinetochore fibers of single autosomal half bivalents the results obtained depended on the relationship of the sex univalents to the half bivalent involved (Sillers & Forer, 1981). Sex chromosome movement was only altered if the irradiated autosomal spindle fiber was adjacent to it, in which case the sex univalent moved to the non-irradiated pole. When both sex univalents were adjacent to an irradiated autosomal half bivalent both moved to the same pole (Fig. 5.12).

Male fleabeetles have giant sex chromosomes of the XY-type. At first metaphase these lie on an independent spindle separated from the main autosomal spindle by a mitochondrial-rich area of cytoplasm. In *Omophoita cyanipennis*, for example, the region of the membrane closest

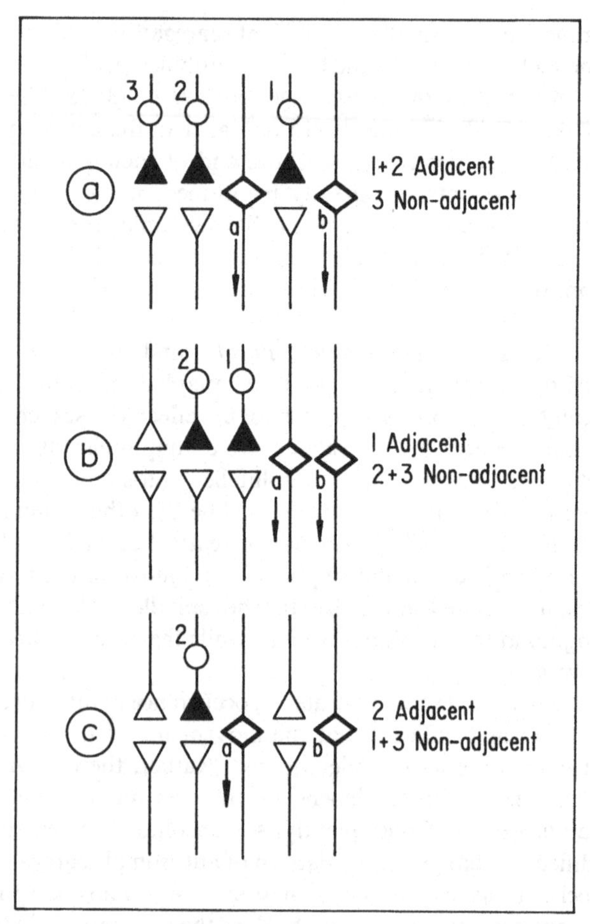

Fig. 5.12. The influence of UV microbeam irradiation of individual autosomal half spindle fibers (-○-) on the anaphase-1 movement of the sex univalents (◇) of the crane fly (after Sillers & Forer, 1981).

to the asters first becomes indented and then begins to undulate under the influence of the growing asters. The nucleus is subsequently pushed to the cell wall opposite the asters and the chromosomes contract into this reduced space. When the nuclear membrane breaks down all the chromosome clump against the cell membrane in a post-diakinetic contraction phase. The autosomal bivalents move out of this clump to form a pro-metaphase-1 group which invades the centre of the cell and gives rise to an equatorial plate on the main spindle system. The unpaired sex chromosomes are left behind at the site of the initial clump where they orient on their own independent mini-spindle at a slightly later stage than the orientation of the autosomes. This specialized sex spindle is separated from the main spindle by a well defined zone of cytoplasm. The physical

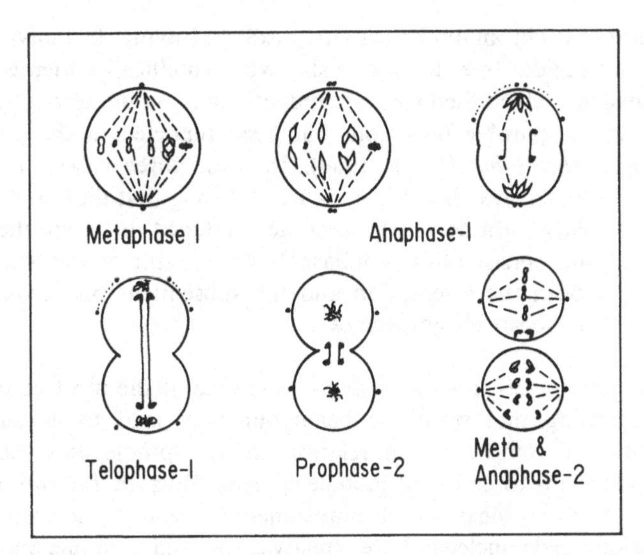

Fig. 5.13. Behaviour of the hereditary X- and Y-univalents in the beetle *Hyperaspis* sp. Counter-rotation of the daughter centrosomes at telophase-1 is followed by a reorientation of the lagging sex univalents resulting in their inclusion in distinct second division cells and their subsequent equational division at meiosis-2. (after Smith & Virrki, 1978 and with the permission of Gebrüder Borntraeger).

basis of the interdependency between the unconnected sex chromosomes within their own mini-spindle has yet to be determined. Here they either immediately become synoriented to opposite poles or, in cases where they initially orient to the same pole, one of them reorients. Although Virrki (1971, 1972) describes this as a distance pairing system, the X- and Y-chromosomes lie close to one another at metaphase-1, quite unlike conventional distance pairing, and do not move apart until anaphase-1. This case is thus better treated as one involving the preferential movement of unassociated sex univalents. In occasional meiocytes, an autosomal bivalent may trap a sex chromosome. In this event, the autosomal bivalent involved remains in the sex spindle where, however, it congresses to the same equatorial plate as the autosomal bivalents of the main spindle (Virrki, 1989b).

Post-reduction following persistent anaphase-1 bridges. In beetles of the genus *Hyperaspis*, the X- and Y-univalents congress independently and show amphiorientation on the first division spindle. The two chromatids of each univalent then begin an equational separation at anaphase-1 which they do not complete because of the development of persistent chromatid bridges (Fig. 5.13). These bridges retract during prophase-2 drawing the two univalents into juxtaposition at the mid-spindle region.

At this point the chromatid bridges are resolved leaving the individual X- and Y-chromatids free to establish two syntellically-oriented XY-pseudobivalents which then separate reductionally on the second division spindles. A comparable behaviour has been reported in the dipterans *Phryne* and *Mycetobia* (Wolf, 1950). In these three cases, the distal 'collochores' of the sex chromosomes are so strong that they do not yield at first anaphase. Rather, they become stretched between the sister chromatids and persist until prophase-2. As a result of this behaviour they control both the orientation and the subsequent post-reductional segregation of the sex chromosomes.

Persistent joint polarization to a single pole. One of the most convincing cases confirming that prophase behaviour may lead to a particular placement of chromosomes in relation to the spindle axis that subsequently forms is to be found in male spiders. Thus, in *Tegenaria atrica* ($2n = 20+X_1X_2$, ♂) the two X-chromosomes are positively heteropycnotic in pre-meiotic male nuclei and are already associated with one another at this stage. This side-by-side association persists into leptotene and is polarized in relation to the centrioles, like the precondensed ends of the autosomes. This polarization continues throughout the remainder of prophase-1. Consequently, when sister centrioles move apart in anticipation of spindle bipolarity, 6–9 of the autosomal bivalents move with the migrating centriole so that the diakinesis nucleus becomes bipolar. Bivalent bipolarization is retained after spindle formation, so delaying metaphase-1 congression which is attained progressively by individual bivalents. When the nuclear membrane breaks down the two X-univalents remain jointly polarized by their centromeres to the initial polar centre and are still aligned in parallel. They retain this relationship throughout first metaphase and first anaphase so they are both automatically included in the same daughter nucleus (Fig. 5.14). This behaviour applies also to $X_1X_2X_3$ species of *Tegenaria* (Revell, 1947).

More recently Wise (1983), using the microspreading technique, has shown that the two X-chromosomes of the male wolf spiders *Lycosa georgicola* and *L. rabida* are also closely aligned during the first prophase. Though they never synapse, a distinct axial core is present in each of them. Using conventional electron microscopy, Benavente & Wettstein (1977, 1980) had earlier reported the existence of a special junction lamina between the two univalent X-chromosomes which was assumed to account for their parallel alignment at first prophase and their subsequent conjoint behaviour. This structure was not evident following microspreading but its presence has been subsequently confirmed by Wise & Shaw (1984).

In the jumping spider *Pellenes* (Aranae: Salticidae) there is an $X_1X_2'X_3'Y$-male system, evidently derived from an X_1X_2-male progenitor,

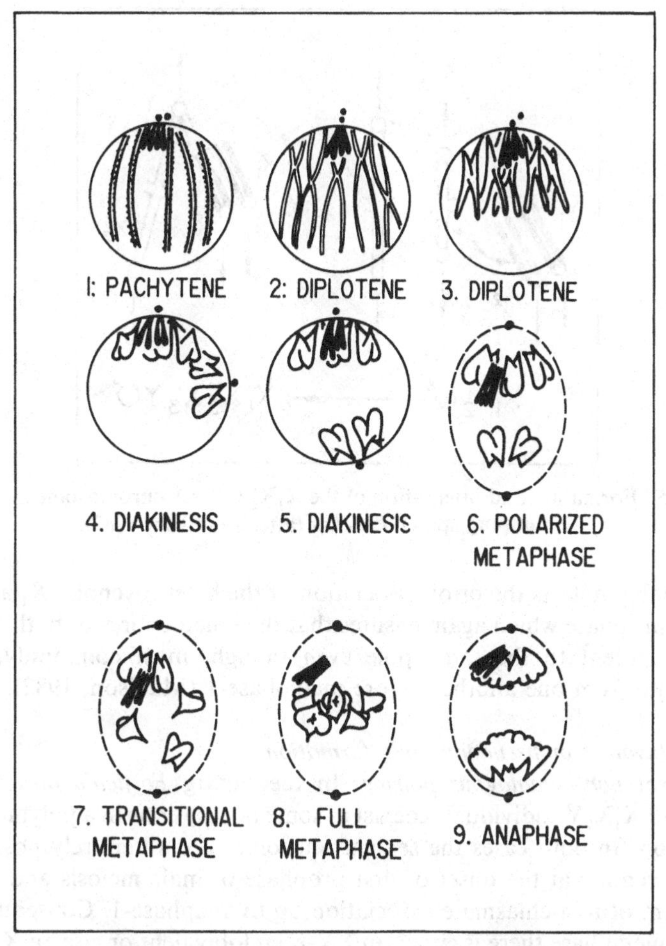

Fig. 5.14. Chromosome polarization at the first male meiotic division in the spider *Tegenaria atrica*. All chromosomes are initially polarized by their centromeres to the centriolar end of the nucleus (persistent Rabl polarization). The two sex chromosomes retain this polarization throughout the entire first division. From 11–14 of the 20 autosomal bivalents also initially retain their initial polarization throughout diakinesis and into pro-metaphase-1. The remainder actively move to the opposite side of the nucleus during diakinesis in conjunction with the passage of the distal centriole. This leads to a polarized metaphase-1 in which all the bivalents exhibit monosyntelic orientation. This is then followed by the reorientation and congression of the autosomal bivalents (after Revell, 1947).

a condition which is still found in some species within the genus (Fig. 5.15). Here, a V-shaped multiple of three is formed with the Y lying between the X_2' and the X_3' and oriented towards the opposite pole to them. The X_1 may be parallel to, or at least close to, the X_2' but more frequently it lies separate but still oriented to the same pole as the X_2'. This

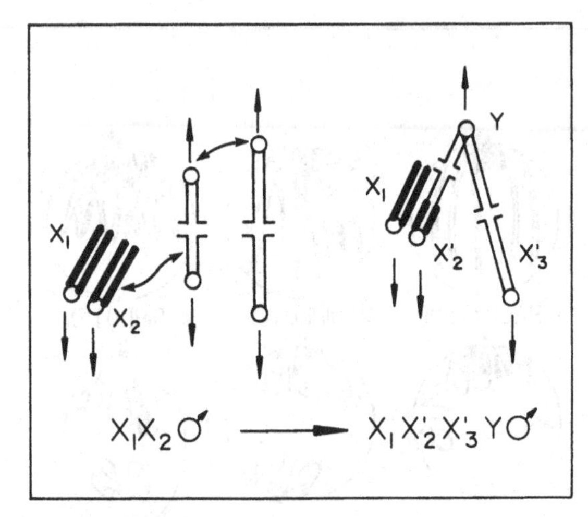

Fig. 5.15. Formation and orientation of the $X_1X_2'X_3'Y$ sex chromosome system of the jumping spider *Pellenes* (after Madison, 1982).

presumably reflects the prior association of the heteropycnotic X_1 and X_2' at first prophase which again ensures that the kinetochores of both the X_1 and X_2' orient to the same pole even though, most commonly, they dissociate from one another before metaphase-1 (Madison, 1982).

Pseudobivalent or pseudomultiple formation

Persistent non-homologous pairing. In the earwig *Forficula auricularia*, XY- and X_1X_2Y-individuals coexist in some populations in a polymorphic condition. In both cases the sex chromosomes are all entirely positively heteropycnotic at the onset of first prophase of male meiosis and retain this form of non-chiasmate association up to anaphase-1. Consequently, at first metaphase there is either an XY-pseudobivalent or else an X_1X_2Y-pseudotrivalent present which invariably results in regular segregation (Callan, 1941; Ortiz, 1969; Henderson, 1970a).

On the assumption that only chiasmata could provide a mechanism for the segregation of homologues, Darlington (1934) explained the demonstrable absence of genetic crossing-over in male *Drosophila melanogaster* by claiming that the bivalents invariably formed reciprocal double chiasmata which thus cancelled one another out genetically so that crossing-over would not be detected. While he later abandoned this claim for the autosomal bivalents he, quite irrationally, persisted in this belief in the case of the sex chromosomes (Darlington, 1973), despite the formidable evidence to the contrary presented by Cooper (1944a, b, 1949) and even after it was clear that neither the autosomes nor the sex chromosomes in the male of this species formed an SC.

The Y-chromosome of *D. melanogaster* is an almost exclusively

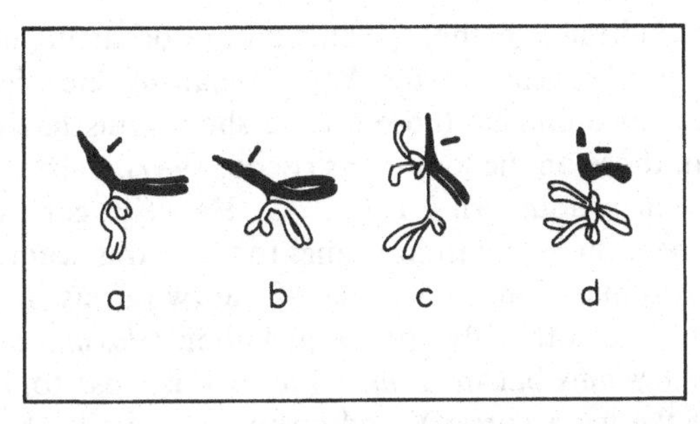

Fig. 5.16. Conjunction of the X- and Y-chromosomes of *Drosophila melanogaster* at first male meiosis. The short bar defines the position of the nucleolar organizer. In the Y the pairing site may be in (*a*) the short or (*b*) the long arm and equivalent pairing occurs in males with (*c*) two or (*d*) three Y-chromosomes. Comparable conjunction patterns (not shown) may also take place proximal to the nucleolar organizer region on the X (after Cooper, 1964).

heterochromatic chromosome and has no equivalent to the euchromatic segment present in the X. Moreover, as Cooper (1964) showed, pairing between the X and Y occurred only at short specific sites within the heterochromatic regions. These were located proximally in the X-heterochromatin (Xh) on either side of the nucleolar organizing region and on the long (L) and short (S) arms of the Y. These specialized interstitial sites Cooper referred to as *collochores*. He noted that, at any one time, only one collochore on each of the two chromosomes was involved in pairing. That is, one of the Xh sites conjoins with either the Y^S or the Y^L but never with both. In experimentally-produced XYY and XYYY individuals all the sex chromosomes conjoin to form a single multiple in which all the Ys are associated at the same site on the X (Fig. 5.16). By examining thin serial sections of metaphase-1 XY-bivalents of *D. melanogaster* with the electron microscope, Ault, Lin & Church (1982) identified a unique fibrillar material at the site of XY-association which they suggested might represent the cohesive elements or collochores which function as pairing sites. Its precise, composition, however, remains unknown. In *D. hydei*, Kremer, Hennig & Dijkhof (1986) comment that at male first metaphase, one or both ends of the Y are associated with the end of the heterochromatic arm of the X. These regions are not homologous, as defined by DNA-hybridization experiments, and here, as in *D. melanogaster*, it appears that pairing is restricted to limited regions of the sex chromosomes and is of non-chiasmate kind. If these pairing sites in *D. hydei* are equivalent to those of *D. melanogaster* this would strengthen Virrki's (1988a) contention that collochores may be interstitial or terminal in location.

Attachment of the X and Y by localized, non-chiasmate, sites is not confined to *Drosophila*. In males of the fly *Lucilia cuprina*, pairing sites are located distally in both long (l) and short (s) arms of both sex chromosomes. Pairing is highly specific, always occurring either between X^S and Y^S or else between X^L and Y^L. The pairing sites, however, show markedly different affinities, those on the short arms having a stronger attraction than those on the long arms (Bedo, 1987). Pairing sites are also found in Tsetse flies (Southern & Pell, 1973). Here too, contact is restricted to a small segment on each chromosome. In the X this segment lies to one side of the centromere, on the shorter of the two arms. It has a similar location in the Y of other fly species including *Glossina austeni* and *G. morsitans submorsitans* but in *G. m. centralis* it is close to, or at, the end of one arm of the metacentric Y and occurs in a similar location on the very short arm of the Y in *G. m. morsitans*. In these two latter taxa, although the X and Y generally remain together throughout first prophase they sometimes separate, even though they are still connected by a fine chromatin thread. In *G. m. submorsitans* and *G. austeni*, separation of the X and Y is frequently exaggerated with no visible form of attachment between the two sex univalents while in *G. pallidipes* the sex chromosomes are never associated, even at very early prophase (Southern, 1980). Despite this they still co-orient and segregate regularly in all species.

Pseudobivalent formation at second division. In the homopteran *Oncopeltus fasciatus* the X- and Y-sex chromosomes are positively heteropycnotic at the outset of meiosis and may, or may not, be associated. At zygotene the nucleus enters a diffuse stage during which the X and Y remain condensed and are frequently non-homologously associated. The autosomes then recondense at diakinesis and by the end of this stage the sex chromosomes are invariably separate and isopycnotic. During first metaphase they lie side-by-side in the centre of the spindle surrounded by a ring of autosomal bivalents (Fig. 5.17). The sex univalents divide equationally at first anaphase and the products of this division then associate as a pseudobivalent during telophase-1. Consequently the X and Y separate reductionally from one another at second anaphase (Wolfe & John, 1965). An equivalent behaviour has been noted by Ueshima (1979) in a variety of other homopterans.

In the male cotton stainer, *Dysdercus intermedius* (Hemiptera: Heteroptera), there are two X-chromosomes ($2n = 16, X_1X_20$). These are heteropycnotic at first meiotic prophase but at zygotene are frequently unpaired. In most pachytene meiocytes they are associated with one another and, by diplotene, pairing of the two Xs is complete in all cells. EM observations fail to reveal any SC between them and at diakinesis they separate from another. By first metaphase, however, they are again associated but this time side-by-side, like the X and Y of *Oncopeltus*. Precisely how such lateral pairing is established is not clear but EM observations indicate that they are surrounded by a sheath formed from the smooth endoplasmic reticulum. At first anaphase both sex univalents

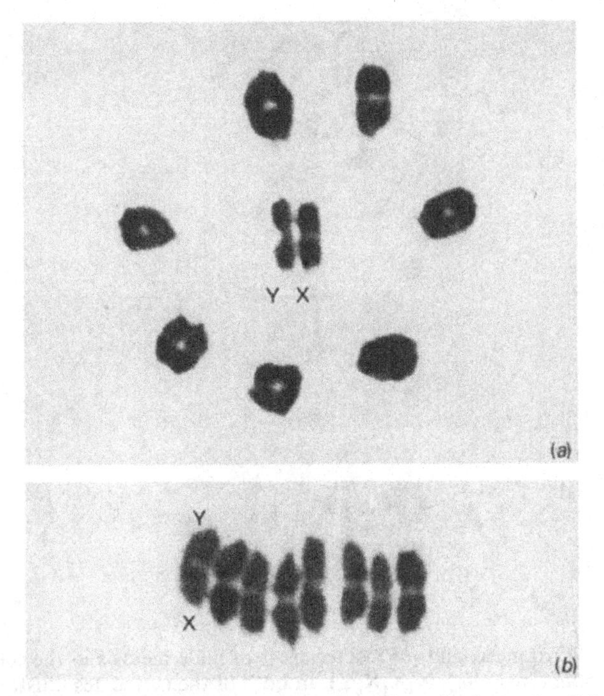

Fig. 5.17. (*a*) First and (*b*) second meiotic metaphase in the male of the milkweed bug *Oncopeltus fasciatus* (2n = 16♂, XY). The X- and Y-chromosomes are unpaired, but lie adjacent to one another and autoriented in the centre of a hollow spindle, at metaphase-1. At metaphase-2, following secondary pairing during interkinesis, the X and Y are co-oriented.

divide equationally. The products remain in close association and at second division they pair end-to-end but with a kinetochore attachment to only one pole from only one of the two chromosomes. Consequently at second anaphase they both move to the same pole (Ruthmann & Permantier, 1973; Ruthmann & Dahlberg, 1976).

Development of a specialized segregation body. In polyphagous beetles there is a large metacentric X and a very small y-chromosome. These associate at meiosis-1 in the form of a distinctive sex parachute (Xy_p) – an association that is unique among animals. Initially this association was interpreted as chiasmate, a view maintained by White (1973) despite the fact that it had already been shown to involve a specialized segregation body which formed between the X and y following their non-specific and heterochromatic association at leptotene–zygotene. This segregation body, which persists until first metaphase, guarantees the proper orientation of the two sex chromosomes in the absence of chiasmata (John & Lewis, 1960). An equivalent body is present in tenebrionid beetles of the

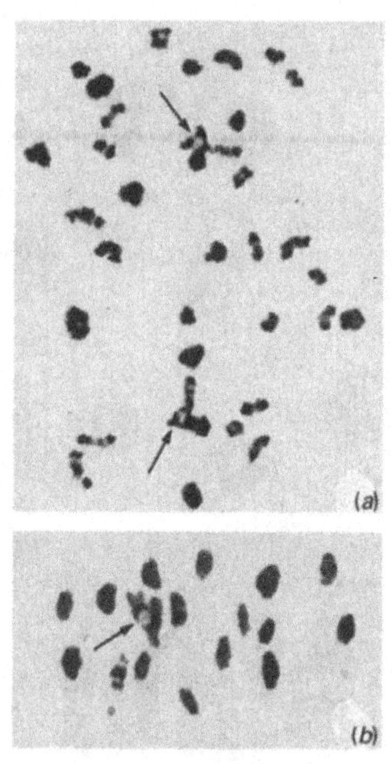

Fig. 5.18. (*a*) Diakinesis and (*b*) metaphase-1 of male meiosis in the cellar beetle *Blaps mucronata* (2n = 36♂, $X_1X_2X_3Y$). In both of the meiocytes illustrated in (*a*) the X_1-chromosome is paired terminally with the Y, and all three Xs are associated with a specialized segregation body (arrow). One of the autosomal bivalents is also associated with this body. By metaphase-1 (*b*) the autosomal association has lapsed but all three X-chromosomes remain associated with the segregation body and are now oriented to a common spindle pole, with the Y facing the opposite pole (arrow).

genus *Blaps* (Fig. 5.18), all known species of which are characterized by multiple sex mechanisms of the X_nY- or X_nY_n-type (Lewis & John, 1957; Wahrman, Nezer & Freund, 1973; Wahrman & Nezer, 1976; Smith & Virrki, 1978). This reaches an extreme in *Blaps cribosa* where there are 6 Ys and 12 Xs (= $[X_1X_2Y]_6$). Here the 18 sex chromosomes include six X-univalents and six XY-bivalents. All 12 Xs are attached to different regions of the segregation body while the six Ys are each individually associated, end-to-end, with six of the Xs. Drets *et al.* (1983) have described a variant form of Xy_p in the coccinelid *Epilachna paenulata*. They claim this involves a non-chiasmate association of the procentric heterochromatic segments of the two chromosomes together with a terminal, non-chiasmate, association of the ends of their euchromatic long arms. However, their description of the mode of formation of this parachute is not entirely convincing.

Combinations of paired and unpaired sex chromosomes

Most Crustacea do not possess differentiated sex chromosomes. Cyprid ostracods are an exception and are characterized by a striking range of

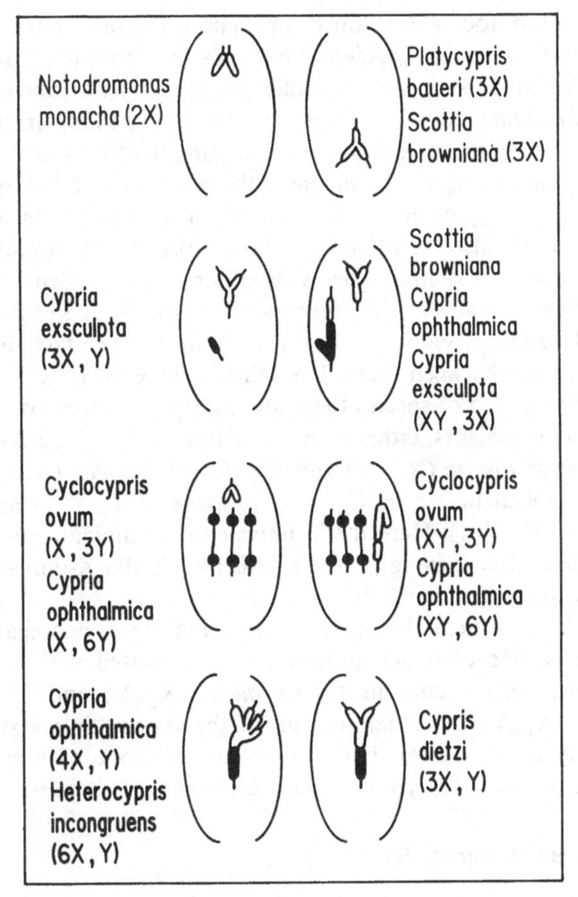

Fig. 5.19. The orientation behaviour of the eight distinct kinds of multiple sex chromosome systems that have been described in male cyprid ostracods (after Dietz, 1954, 1958).

different mechanisms (Dietz, 1954, 1955, 1958). In addition to conventional XY- and X0-systems there are three distinctive modes of segregation (Fig. 5.19). The first of these involves multiple and non-chiasmate aggregates of X-chromosomes. Such aggregates may occur alone, as in the crustaceans *Notodromonas monacha* ($X_{1\text{-}2}$) or *Scottia browniana*, *Physocypria kliei* and *Platycypris baueri* (all with $X_{1\text{-}3}$). Alternatively, X-aggregates may be present in conjunction with a Y-univalent, as in *Cypria exscultpta* ($X_{1\text{-}3}Y$), or with an XY-bivalent, as in *Scottia browniana* and *Cypria ophthalmica* (both with $X_{1\text{-}3}XY$).

In the case of *Notodromonas monacha*, Dietz (1954) showed that the two telocentric X-chromosomes were heteropycnotic, and loosely associated, at the onset of meiosis. They retained this association throughout the entire first division. By pre-diakinesis the centromeres of the two X-chromosomes were both polarized to the extranuclear centrosome and this polarization, too, was retained throughout the remainder of meiosis-1. Segregation in this case evidently parallels the continuous polarization system of X_n0 spiders. A more complex pattern of behaviour was present in *Scottia browniana* and *Platycypris baueri* (Dietz, 1958). In these cases, the aggregates of three X-chromosomes consistently oriented in a 2:1 fashion but then segregated *en masse* to the pole to which two of the three Xs within the aggregate were directed. Presumably an equivalent behaviour must also characterize those species where multiple X-aggregates occur in conjunction with either a Y-univalent or an XY-bivalent. Here, however, the movement of the X-aggregate must be regularized so that it segregates from the Y despite the fact that there is no direct contact between them. How this is achieved is not known.

In a second group of species there are multiple Y-chromosomes which all appear as univalents, either in conjunction with a single X-univalent, as in *Cyclocypris ovum* (X,Y_{1-3}) and *Cypria ophthalmica* (X,Y_{1-6}) or else with an XY-bivalent, as in *Cyclocypris ovum* (XY, Y_{1-3}) and *Cypria ophthalmica* (XY,Y_{1-6}). Here the Y-univalents all divide equationally at first anaphase, after lagging on the spindle, while the X-univalent, when present, moves reductionally to one pole.

Yet a third system involves a non-chiasmate aggregate of X-chromosomes, one of which is terminally associated with a single Y-chromosome. This occurs in *Cypris dietzi* ($X_{1-3}Y$) and *Heterocypris incongruens* ($X_{1-6}Y$) and is reminiscent of the sex multiple system of the beetle genus *Blaps* though it differs in the absence of any apparent segregation body of the type that characterizes these beetles.

B.2 Preferential segregation

B.2.1 Preferential segregation at male meiosis

Segregation on a monopolar spindle

Conventional meiosis is based on the formation of a bipolar spindle and on the bipolar orientation of individual bivalents within that spindle. The polar complex that forms at the breakdown of the nuclear membrane at the first meiotic division in males of the fungus gnat, *Sciara coprophila*, however, defines the acuminate end of a structurally asymmetrical and monopolar spindle (Metz, 1933). Despite the unique nature of the first division spindle, homologues still segregate on it, though they do not pair. Additionally, the 10 chromosomes, comprising four paternal and four homologous maternal members, together with two germ-line-limited (L) chromosomes which may be of either maternal or paternal origin, do not form a metaphase plate. Subsequently, at late meiosis-1, the four paternal chromosomes are cut off in a cytoplasmic bud and degenerate. Only the maternal chromosomes and the L-chromosomes remain at the pole.

From an ultrastructural analysis of this sequence, Kubai (1982) showed that at early prophase-1 the nuclear envelope was irregular in contour with the chromosomes located in protuberances of the nucleus. She found that the two L-chromosomes and the maternal X, which could be individually distinguished, lay close to the polar organelle present in the cytoplasm and already surrounded by microtubules. The paternal X, however, was not so affiliated. That is, there was at least a partial segregation within the prophase-1 nucleus which foreshadowed the distribution of the chromosomes in the metaphase-1 nucleus. Whether the remaining chromosomes were similarly segregated relative to the polar organelle was not clear but appeared likely.

Within the monopolar spindle itself Kubai found that neither the maternal chromosomes nor the L-chromosomes, both of which segregate to the single pole, actually formed kinetochore microtubules. The retention of these chromosomes at the polar region thus takes place in the complete absence of any direct connection with that region and presumably as a consequence of a persistence of the polarization to the polar centre which was evident prior to the breakdown of the nuclear membrane. While the kinetochores of both the maternally-derived and the L-chromosomes are constrained from developing microtubules, and have no recognizable orientation, those of the paternal chromosomes which directly face the pole develop microtubules that run towards it. Paradoxically, however, they do not move to the pole. That is, while the paternally-derived chromosomes develop kinetochore microtubules these do not generate chromosome movement as they conventionally do in bipolar spindle systems.

In the second meiotic division, as in all of the mitotic divisions in *Sciara*, a bipolar spindle forms and all the kinetochores of the maternal autosomes and the L-chromosomes develop microtubules connected to both poles. The maternal X, however, remains oriented to only one of the two poles and so undergoes directed non-disjunction. This appears to represent a carry-over of the polarization established during the first division (Abbott & Gerbi, 1981).

In sum, the monopolar spindle system, and the differential reaction of maternal and paternal chromosome sets to this spindle, evidently function to ensure the physical separation of chromosomes which are already spatially segregated in the prophase-1 nucleus. Moreover the chromosomes that segregate to the single pole fail to form kinetochore microtubules whereas the chromosomes which do not move to the pole do form them. Consequently, segregation is achieved by persistent polarization and not by microtubule-initiated movement.

Monocentric movement on a structurally bipolar spindle

In the gall midge *Mycophila speyeri* (Diptera: Cecidomyidae) spermatogenesis is unorthodox. There is no meiotic pairing and the

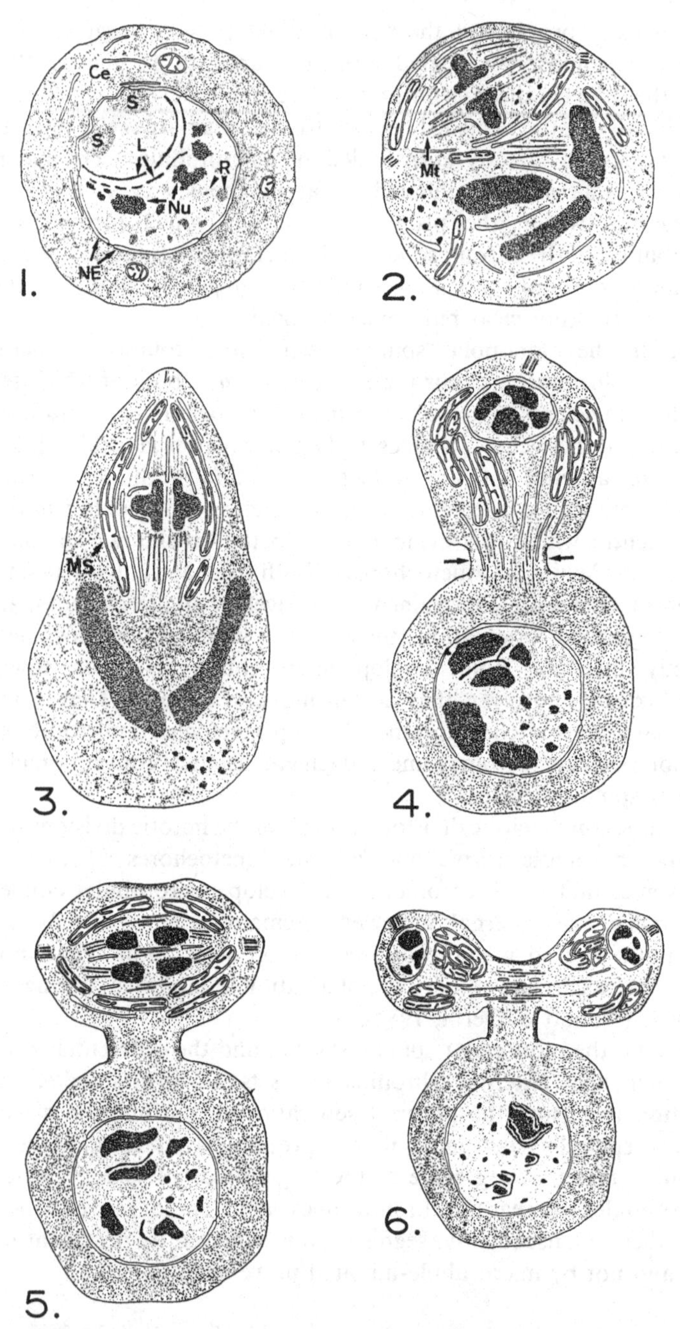

Fig. 5.20. Meiosis in the male of the midge *Mycophila speyeri*. (1) At prophase-1 there are two groups of unpaired chromosomes (S and R), separated by a lamella (L) of electron-dense material, within the nuclear envelope (NE). The S-chromosomes lie polarized to the centrioles (Ce) and a series of nucleolar masses

chromosomes are separated into two groups at the onset of prophase-1. Electron microscopy indicates that a lamella of electron-dense material separates the two and that a pair of centrioles is located near the haploid set which is subsequently incorporated into the sperm, the so-called *S-group*. The other group includes a *residual haploid set* (*R*) which is segregated from the S-group at anaphase-1. Unequal cytokinesis then produces a small secondary spermatocyte containing the S-chromosomes and a large residual cell which includes the R-chromosomes. Only the S-group undergoes a second meiotic division and so only two spermatids form (Camenzind & Fux, 1977).

During prophase-1 the lamella separating the R- and S-groups disintegrates and a spindle precursor develops between the two centrioles. When the spindle forms, only S-chromosomes orient on it (Fig. 5.20). The R-chromosomes gather together in the vicinity of the broad inner spindle pole. At this stage centrioles are present only at the outer pole, and at anaphase-1 the S-chromosomes move intact to the outer pole despite the fact that they maintain a bipolar orientation. This monocentric movement is accompanied by a spindle shift away from the prospective residual cell and the development of a constriction at the inner spindle pole. The R-chromosomes then clump to form a residual nucleus while the S-chromosomes form the nucleus of a secondary spermatocyte. Cytokinesis is incomplete so that the two nuclei remain in a single cytoplasmic area. Only in the secondary spermatocyte nucleus does a second division spindle form.

The X_1X_2Y system of Neocurtilla hexadactyla

The independent assortment of non-homologous chromosomes in the first meiotic division was first confirmed by the pioneer work of Carothers (1917). She showed that when heteromorphic (unequal) bivalents were present in male (X0) grasshoppers each heteromorph segregated at random relative to every other heteromorphic pair as well as with respect to the X-univalent. Just prior to this study Payne (1916) had reported on a heteromorphic pair in the mole cricket (*N. hexadactyla*) which, unlike

---

Caption for Fig. 5.20 (*cont.*)

(Nu) are associated with the R group of chromosomes. (2) At pro-metaphase-1 only the univalent S-chromosomes orient on the developing spindle. (3) By metaphase-1 this spindle has become visibly asymmetrical with the centrioles now restricted to the pointed outer pole. The R-chromosomes accumulate at the broad inner spindle pole where the centrioles have aborted. (4) At anaphase-1 the S-chromosomes move to the pointed outer spindle pole despite the fact that they are oriented to both poles. (5) Cytokinesis is incomplete and the second meiotic division is restricted to the outer nucleus so that only two spermatids form. (6) The residual nucleus, containing the R-chromosomes, plays no further part in meiosis and eventually degenerates (after Camenzind & Fux, 1977 and with the permission of *Caryologia*).

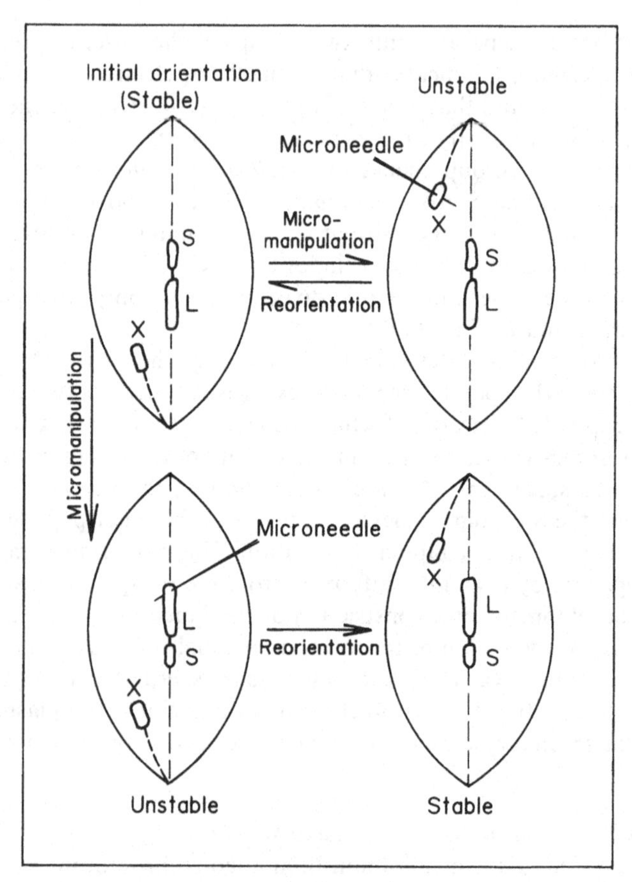

Fig. 5.21. Reorientation behaviour of the X_1X_2Y male sex multiple of the mole cricket *Neocurtilla hexadactyla* following micromanipulation (after Camenzind & Nicklas, 1968).

Carother's cases, segregated non-randomly with respect to the univalent-X. In the mole cricket, the larger heteromorph consistently moved to the same pole as the X at first anaphase despite the fact that the two chromosomes concerned showed no apparent physical association.

The validity of this case was subsequently confirmed in living meiocytes by Camenzind & Nicklas (1968). They also showed that when an incorrect orientation was induced by micromanipulation, with the X oriented to the same pole as the small heteromorph, a reorientation of the X invariably occurred (Fig. 5.21). In effect, therefore, this system behaves in the manner expected of an X_1X_2Y mechanism where X_1 designates the univalent sex chromosome while X_2 designates the larger and Y the smaller member of the heteromorphic pair. These micromanipulation experiments also demonstrated that, while the X_1-univalent responded actively to an

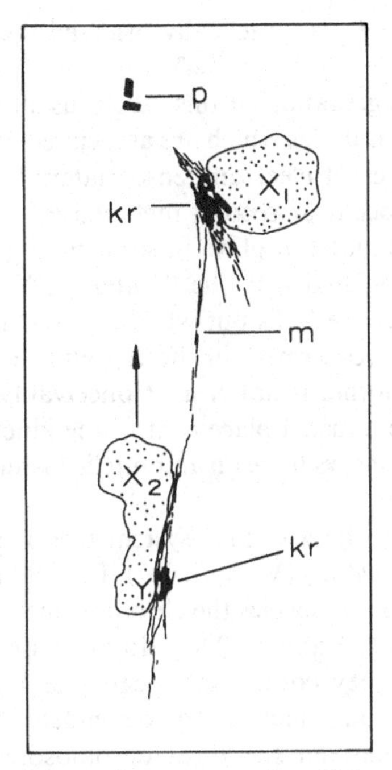

Fig. 5.22. An electron micrograph reconstruction of the orientation of the X_1X_2Y sex chromosome system of *Neocurtilla hexadactyla* at first metaphase of male meiosis. The kinetochore regions (kr) of the X_1- and Y-chromosomes are interconnected by microtubules (m). The direction of the spindle microtubules of the X_2-chromosome is indicated by an arrow and P marks the position of the spindle pole to which the two X chromosomes are oriented (after Kubai & Wise, 1981 and with the permission of *The Journal of Cell Biology*).

alteration in the orientation of the X_2Y-bivalent, the reverse was not true; the X_2Y-bivalent appeared to be insensitive to the orientation of the X_1-univalent. By using UV microbeam irradiation, Wise, Sillers & Forer (1984) were able to confirm that the X_1 is the active chromosome in determining the orientation of the two Xs to the same pole. They found that, in about one-third of the treated cells, irradiation of the X_1, X_2, Y or X_1Y spindle fibers resulted in a movement or a rotation of the X_1-univalent. By contrast the X_2Y bivalent did not respond to such irradiation. Neither did the irradiation of autosomal spindle fibers have any effect on the X_1.

This system has also been examined using electron microscopy by Kubai & Wise (1981). They report that, at metaphase-1, the kinetochore regions of the X_1 and the Y, both of which are located laterally on the spindle, are marked by massive aggregates of electron-dense material associated with microtubules running to both poles. Thus, although they lie in opposite half spindles, they are connected by the bundle of microtubules running between their kinetochores (Fig. 5.22). The net result is that while the X_2 is syntelically-oriented the X_1 and the Y are in effect amphitelic.

The non-random segregation in this case thus stems from the unusual properties of the microtubules which are associated with the kinetochores. The massive aggregates of electron-dense material which are present at metaphase-1 do not occur at second metaphase or at mitosis. Precisely what role, if any, this material plays in securing segregation is not clear. The microtubules associated with the X_1 and the Y are certainly distinct from ordinary kinetochore MTs but whether these atypical fibers, which link the two kinetochores, constrain the X_1- and the Y-chromosomes to segregate from one another is not clear. Conceivably, their presence may be associated with the unusual placement of the kinetochore faces, which are oriented in what appears to be an amphitelic fashion reminiscent of the X and Y of crane flies.

What could prove to be a similar system operates in the male of the coleopteran *Omphoita clerica* (Virrki, 1967). Because the female condition has not been defined in this species the chromosomes involved are referred to simply by shape as J, I and V. They do not pair but form a distance trivalent from which they consistently segregate J to I+V. Here, too, some kind of interaction which is not dependent on obvious physical contact must occur between the three chromosomes to regulate their orientation and segregation.

B.2.2 Preferential segregation at female meiosis

Female meiosis is most usually polarized such that meiotic products in different locations within the oocyte differentiate into distinctive structures, not all of which become gametes. In animal oocytes, three of the four meiotic products become polar bodies which most commonly degenerate (but see section 7B). In a number of 'female' plants there is a linear tetrad in which only the terminal member becomes a functional megaspore. Both these situations depend on the directed orientation of the meiotic spindles in conformity with an established cytoplasmic gradient operative within the female meiocyte. Particular chromosomes may utilize this gradient to ensure their preferential segregation into the functional female gamete and so produce systems of meiotic drive.

One of the clearest cases of this behaviour is to be found in the supernumerary, or B-chromosomes, which are sometimes present in polymorphic situations in a proportion of individuals within a given population. From controlled crosses involving parental plants with B-chromosomes, Kayano (1957) was able to show that in *Lilium callosum* such supernumeraries were preferentially included in egg nuclei. By direct cytological observation of EMCs he found that univalent Bs lay preferentially on the micropylar side of the first division spindle and that

Table 5.4. *Preferential segregation of supernumerary univalents (B-Is) in the EMC of Lilium callosum*

Stage	% B-Is at micropylar pole
Metaphase-1	75.4
Anaphase-1	84.1
Metaphase-2/Anaphase-2	80.0
Post-meiotic mitoses	73.1
% Eggs with Bs	83.8

After Kayano, 1957.

Table 5.5. *Behaviour of univalent B-chromosomes in 57 metaphase-1 oocytes of the grasshopper Myrmeleotettix maculatus*

		M-1 distribution of B univalents			Transmission frequency	
Number of Bs per individual	Number of eggs examined	Polar end	Equa-torial	Egg end	Inferred	Observed
1	48	8	14	26	0.77	0.78
2	9	4	1	13	0.79	

After Hewitt, 1976.

this behaviour was repeated not only at the second meiotic division but also in the post-meiotic mitosis (Table 5.4). Since it is the micropylar megaspore that gives rise to the haploid egg nucleus, while the three other meiotic products migrate to the chalazal end where they fuse to form a triploid chalazal nucleus, this polarization of B-chromosome movement leads to the preferential inclusion of Bs in the egg nucleus.

Hewitt (1976) reported comparable behaviour in the grasshopper *Myrmeleotettix maculatus*, where again B-chromosomes have been inferred, from single pair crosses, to accumulate during egg meiosis. Here, the egg spindle is markedly asymmetrical with the external pole, which is destined to give rise to the polar bodies, being characteristically shorter, flatter and wider. By contrast, the internal pole, which produces the definitive egg nucleus, is pointed. Relative to this asymmetry, a majority of B-univalents in individuals with either one or two supernumeraries lie on the egg-nucleus side of the first division spindle. The inferred transmission frequency obtained from these data agrees well with the observed transmission frequency provided from controlled crosses (Table 5.5). Thus here, as in *Lilium*, the preferential distribution of Bs must be

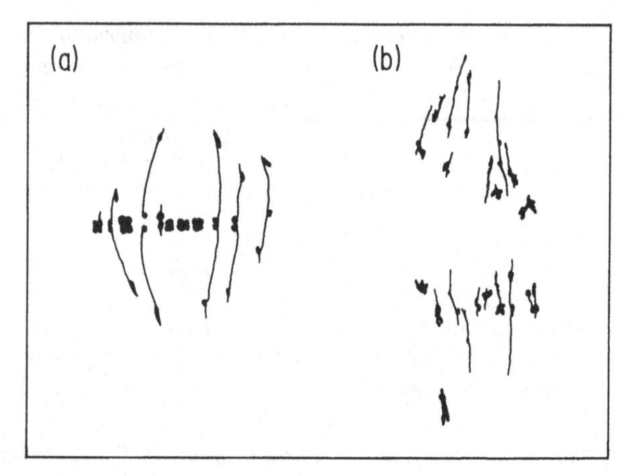

Fig. 5.23. (*a*) Metaphase-1 and (*b*) anaphase-1 of meiosis in PMCs of *Zea mays* homozygous for a supernumerary heterochromatic segment (knob) on chromosome-10. In the first metaphase cell, this bivalent, and four others which are similarly knobbed, show homozygosity for neocentromere activity. These active neocentric ends then lead the way to the poles at anaphase-1 (after Rhoades, 1952).

repeated at the second meiotic division though no observations were actually made at this stage.

Yet a third example of this form of preferential segregation has been described in the EMCs of individuals of *Zea mays* (2n = 20) heterozygous for supernumerary segments, referred to as knobs (K), on the shortest, 10th, chromosome. In such heterozygotes, crosses indicate that approximately 70% of the functioning megaspores transmit a segment-bearing K10-chromosome. The same is true of heteromorphic pairs other than 10, provided that 10 itself is also heterozygous (Longley, 1945).

PMCs carrying one or two K10s produce supernumerary half-spindle microtubules at distal sites in the vicinity of the extra segment, referred to as *neocentromeres*, in addition to those formed by the true centromere, at both first and second meiosis.

Comparable neocentromeres are assumed to occur in EMCs though this has not been confirmed by direct observation. Neocentromere formation takes place in plants which are either heterozygous or homozygous for K10 but is not restricted to this chromosome. Neocentromeres also develop in other heteromorphic pairs carrying supernumerary segments, provided one or two K10 chromosomes are also present. No attempt has been made to analyse these neocentromeres by electron microscopy though Bajer (1968) has identified a neocentromere in the plant *Haemanthus* by this technique. His studies indicate that typical details of kinetochore structure are lacking and microtubules simply insert directly into the chromatin.

Neocentromeres are drawn poleward ahead of the true centromeres. Consequently, the distal ends lead the way to the pole (Fig. 5.23) rather than trailing in the spindle. When coupled with crossing-over, the net result of this behaviour, and its persistence into the second meiotic division, is the non-random orientation of half bivalents on the second division spindle. Maize forms a linear tetrad, with the basal megaspore giving rise to the female gametophyte, the other three aborting. Consequently, any mechanism that preferentially directs a chromatid into the basal megaspore will automatically lead to its preferential segregation.

6

The genetic control of meiosis

The typical mutant deficient in general recombination is either sick or dead.
Alan Campbell

All the events of meiosis are under some form of genetic control and mutations which are defective in functions specific to meiosis have been identified in a variety of eukaryotes (Rees, 1961; Baker *et al*., 1976; Sears, 1976; Golubovskaya, 1979). Table 6.1, for example, summarizes the principal categories of mutations identified in plants. From such cases it has been inferred that the normal alleles of such mutations produce gene products that play crucial roles in regulating the meiotic activities of chromosomes in space and time. Like mutations generally, those which influence chromosome behaviour most commonly have deleterious consequences and lead to an impairment of the efficiency of meiosis. They have been most thoroughly characterized in the fruit fly *Drosophila melanogaster* and in the yeast *Saccharomyces cerevisiae*.

A MEIOTIC MUTATIONS IN *DROSOPHILA*

A majority of mutations in *Drosophila* affect only one sex. This is not surprising when one recalls that SC formation and crossing-over is restricted to the female.

A.1 Male mutations

Microtubules, as we have seen, are dimers composed of equimolecular amounts of two 50000 dalton subunits, α- and β-tubulin. While there are other microtubule-associated proteins, the tubulins are the only ones that have been shown to play a role in chromosome movement since they are the principal functional components of the spindle microtubules. A specific β_2-tubulin subunit has been identified which is expressed only at male meiosis in *D. melanogaster* and so functions in the production of the meiotic, but not the mitotic, spindle as well as in the differentiation of the axoneme of the male sperm (Kemphues *et al*., 1982). The β_2-tubulin gene

Table 6.1. *Mutations affecting meiosis in angiosperm plants*

Effect	Mutation	Species
Entry into meiosis blocked	*am* (ameiotic)	*Zea mays*
Equational division substituted for meiosis-1	*afd-W23*	*Zea mays*
No synapsis	*as* (asynaptic)	*Triticum durum*
Failure to maintain pairing	*ds* (desynaptic)	*Avena strigosa* *Hordeum vulgare* *Lolium perenne*
Desynapsis with abnormal synaptonemal complex	*dsy-A344*	*Zea mays*
Desynapsis with precocious sister chromatid separation at anaphase-1	*a*	*Triticum monococcum*
	2982	*Pisum sativum*
Impaired separation of half bivalents at anaphase-1	*dv* (divergent)	*Zea mays*
Precocious separation of sister chromatids at interkinesis	*pc*	*Lycopersicon esculentum*
Absence of division-1 cytokinesis	*va*	*Zea mays*
Loss or impairment of second division	*dy* (dyad) *el* (elongate)	*Datura stramonium* *Zea mays*

After Golubovskaya, 1979.

Table 6.2. *Male sterile mutations affecting β_2-tubulin in male Drosophila melanogaster*

β_2-tubulin variant	Effect
Dominant stable (*B2t*[d])	Formation of multipolar spindles at meiosis-2 in heterozygotes. Complete failure of spindle formation in homozygotes.
Recessive class I, unstable	Unable to form 2-β dimers, consequently both mutant β and wild type α tubulins are degraded. No meiosis occurs.
Recessive class II, stable	Specific defects in microtubule functions which proceed abnormally

After Raff, 1984.

Table 6.3. *Mutations affecting non-disjunction at meiosis-1 in male Drosophila melanogaster*

Mutation	Affected chromosomes	% non-disjunction
mei-269 (X)	X, Y	1–10% (70–95% of recovered sex chromosome exceptions are *nullo*-XY).
mei-S8 (2)[a]	4	
mei-081 (3)[a]	X, Y and 4	X, Y – 5% 4 – 7.8%
mei-G17 (2)	X, Y and 2	
pal (= *mei-W5*) (2)[b]	X, Y, 2, 3 and 4	4 – 17% X, Y – 5%

After Baker & Hall, 1976; Baker *et al.*, 1976.
[a] Disjunction of major autosomes not yet examined.
[b] Also causes frequent somatic loss of paternal X, Y and 4.

is expressed shortly before the onset of male meiosis and is not expressed at all in females or elsewhere in males. Raff (1984) defines three classes of mutations in the β_2-tubulin locus that produce different effects on microtubule function in *Drosophila* (Table 6.2).

Ripoll *et al.* (1985) have described a semi-lethal mutation, abnormal spindle (*asp*), which, when homozygous, affects both mitosis and male meiosis. Heterozygotes are normal but in *asp* homozygotes mitosis is arrested leading to the production of polyploid cells. In homozygous *asp* males sex chromosome segregation at meiosis-1 parallels the behaviour of individuals in which the X lacks a collochore. Consequently, Y-chromosome gametes constitute only 16–19% of the progeny compared to the 46% which carry an X-chromosome. Homozygous females are sterile.

In the absence of SC formation and crossing-over, most other male mutations influence only chromosome disjunction (Table 6.3). Moreover, a majority affect only a subset of chromosomes. In the case of the sex chromosomes this is easy to understand since they possess special pairing sites or collochores but why, and how, specific autosomes are also sometimes affected is not so simply explained.

A.2 Female mutations

Mutations concerned with female meiosis are far more numerous and fall into two categories, those which influence recombination and those which disturb disjunction.

A.2.1 Recombination defective mutations

Mutations affecting recombination at female meiosis fall into several distinct categories (Table 6.4). All are restricted to modifying recombination within euchromatic segments. None induce crossing-over within the heterochromatin (Carpenter, 1984b). Some decrease the frequency of crossing-over (*mei-9*). Others produce a more random distribution of crossover sites (*mei-352*), while still others both decrease exchange frequency and lead to an alteration in the distribution of exchanges along chromosomes (*mei-41, 218, 251, S282, mei-B, abo* and *mei-W68*^{L1}). In the latter class, the reduction is most extreme in the distal regions and less pronounced in the proximal regions. Indeed in the case of *mei-B, 195* and *251* the recombination frequency in the proximal region is significantly higher than in the wild type. This increase in the number of proximal exchanges cannot be explained by any alteration in the structure of the SC since the location of the zone of morphological transition between the eu- and heterochromatin remains unchanged in these polar mutations. The behaviour of such recombination-defective mutations suggests that exchange involves two distinct events. One of these determines the distribution and frequency of the sites at which crossing-over occurs, including their distribution along individual chromosome arms and their interference relationships, with the implication that the normal alleles of the mutations in question determine where exchanges can occur within the euchromatin. A second event then influences the frequency of crossing-over at these sites. The greater majority of the well characterized mutations appear to influence site choice rather than the exchange mechanism itself, though these two events may well share some processes in common. Whereas most of the mutations influencing female recom-

Table 6.4. *Mutations affecting recombination in female Drosophila melanogaster*

Mutation	Effect	Presumed wild-type function
mei-9, *mei-9*[b](X)	Reduce recombination in 2 and X without altering distribution of exchanges.	Specifies a product that functions directly in the exchange process.
mei-352(X)	Affects distribution but not frequency of crossing-over leading to a more random distribution	Specifies spatial restriction of exchange sites.
$c(3)G^{17}$, $c(3)G^{68}$(3)	Greatly reduces exchange in X, 2 and 3.	Specifies condition for exchange, acting to promote synapsis.
mei-B(2) *mei-S282*(3) *mei-41*(X)[a] *mei-218*(X) *mei-251*(X) *abo*(2) $mei\text{-}W68^{L1}$(2)	All seven both reduce recombination in X, 2 and 3 and alter the distribution of exchanges. Reduction is most extreme in the distal regions. In *mei-B* and *251* there is increased recombination proximally.	Specify preconditions for exchange.
mei-1(3)	Affects restricted to X and are both chromosome and region specific.	Specify conditions affecting specific sites on X.

After Baker & Hall, 1976; Baker *et al.*, 1976.

[a] *mei-195* is allelic with *mei-41*.

[b] Requires homozygous factors on 2 and 3 for its expression.

Table 6.5. *Influence of the c(3)G mutation on recombination in female Drosophila melanogaster*

Genotype of parent	Total progeny analysed	% Recombination Inter-genic y/w^{a}w	Intra-genic w^{a}w	Inter-genic w^{a}w/spl
+/+	96928	1.34	0.01	1.42
c(3)G/c(3)G	216420	0.10	—	0.21
c(3)GG/+	84812	5.10	0.03	4.41

After Sen *et al.*, 1981.

bination in *Drosophila melanogaster* affect all chromosomes in a similar way, *mei-1* is unique. It causes a reduction of recombination specifically in the X-chromosome and especially in its central region.

More severe results follow from *c(3)G*^{17}, *c(3)G*^{68} and *mei-W68*^{L1} in which there is an especially marked reduction in crossing-over in female homozygotes. Intergenic recombination in *c(3)G* homozygotes, for example, is reduced some seven-fold compared to wild type (Table 6.5), while intragenic recombination is virtually abolished. In heterozygotes, by contrast, both types are significantly increased. These increases are more pronounced for *c(3)G*^{17} than for *c(3)G*^{68} heterozygotes (Hinton, 1966; Hall, 1972; Sen *et al.*, 1981). In flies heterozygous for a deficiency of the *c(3)G* locus, crossing-over is decreased by about one-third, with the most marked changes again occurring in the distal regions. This pattern is reminiscent of those mutations which appear to be defective in preconditions for exchange and imply that *c(3)G* also belongs to this category.

Where crossing-over is virtually absent, as in homozygotes for *c(3)G*^{17} and *mei-W68*, no synaptonemal complex forms (Meyer, 1964; Carpenter unpublished, referred to in Baker *et al.*, 1976) and pachytene chromosome condensation also fails. Rasmussen (1975) reported that, while the central elements of the SC were present in homozygous *c(3)G* females, no lateral elements formed. In heterozygous females, by contrast, he found normal complexes. SC formation was also shown to be normal in mutations, like *mei-9*^{a}, *mei-218* and *mei-41*, which show some residual recombination (Carpenter, 1975a).

Since recombination was not completely abolished in *c(3)G* homozygotes, Sen *et al.* (1981) concluded that crossing-over cannot be excluded even in the absence of SC formation. There remains the possibility that very short stretches of SC are formed which would be difficult to detect using EM sections and until microspreads have been examined it would be safer to suspend judgement on the matter.

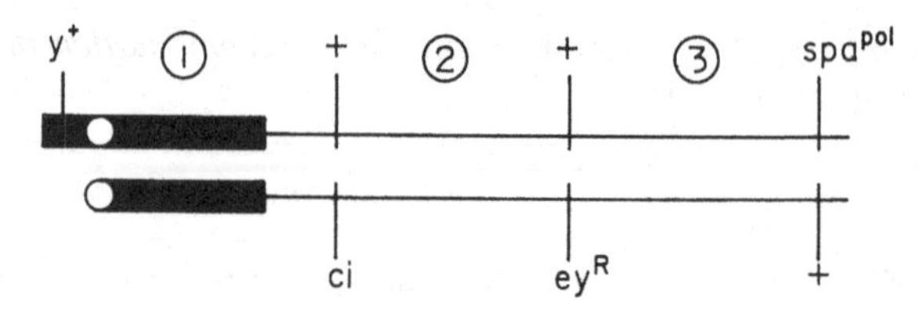

Fig. 6.1. Structure of the heterozygous bivalent used by Sandler & Szauter to measure recombination in chromosome-4 of *Drosophila melanogaster* in the presence of four recombination-defective meiotic mutations.

While a substantial number of recombination-defective mutations have now been identified in *D. melanogaster* we know virtually nothing about their probable molecular activity. Nguyen & Boyd (1977) showed that the *mei-9* mutation was defective in the repair replication of DNA damaged by either UV- or X-ray treatment. From this they suggested that the wild-type function of the *mei-9* locus was, in some way, necessary for the excision repair process. Beyond this we have no additional insight into the molecular biology of this class of mutations.

In normal female meiosis there is only one pathway to the regular disjunction of homologues in the case of the X-chromosome and autosomes-2 and -3; namely, homologous pairing followed by crossing-over. The small fourth pair of homologues, however, never recombine under normal circumstances. They are consistently achiasmate, yet they disjoin regularly. Despite the fact that crossing-over does not normally take place in chromosome-4 it is possible to induce it under special circumstances. These include diplo-4 triploids as well as through the influence of meiotic mutations that alter the spatial distribution of exchanges in the major chromosomes. This was clearly demonstrated by Sandler & Szauter (1978) who measured the extent of recombination in three regions of chromosome-4 (Fig. 6.1) using females which were homozygous for three of the recombinant-deficient mutations (*mei S-282*, *mei-218* and *mei-9*^{b}) as well as for a fourth (*mei-352*) which affects the distribution but not the frequency of exchanges. Region 1 was largely heterochromatic whereas regions 2 and 3 were exclusively euchromatic. Despite the fact that chromosome-4 is normally achiasmate in the female it was affected by these mutations. Specifically: there was no recombination in the two euchromatic regions in the presence of *mei-9*^{b} but reciprocal recombinant classes were recovered with the other three mutations. Recombination was also detected in the predominantly heterochromatic region 1 but here it was invariably non-reciprocal with only one recombinant class being recovered in all four mutations, namely *y pol*.

Because a mitotic exchange event will be multiplied mitotically, its occurrence is expected to produce clusters of recombinants in the progeny of single females. Meiotic recombination, on the other hand, leads to a binomial distribution of recombinants among the progeny of single females. While clustering was evident for exchanges in region 1 it was not present in either of the two euchromatic regions. For this reason, Sandler & Szauter concluded that the non-reciprocal recombination in the heterochromatic region must have resulted from pre-meiotic exchange and was mitotic in character whereas the reciprocal recombination observed in the two euchromatic regions was due to meiotic crossing-over. This was supported by the fact that mitotic recombination was shown to be either absent, or else very infrequent, in males carrying a mutation that induces substantial numbers of meiotic recombinants in chromosome-4 at female meiosis. Their results also imply that while the fourth chromosome is physically capable of crossing-over it must ordinarily be restrained in some way from doing so. A similar form of constraint may also explain the marked regional differences in the frequency of crossing-over per unit length of the major chromosomes (X, 2 and 3) under normal circumstances. Thus, it may be significant that meiotic crossing-over can be induced in chromosome-4 by mutations which, like *mei-218, S282* and *352*, attenuate these localized constraints, but not by *mei-9*^{b} which does not do so.

A.2.2 Disjunction defective mutations

Mutations which lead to decreased levels of recombination also lead to increased levels of non-disjunction at the first meiotic division. In the case of the major chromosomes there is no reason to doubt that they fail to disjoin because of their failure to crossover. Thus, most of the female recombination-defective mutations do not influence disjunction in the achiasmate males. The mutation *mei-S332* is unusual since, when homozygous in either males or females, it leads to a very high frequency of non-disjunction for all chromosomes. Non-disjunction is also substantially higher than normal in male and female heterozygotes, since *mei-S332* is semidominant. Additionally, the frequency of non-disjunction in autosomes-2 and -3 is correlated with the frequency of production of non-exchange chromosomes (Baker & Hall, 1976). This argument cannot, however, be applied to chromosome-4 which, as we have seen, does not normally recombine at female meiosis.

Non-homologous chromosomes are expected to segregate independently of one another. However, in a number of experimentally-constructed and chromosomally-aberrant situations, where crossing-over is abolished between two different pairs of homologous, non-homologous chromosomes may disjoin from one another at female meiosis in *Drosophila melanogaster* (Anderson, 1929; Grell, 1976). To accommodate such cases, Grell has consistently argued (see Grell, 1985 for references)

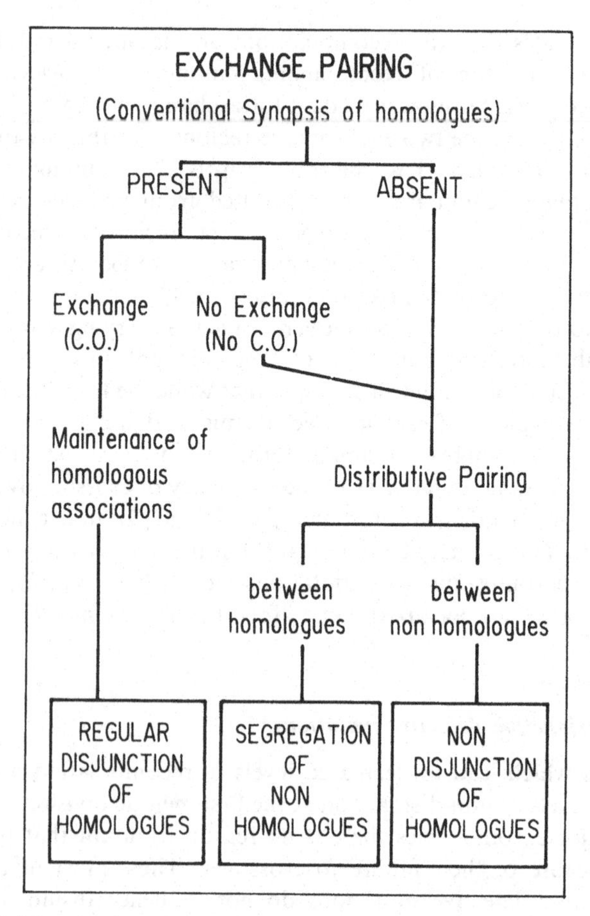

Fig. 6.2. Pairing and segregation sequences in the female of *Drosophila melanogaster* as proposed by Grell (1985).

that when homologues fail to exchange by crossing-over, or else when a chromosome lacks a homologue for normal exchange, then such chromosomes may participate in a second, discrete and non-homologous, pairing event. This event, which occurs subsequent to, and independent of, conventional synapsis, she refers to as *distributive pairing* and emphasizes that such pairing may occur either between non-exchange homologues or between non-homologues. Consequently, distributive pairing may lead both to the regular segregation of non-homologues or to the non-disjunction of homologues (Fig. 6.2). In this scheme, whether non-exchange homologues segregate from one another of from non-exchange heterologues is assumed to depend on chromosome size. The more similar two chromosomes are in size, the more likely they are to disjoin from one another. Size recognition is not, however, a necessity for non-homologous segregation. If only two non-exchange chromosomes are present in a female meiocyte they will still disjoin despite a disparity in size (Baker & Hall, 1976). Likewise in *mei-S51* there is no evidence that non-homologous pairing is governed by size distribution. Grell also argues that the disjunction of the small fourth pair of homologues, which do not recombine at female meiosis, regularly involves the distributive system.

It is important to emphasize that there is no cytological evidence to support the occurrence of any form of secondary pairing of the kind implied by Grell's hypothesis. Rather, distributive pairing is inferred from genetic data – data which are not in dispute. Cytological evidence does, however, confirm that, as predicted by Schultz & Redfield (1951), a single compound chromocentre is present at the onset of female meiosis. This chromocentre involves the fusion of the paracentromeric heterochromatic regions of all the chromosomes and persists until the breakdown of the nuclear membrane (Dävring & Sunner, 1973, 1976, 1977; Nokkala & Puro, 1976).

Novitski and Puro (1978), in a careful and thoughtful analysis of female meiosis in *D. melanogaster*, argue cogently against the distributive pairing hypothesis. Instead, they propose that what Grell has interpreted as a distinct and novel form of pairing in reality involves the preferential separation of non-exchange chromosomes from the common chromocentre. They point out that this chromocentre provides a means of maintaining a pairing relationship between chromosomes, even if they fail to undergo exchange, until the first division spindle is organized at prometaphase-1, This, they believe, adequately explains the lack of independence of the two achiasmate pair-4 homologues (Puro, 1978) which are largely heterochromatic in character. Additionally, in homozygous *c(3)G* females, Puro & Nokkala (1977) found extensive asynapsis of the euchromatic regions of all homologues accompanied, however, by the maintenance of a normal chromocentral association. This follows from the fact that the heterochromatic regions, which do not crossover, associate prior to the synapsis of the euchromatic regions and so are not influenced in the same manner by the *c(3)G* mutation. While, therefore, Novitski & Puro (1978) confirm the existence of a distributive system of segregation they conclude that this system involves the maintenance of heterochromatic associations established as a consequence of chromocentre formation at the outset of meiosis and not from a separate system of distributive pairing of the type proposed by Grell.

Robbins (1971) identified and analysed a mutation, *S51*, which resulted in decreased exchange accompanied by non-homologous segregation. This mutation was shown to involve two recessive genes located near the centromeres of chromosomes-2 and -3, respectively, both of which were required for the expression of the mutant phenotype. The behaviour of *S-51* indicates that the occurrence of non-homologous segregation is not

Table 6.6. *Disjunction-specific defective mutations affecting meiosis in Drosophila melanogaster*

Time of non-disjunction	Mutations	Effect
Meiosis-1	ca^{nd}, claret non-disjunction and $1(1)TW\text{-}6^{cs}$	Very high non-disjunction frequencies in all chromosomes, both exchange and non-exchange. Also frequent chromosome loss, especially of 4.
	nod, no distributive disjunction (X)	Increased non-disjunction of all chromosomes plus loss, especially of 4.
	mei-38, mei-99 and *mei-160* (X)	Increased non-disjunction of all chromosomes and some chromosome loss. At least for *mei-38* and *mei-99* some non-disjunction of recombinant chromosomes occurs. Occasional second meiotic non-disjunction observed.
Meiosis-2	semi-dominant *mei-S332* (2)	Increased non-disjunction of all chromosomes and to the same extent in both sexes. Sister chromatids separate precociously before metaphase-2.
	mei-G87 (2)	Increased non-disjunction in both sexes but only of the chromosome (2) that carries it.
Meiosis-1 and 2	*ord* = *mei 312* (2)	Both recombination defective and disjunction defective in both sexes.

After Baker & Hall, 1976; Baker *et al.*, 1976.

simply the result of structurally aberrant chromosomes and, presumably, it is the wild-type alleles of the two genes in question which prevent non-homologous disjunction during normal meiosis. That is, in Novitski's terms, they prevent the persistence of heterochromatic pairing. So far as I know this has not been checked cytologically though it would certainly provide an objective test of the hypothesis.

Turning to the known female-specific disjunction mutations which influence meiosis (Table 6.6), two of them, ca^{nd} and $l(1)TW\text{-}6^{cs}$, greatly increase the non-disjunction of both exchange and especially of non-exchange chromosomes, at the first meiotic division. Thus, there is a substantial loss of chromosome-4 with both these mutations. In the case of *nod*, non-disjunction involves only those chromosomes whose disjunction is normally ensured by the distributive system, namely chromosome-4 and other non-exchange and compound chromosomes (Carpenter, 1973).

Unlike the disjunction-defective mutations which affect the first female meiotic division exclusively, the three known mutations which increase non-disjunction at meiosis-2 affect both sexes. In *mei-S332* males, for example, sister chromatid separation within individual half bivalents is frequently precocious. Rarely this may occur as early as anaphase-1 and commonly by metaphase-2, and it is this that leads to non-disjunction at second division. Davis (1971) was of the opinion that the normal allele of *mei-S322* either regulated a component of centromere structure or else influenced sister centromere separation. As long ago as 1956, Lima-de-Faria provided a convincing demonstration that sister chromatids at anaphase-1 are held together by the region of the chromosome arm proximal to the centromere itself. An alternative explanation, therefore, is that the wild-type allele of *mei-S332* controls whatever holds sister chromatids together at these specialized proximal regions. In either event, it is clear that second division segregation depends on genes whose function is to preclude sister kinetochores from becoming functionally independent until the second division. In *ord*, precocious sister chromatid separation occurs even earlier, at prophase-1 or metaphase-1 (Goldstein, 1980). Normal congression does not occur in *ord* spermatocytes despite the fact that there are no differences in spindle ultrastructure compared with wild type. Using electron microscopy, Lin & Church (1982) found that in wild-type males of *D. melanogaster* half bivalents at metaphase-1 are characterized by two kinetochores which are held together and so appear at the ultrastructural level as a single hemisphere. At anaphase-1 the 'single' kinetochore is still a double, but coupled, disc. In *ord* males, however, sister kinetochores are distinct and separate by metaphase-1, as too are sister chromatids . That is, in this mutation, sister kinetochores become functionally independent well ahead of their normal time. The *ord* mutation also causes non-disjunction during gonial divisions in addition

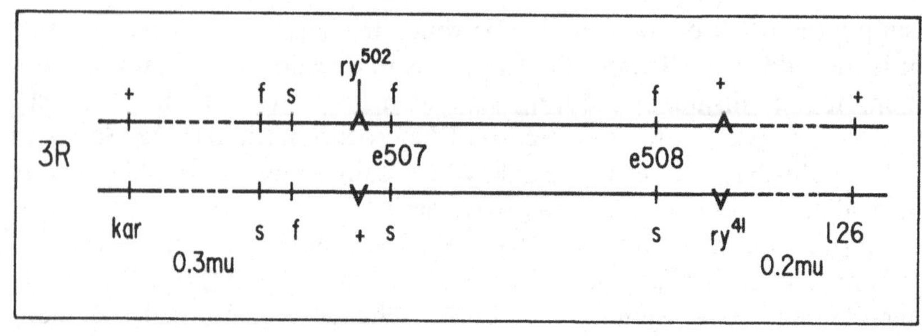

Fig. 6.3. Recombination in the rosy (*ry*) cistron of chromosome arm 3R in female *Drosophila melanogaster*. The symbols ry^{502} and ry^{41} identify ry^-, XDH^- mutant alleles while e^{507} and e^{508} are sites that affect the electrophoretic mobility (f = fast, s = slow), but not the function, of xanthine dehydrogenase (XDH), and *kar* and *l26* are functionally unrelated flanking markers. Recombination events within this cistron, which generate ry^+ chromatids, are detectable by eye colour and by relative resistance to purine since in larvae lacking XDH this substance is toxic (after Carpenter, 1984c and with the permission of Cold Spring Harbor Laboratory).

to a syndrome of other cyst abnormalities. Consequently, a high frequency of primary spermatocytes are either mono-or tri-somic for autosomes-2 and -3.

Mei-G87 is exceptional; it affects only the chromosome that carries it. This is also true of the X-linked mutation *eq* (equational) in which second division non-disjunction is restricted to the male (Valentin, 1984). This implies that at least some aspects of the control of meiosis-2 must differ between the two sexes. The paucity of mutations which affect the second meiotic division in *D. melanogaster* may mean that this division is under the control of the same genes that regulate mitosis. Mutations of such genes might then be expected to be lethal so that the products would not be recovered.

A.2.3 Inter- and intragenic effects

Meiotic recombination events are, as we have seen, recognized by two rather different outcomes, namely by crossing-over (*intergenic recombination*) and by conversion (*intragenic recombination*). The *c(3)G* mutation, which drastically reduces crossing-over also abolishes conversion. Two other mutations, *mei-9* and *mei-218*, have been shown to uncouple crossing-over and conversion. Carpenter (1984c) analysed recombination events within the rosy locus of *D. melanogaster*. Her approach was to employ homologues differing in respect of loss of function rosy (*ry*) alleles and then select for intragenic events involving sites within the rosy cistron which affected the electrophoretic mobility of xanthine dehydrogenase, the protein product of the rosy locus (Fig. 6.3).

Table 6.7. *The influence of two recombination defective mutations on inter- and intra-genic recombination frequencies at the rosy (ry) locus of Drosophila melanogaster*

	Recombination frequency ($\times 10^{-6}$)			
	Conversion		Crossing-over	
Genotype	ry^{41}	ry^{502}	*kar 126*$^{+}$	% *kar-e* (control)
control	5.0	3.9	13.3	100.0
mei-9	6.3	7.6	3.8	11.3
mei-218	7.8	10.6	< 2.6	6.8

After Carpenter, 1984c.

She was able to demonstrate that the frequency of crossing-over was strongly reduced relative to control levels by both mutations whereas conversion frequencies were not (Table 6.7). Indeed, in *mei-218* there was a marked increase in conversion events. In *mei-9* the conversions were as long as, or even longer than, wild type and there was frequent post-meiotic segregation manifested by the production of mosaic progeny. By contrast, conversions in *mei*-218 were shorter than wild type and, as in the controls, there was no post-meiotic segregation.

In *D. melanogaster*, recombination nodules (RNs) come in two morphologically and temporally distinct forms (see Chapter 4D.1). Carpenter (1979a, 1984b, 1987) has argued that the earlier-produced, and more numerous, ellipsoidal forms correspond to non-reciprocal conversion events. The later-produced, and less numerous, spherical forms she assumes to correspond with crossovers. In support of this she found that the average number of spherical RNs is proportionally reduced in mutations which, like *mei-218, mei-41* and *mei-S282*, result in reduced crossing-over. Added to this, half of the reduced number of spherical RNs found in *mei-218* were morphologically abnormal indicating that the wild-type allele of this locus is necessary for normal RN structure as well as for the occurrence and distribution of exchange events. In *mei-9*, on the other hand, there was no reduction in the average number of spherical RNs per nucleus (Table 6.8) and here their frequency, distribution and morphology all correspond exactly with wild type despite the fact that, in this mutation too, recombination events are dramatically reduced compared to control levels. Thus, whereas in the case of the *mei-218* locus, wild-type function appears to be necessary for normal RN formation this is not so in *mei-9*. This implies that completion of exchange is not a necessary consequence of RN formation (Carpenter, 1984b).

The overall picture that emerges from Carpenter's work is that

Table 6.8. *A comparison of recombination nodule (RN) numbers in three recombination-defective mutations of Drosophila melanogaster*

Mutant	Mean number of RNs per nucleus	
	Mutant cyst	Wild-type cyst of same age
mei-9	3.63 ± 0.75	2.81 ± 0.63
mei-218	0.54 ± 0.13	3.37 ± 0.24
mei-41	0.42 ± 0.23	3.37 ± 0.24

After Carpenter, 1979b.

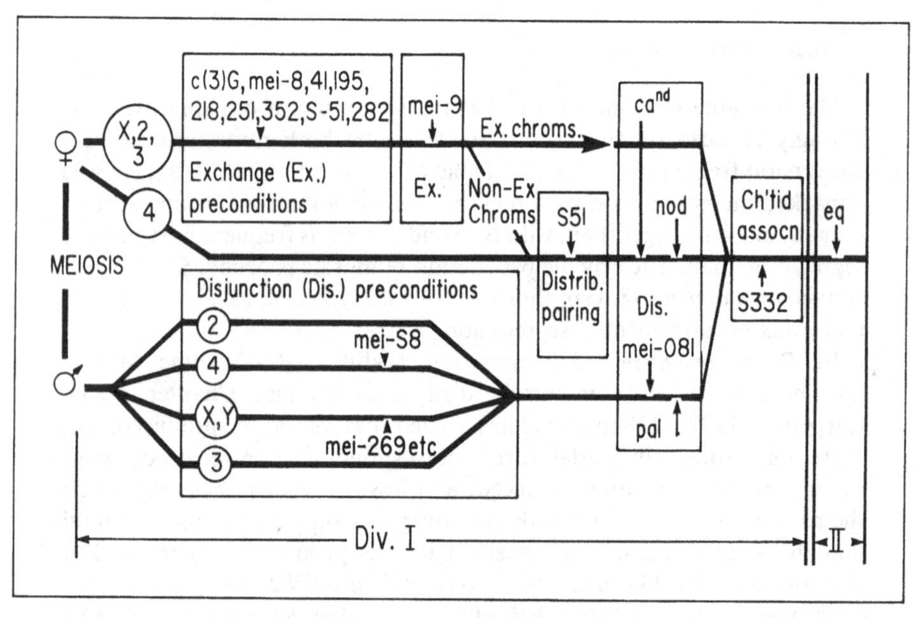

Fig. 6.4. The genetic control of meiotic chromosome behaviour in *Drosophila melanogaster* (after Baker & Hall, 1976 and with the permission of the Academic Press Inc.).

recombination events in *Drosophila* begin at a number of randomly determined sites in the euchromatic portions of chromosomes. These sites are defined by the distribution of ellipsoidal RNs. This randomness is then converted into non-randomness through the assembly of spherical RNs. This is an active process which involves products of *mei-9*$^{+}$. Consequently, while the defect induced by the *mei-9* mutation occurs relatively late, the defective step resulting from the presence of *mei-218* occurs earlier and dramatically affects the number of spherical RNs.

In summary, despite the substantial amount of information now available on mutations affecting meiosis in *D. melanogaster* (Fig. 6.4) the detailed biochemical analysis of these meiotic mutations has not even begun. Yet, this is precisely what is needed to resolve both their role and their regulation. While RNs have been assumed to function as recombination factories, we have no idea of how they are constructed or how they carry out this presumed role.

B MEIOTIC MUTATIONS IN FUNGI

Because of the ease with which meiosis can be induced and manipulated, yeast provides an especially attractive organism in which to analyse both the molecular events of meiosis and their mode of genetic control. The one disadvantage is that it is not possible to directly observe yeast chromosomes, though the 16 SCs present at prophase-1 can be visualized with the electron microscope (Horesh, Simchen & Friedmann, 1979) and these, in conjunction with the kinetic apparatus which is also visible with the EM, provide a morphological guide to the meiotic sequence (Fig. 6.5).

It has been known for some considerable time that mitotic cell cycle (*cdc*) functions, which are required for DNA synthesis and nuclear division, are also required for meiosis, which is arrested in all cases where *cdc* mutations are involved. This is also true for the 'start' mutation, *tra 3*. Simchen (1974, 1978), therefore proposed that meiosis evolved from the nuclear division pathway that characterizes mitosis but with specific meiotic functions having been superimposed on the shared systems. Thus *rad 50-1, rad 52-1* and *spo-11* all abolish meiotic exchange in yeast. The first two of these also affect mitotic exchange whereas *spo-11* is meiosis-specific. On the other hand, there is no apparent involvement of genes in the *RAD 1* group, which are known to be involved in excision-repair processes, in either meiosis or meiotic recombination (Dowling, Maloney & Fogel, 1985). There is also a comparable partial overlap in *D. melanogaster* in the case of genes, such as *mei-9*$^{+}$, which function in both mitotic and meiotic events, though as Carpenter (1984a) points out, this does not necessarily imply that they are involved in identical pathways. Additionally, in *Drosophila* too, there are genes, such as *mei-218*$^{+}$, which function only in meiosis.

In yeast, *cdc* mutations can be used to arrest meiosis at specific intermediate stages. In this way it has been possible to separate recombination commitment, which takes place at pre-meiotic DNA synthesis, from the realization of recombination during prophase-1. All the mutations that block pre-meiotic DNA synthesis thus automatically prevent cells from undergoing recombination. Even so, DNA replication has been found to occur in certain mutant cells without any subsequent commitment to recombination (Kassir & Simchen, 1978). A diploid strain of yeast homozygous for the mutation *cdc 5-1* undergoes normal meiosis

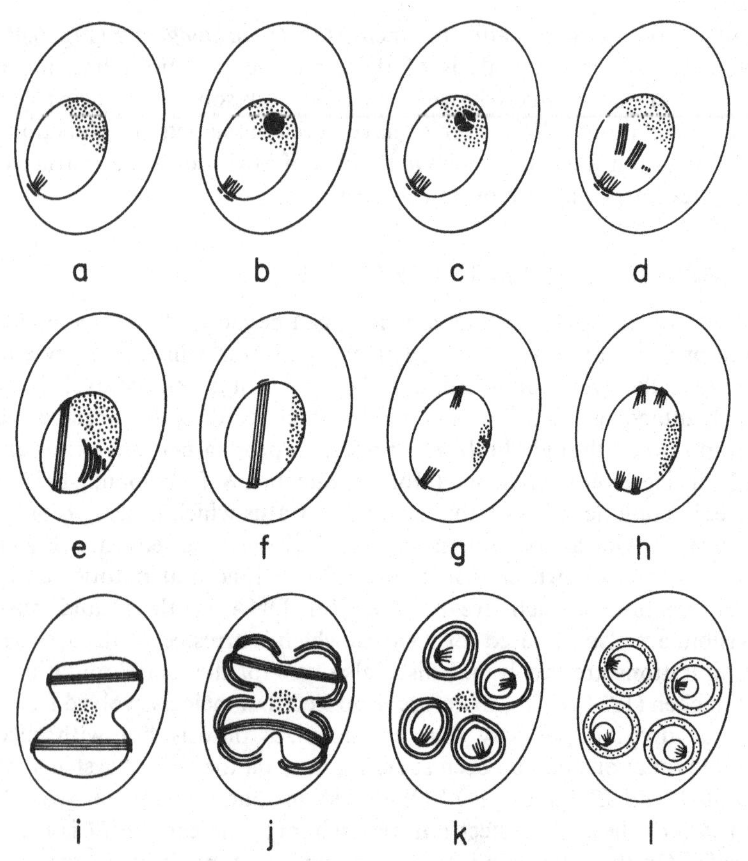

Fig. 6.5. Events of meiosis in the budding yeast *Saccharomyces cerevisiae* identified by electron microscopy. The spindle pole body, with its associated microtubules, is unduplicated in (*a*) but duplicated in (*b*) which also contains a dense body associated with a nucleolus (stippled). Elements of the synaptonemal complex form within this dense body (*c*) and these elements appear in the nucleoplasm after its disappearance (*d*). With the formation of an intranuclear spindle (*e*) the synaptonemal complexes are replaced by a single polycomplex body which subsequently disappears (*f*). After the dissolution of the first meiotic spindle, the outer layer (plaque) of each spindle pole body enlarges (*g*) prior to its duplication (*h*) so that two second division intranuclear spindles form (*i*). The second meiotic division is followed by prospore wall growth, which is initiated at the outer plaque of each spindle pole body (*j*). With the closure of the prospore wall the post-meiotic coenocyte is separated into four haploid nuclei (*k*) which then (*l*) develop into four spherical spores (after Esposito & Klapholz, 1981 and with the permission of Cold Spring Harbor Laboratory).

at 25 °C but is arrested in meiosis-1 at the restrictive temperature of 34 °C after both pre-meiotic DNA replication and recombination commitment have taken place. SCs form but there is a prevalence of modified types in which the central elements are missing. Paradoxically, despite this, prolonged incubation at the stage of arrest leads to unusually high levels of recombination (Simchen *et al.*, 1981).

At 25 °C the *spo-11* mutation has normal DNA metabolism and sporulation. At 34 °C, however, this mutation fails to undergo pre-meiotic S but, nevertheless, continues through the meiotic cycle with unreplicated chromosomes. Axial cores form and a polycomplex is present, usually in association with the nucleolus, but no SCs form. All subsequent stages of meiosis have been observed as, also, have the early stages of spore formation though all the meiotic products eventually degenerate (Klapholz, Waddell & Esposito, 1985). This case is instructive in demonstrating that the formation of axial cores and polycomplexes is not blocked by the inhibition of DNA synthesis. The *spo-11* mutation is thus recombination-deficient. It is also meiosis-specific since Di Domenico *et al.* (1984) have identified a meiosis-specific transcript of the *SPO* 11 gene.

Two additional mutations, *spo-12* and *spo-13*, eliminate the first meiotic division so that meiocytes go through a single equational second division. Events prior to anaphase, including recombination, occur normally but the ascus that is produced has only two spores (Klapholz & Esposito, 1980).

Davidow & Byers (1984) have described a mutation, *pac 1*, in which pachytene was arrested and both gene conversion and post-meiotic segregation were enhanced at the restrictive temperature for this mutation. The enhancement of post-meiotic segregation in this case, coupled with the increased length of the converted segment, parallels the behaviour of *mei-9* in *Drosophila melanogaster* and, like it, is indicative of a long-lived heteroduplex.

A number of additional mutations, identified by their increased sensitivity to ionizing radiation and consequently designated as *rad* mutations, have been shown to be essential for meiosis (reviewed in Resnick, 1987). The wild-type alleles of several of them are evidently required for meiotic recombination since the non-disjunction associated with these mutations leads both to a reduction in the number of asci produced and to the formation of inviable ascospores. This is true for *rad 50* which lacks SCs. Meiotic recombination is, as we have seen, not necessary for correct disjunction in the presence of the *spo-13* mutation because the only segregation that occurs here is equational. By using *spo 13 rad 50* diploids it is possible to produced viable diploid spores which lack recombination, although normal recombination levels occur in *rad*$^+$ *spo-13* mutants.

The *RAD 52* gene, by contrast, appears to control the production of a deoxyribonuclease since *rad 52* mutants completely lack nuclease activity in both mitotic and meiotic cells. Associated with this lack, comparable numbers of single-strand breaks occur in both parental and newly-synthesized DNA strands following the onset of pre-meiotic S. Based on the time of appearance and the frequency of these breaks, relative to the known recombination events in wild-type cells, it has been proposed that the *RAD 50* gene must influence an early step in the recombination process. This is supported by the fact that SCs are not present in *rad 50* mutants and that, also unlike *rad 52*, this mutation is not meiosis-specific. Additional support for this proposal comes from the fact that no single-strand breaks can be identified either in *rad 50* mutants or in the recombination of the silent ribosomal DNA region of the wild-type yeast genome (Høgset & Øyen, 1984).

The genetic dissection of meiosis, based on an analysis of the major mutations which affect the process in both *Drosophila* and yeast, indicates that these two organisms share equivalent systems of genetic control. There remains the formidable task of determining the products of the genes involved and evaluating the precise roles that these products play in regulating the events of meiosis. As Magee (1987) points out, this task may well be complicated by the occurrence of gene products which persist from pre-meiotic stages but which are nevertheless required during meiosis, as well as by functional redundancy in the meiotic transcription process itself.

C SPECIALIZED FORMS OF MEIOTIC CONTROL

Apart from the general gene control systems identified in *Drosophila* and yeast, there are at least four rather more specialized forms of meiotic control which, while as yet less understood, merit some comment.

C.1 Genes with local effect

Differences in recombination frequency between different regions of the same bivalent can sometimes be attributed to differences in the primary structure of the regions concerned. This is most evident in heterochromatic regions which, unlike euchromatic regions, do not undergo crossing-over. Alternatively, regional differences in recombination may depend on the position of a given locus relative to that of *cis*- or even *trans*-acting regulatory factors which may then either increase or decrease recombination frequencies. For example, *M26* in *Schizosaccharomyces pombe* stimulates meiotic gene conversion at *ADE6* (Gutz, 1971) and equivalent *cis*-acting genes are essential for high frequencies of meiotic gene conversion at the *ARG4* and *HIS2* loci of *Saccharomyces cerevisiae* (referred to in Lambie & Roeder, 1988). In several fungi, mutations are known which act on recombination but only at specific and very localized sites. The most thoroughly documented of these are the repressor (*rec*) genes of *Neurospora crassa* (reviewed in Catcheside, 1977). Three such genes have been uncovered, located respectively on linkage groups 1 (*rec 3*) and V (*rec*

1 and *rec 2*). The dominant allele of each of these genes acts as a repressor of local recombination but is not closely linked either to the specific locus it affects or, in the case of *rec 1* and *rec 2*, to each other. In the absence of the gene products encoded by the wild-type allele (rec^+) of these genes the recombination rates rise by several-fold in limited regions of the genome. No comparable genes have yet been identified in other organisms, though there are indications that they may well occur in higher eukaryotes too. Hawley (1980) claimed that the X-chromosome of *D. melanogaster* could be subdivided into five exchange intervals, whose boundaries were defined by blocks of intercalary heterochromatin, and that exchange suppression in these intervals was mediated by the boundary regions, which he speculated were also meiotic pairing sites. There is, however, no hard evidence to support either of these claims. Szauter (1984) subsequently concluded that the non-uniform distribution of crossovers along the euchromatin of *D. melanogaster* was mediated by a genetically-controlled system of regional constraints on exchange, each operating over a limited domain. The components of this system were held to consist of genes whose products specified the distribution of exchanges. Mutations of these genes would then either attenuate or else eliminate these constraints, so producing a more uniform distribution of crossovers.

Where genes with local effect do operate, the responding elements in the euchromatin must, presumably, correspond to features that are capable of being recognized by the products of these genes and which then initiate either meiotic pairing or else recombination itself. This might involve the degree of chromosome condensation, the portion of chromosomal DNA associated with the SC, the regional nature of chromosomal proteins or some other structural property that varies along the chromosome.

C.2 Chromosome-specific control of chiasma formation

With the exception of the non-chiasmate 4th-chromosome of *Drosophila melanogaster*, most of the recessive-defective mutations in this species affect all the other chromosomes in the diploid set to some extent. The *mei-1* mutation is exceptional; its effects are confined to the X-chromosome. Chromosome-specific and recessive desynaptic mutations, that is, mutations which lead to a secondary failure of chiasma formation in individual chromosomes following normal synapsis, have also been described in two members of the plant family Compositae, namely *Hypochoeris radiata* (Parker, 1975) and *Crepis capillaris* (Tease & Jones, 1976). In *H. radiata* (2n = 8), univalency occurs in over 90 % of the PMCs in mutant plants and always involves the same bivalent (no. IV). The degree of univalency is less marked in PMCs of *C. capillaris* (2n = 6) ranging from 2–40 % with an average of 20 %. Moreover, here, univalency was found to affect any one of the three (A, C, D) chromosome pairs though, most commonly, only one pair of univalents was present in affected PMCs. Out of 31 plants sampled at random, 12 were considered to display chromosome-specific univalency including five A-specific, four C-specific and three D-specific types.

C.3 The genetic control of homoeologous pairing in plant hybrids

A number of naturally occurring allopolyploid plants are known in which pairing is confined to strict homologues despite the co-existence within the hybrid genome of homoeologous chromosomes sharing considerable partial genetic homology. The best known of these systems is that which characterizes bread wheat, *Triticum aestivum* (2n = 6x = 42), though comparable control systems have been claimed for other naturally occurring allopolyploid plants (see Golubovskaya, 1979). Wheat is an allohexaploid in which three genomes (A, B and D), derived from three different diploid progenitors characterized by partial genetic equivalence, have been combined into one complex polyploid genome. In *T. aestivum*, and by circumstantial evidence in *T. turgidum*, *T. timopheevii* and *T. zhukovskyi* also, there is a major gene regulating the normal pattern of bivalent pairing. In *T. aestivum* this gene is located on the long arm of chromosome-5B (Sears, 1976). When this locus is present, even in a hemizygous condition, chromosome pairing is confined to strict homologues. The pairing suppressor responsible for this restriction has been named pairing homoeologous (*Ph 1*). In aneuploid conditions in which no homologous chromosomes are present, as in nullihaploids or in hybrids deficient for 5B, homoeologues pair at considerably higher frequencies than when 5B is present in comparable types. Alternatively, in aneuploid situations where homologous chromosomes are present but 5B is absent, as in nullisomic 5B, both homologues and homoeologues pair, often in very complex configurations, sometimes larger than the expected hexavalents.

A homeoallele of *Ph 1*, which has an opposite effect and which promotes homoeologous pairing, is known on 5A. Additionally, a complex system of control elements, involving several promoters and suppressors of pairing are known to occur elsewhere in the wheat genome. Pairing promoters are present on chromosomes-5BS and -5D as well as on 3BL, 3AL and 2AS while 3DS and 3AS both carry suppressors of pairing (Sears, 1976). There are also at least two loci in both *Aegilops speltoides* and *A. mutica* which suppress the action of *Ph 1* in interspecific hybrids with wheat. Allelic variation at these two loci generates a series of mutations in which the pattern of pairing ranges from almost complete asynapsis to full homoeologous pairing of all chromosomes as multivalents (Dover & Riley, 1977). A not too disimilar situation applies to certain experimentally-constructed hybrid polyploids in the genus

Table 6.9. *The influence of B-chromosomes on homoeologous pairing in Lolium hybrids*

Hybrid combination	Mean cell chiasma frequency		
	High line (Lp10)	Low line (Lp 19)	Difference
L. temulentum × *L. perenne*			
OB	11.85	8.47	3.38
2B	7.45	2.07	5.38
L. temulentum × *L. rigidum*			
OB	11.94	11.39	0.55
2B	6.89	2.01	4.88

After Evans & Taylor, 1976.

Triticum. Here, however, the control is exercised not by genes on standard members of the chromosomes represented in the hybrid but by supernumerary or B-chromosomes. Thus *Triticum tripsacoides* and *T. speltoides*, both diploid species, carry B-chromosomes which have a suppressive effect on pairing which parallels that of the 5B-chromosome of *T. aestivum*. In hybrids between *T. aestivum* lacking 5B but carrying B-chromosomes from either of these diploids, pairing is again diploidized (Dover & Riley, 1972). The presence of supernumerary chromosomes in *A. speltoides* and *A. mutica* does not directly affect the levels of pairing determined by the pairing control genes. However, hybrids which normally have very high levels of pairing in the absence of chromosome-5B of *Triticum aestivum* are asynaptic in the presence of B-chromosomes from *Aegilops*. The supernumeraries evidently compensate for the absence of 5B (Dover & Riley, 1977).

An equivalent, though by no means such an extreme, situation is known also in hybrids between inbreeding and outbreeding species of the genus *Lolium*. For example, *L. perenne* and *L. rigidum* are closely related species and the B-chromosomes of both exert a marked effect on homoeologous pairing in hybrid combinations with *L. temulentum* (Table 6.9). Moreover, B-chromosomes of different origin and size are equally capable of reducing homoeologous associations which are easily detected on account of the differences in DNA content, and hence also in chromosome size, which exists between the parental species (Evans & Taylor, 1976).

C.4 Disjunction regulator regions

Goldstein (1985) has identified granular structures, in the form of partially decondensed clumps of chromatin, present in association with the central

Table 6.10. *Correlation between the frequency of X-chromosome non-disjunction and the occurrence of disjunction regulator regions (DRRs) in Caenorhabditis elegans*

Strain	Average number of DRRs	% X-chromosome non-disjunction
Triplo X	0	51.0
him-8	0	37.0
mn T6 (II:X)	0	37.0
him-4	3.25	6.0
him-7	3	3.0
Dpl	3	1.0
F4	5	0.3
(W_2 Bristol)	6	0.3
rad-4	7.25	0.03

After Goldstein, 1985.

elements of the SCs in *Caenorhabditis elegans*. He refers to them as disjunction regulators since their presence and frequency is negatively correlated with the frequency of X-chromosome non-disjunction (Table 6.10).

7

Sequences and consequences of meiosis

Logic is the art of going wrong with confidence
Anon

A TWO-TRACK HEREDITY

That differences exist between the two sexes within a given species in respect of sex chromosome behaviour has long been recognized. This is most obvious in X0, or Z0, systems. Here, genes carried by the X- or Z-chromosome pass unrecombined through the heterogametic sex because of the lack of a homologous partner. The same is also true for many, in some cases probably all, of the genes in the X- or Z-chromosomes of XY, or ZW, systems, respectively, since here too the X or Z is paired with a Y or W with which at best it has only partial homology (see Chapter 4C.6).

Differences of this kind are, of course, expected and easily explained. What is surprising, however, is that the autosomes may also behave differentially in the two sexes in respect of their meiotic behaviour.

A.1 Sex differences in recombination

In a majority of species, meiosis is without doubt chiasmate in both sexes. In most of these, no detailed analysis has been undertaken of either chiasma frequency or chiasma distribution. In a growing number of cases where both sexes within a given species have been analysed meiotically there are significant differences between them in respect of the occurrence, the frequency or else the distribution of chiasmata within the autosomal sets (Table 7.1). The most overt examples concern species where meiosis is achiasmate in one sex but chiasmate in the other. No less impressive are those cases, like the grasshopper *Stethophyma grossum* (Perry & Jones, 1974), where chiasmata are proximally localized in one sex and distally localized in the other. In such situations it is unlikely that there is any significant recombination in either sex since, as Zarchi *et al.* (1972, 1974) point out, both of these chiasma classes are likely to be genetically meaningless. If this is indeed the case then these organisms represent an even more extreme situation than those which are achiasmate in only one

Table 7.1. *Recombination differences between the sexes*

Species	Number of bivalents	Xa frequency ♀	Xa frequency ♂	Xa pattern[a] ♀	Xa pattern[a] ♂	Reference
Animals						
Dendrocoelum lacteum	7⚥	20.4	11.8	D–I	D	Pastor & Callan, 1952
Mesostoma ehrenbergii ehrenbergii	5♀/3♂	—	3.0	AXte	Xte	Oakley & Jones, 1982; Oakley, 1982
Bolbe nigra	13♀/12♂	?	—	Xte	AXte	Gassner, 1969
Bombyx mori	28	58.0	—	Xte, D	AXte	Rasmussen & Holm, 1982
Panorpa communis	12♀/11♂	13.0	—	D–I	AXte	Ullerich, 1961
Parapleurus alliaceus	12♀/11♂	13.6	12.3	R	P	Fletcher & Hewitt, 1980b
Stethophyma grossum	12♀/10♂	15.0	11.3	D–I	P	Perry & Jones, 1974
Triturus cristatus	12	25.0	37.0	P	R	Watson & Callan, 1963;
T. helveticus	12	25.0	22.0	I	D	Callan & Perry, 1977
Odontophrynus americanus	11 (2x)	29–31	16–20	D–I	D	Rahn & Martinez, 1983
	22 (4x)	43–44	37–44	D	D	
Ceratophrys cranwelli	13 (2x)	37	22	D–I	D	
C. ornata	52 (4x)	21–23	26–30	D	?	
Sminthopsis crassicaudata	7	10.2	13.6	D	D–I	J. H. Bennett *et al.*, 1986
Mus musculus	20	27.5	22.5	DI or PD	D or PD	Polani, 1972
Plants						
Fritillaria longiflorum	12⚥	31.4	27.3	—	—	Fogwill, 1958
F. meleagris		37.8	24.8	—	—	
F. martagon		41.0	36.3	—	—	
Fritillaria japonica group	11 or 12⚥	?	—	Xte	AXte	Noda, 1975
Allium nigrum	7⚥	16.9	21.9	R	R	Gohil & Kaul, 1980
A. consangineum		17.5	21.9	R	R	
A. cepa		17.9	22.4	R	R	
A. kachrooi		15.0	12.9	P	R	

[a]P = proximal, I = interstitial, D = distal, R = random chiasmata, Xte = chiasmate, AXte = achiasmate meiosis.

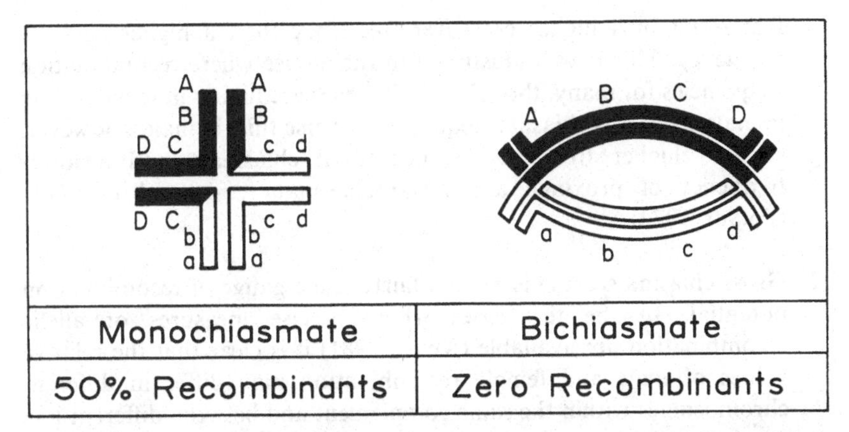

Fig. 7.1. The chiasma frequency of a bivalent is not necessarily positively correlated with recombination frequency. Thus, a monochiasmate bivalent with an interstitial chiasma may show recombination whereas a bichiasmate bivalent with two near terminal chiasmate does not (after Henderson, 1970b).

sex, and they approach the behaviour of hermaphrodite enchytraeid worms in which both male and female meiocytes are achiasmate (Christensen, 1961).

In a majority of species it is the heterogametic sex which either lacks chiasmata or else has a reduced chiasma frequency. Equivalent differences have also been noted between 'male' and 'female' meiosis in hermaphrodite plants and animals. There are, however, a few notable exceptions to this generalization. In the marsupial *Sminthopsis crassicaudata*, paralleling the lower mean cell chiasma frequency of the female there is also a much closer linkage in females with respect to the six electrophoretic loci that have been studied to date in two of the linkage groups (Bennett, Hayman & Hope, 1986). Equally, the mean cell chiasma frequency is higher in the male of the amphibian *Triturus cristatus* compared to the female, though the reverse is true in *T. helveticus*.

Since there is no longer any reason to doubt that chiasmata are indeed the cytologically visible manifestations of genetic crossing-over (Jones, 1987) the frequency of chiasmata provides one, though admittedly crude, measure of the recombination potential of a species. Using this criterion, the most general statement that can and has been made is that recombination rates are lower in the heterogametic sex. Darlington (1973) coined the term *two-track heredity* to describe this type of situation, on the assumption that these differences in recombination between the two sexes were likely to have, or to have had, significant evolutionary consequences. This argument needs, however, to be tempered by three additional considerations:

(1) As emphasized by Henderson (1970b), depending on the actual distribution of chiasmata (Fig. 7.1), a low chiasma frequency may in

fact result in a higher crossover frequency than a higher chiasma frequency. This is well illustrated in the mouse where recombination frequencies for many, though not all, genes are higher in females than in males, whereas chiasma frequencies reverse this. Females, however, show a higher frequency of interstitial chiasmata and a lower frequency of proximal and distal chiasmata compared to males (Polani, 1972).

(2) Gross chiasma scoring is often a fairly crude gauge of recombination potential. In the few cases where precise measures of allelic recombination are available (Keats, 1981) it is clear that the relative values of male and female recombination may differ in different chromosomes within the same complement and between different loci on the same chromosome. One needs, therefore, to exercise some caution before arguing in terms of broad generalizations.

(3) It is also important to emphasize that there are certainly species, like rye (see Chapter A2.1), in which no differences appear to exist between the chiasma characteristics of the two sexes. If one ignores the X-chromosome itself, this applies equally to the grasshopper *Chorthippus brunneus* where again there is no overt difference between the two sexes in either chiasma frequency or chiasma location in the autosomes (Jones, Stamford & Perry, 1975).

There is a temptation to assume that, when recombination differences do exist between the sexes, these differences are necessarily genetically determined. Indeed, this situation might be accommodated in terms of a general biochemical mechanism regulated by a gene tightly linked to the sex determining locus which, initially, inhibited crossing-over between the two sex chromosomes in the heterogametic sex but whose effect then spread to all other chromosomes. This, however, does not account for the behaviour of 'male' and 'female' meiocytes in plants which appear to be primary hermaphrodites and to lack sex chromosomes. Cases like this suggest that either the genetic switch mechanism that governs basic sexual differentiation in the absence of distinguishable sex chromosomes also, in some way, determines the recombination difference between the two classes of sex cells, or else that environmental factors play a key role. Reddy, Reddy & Rao (1965) claimed to have induced crossing-over in a proportion of the progeny of males of *Drosophila melanogaster* that had been injected in the testis region with fresh ovarian extract obtained from 300 females. The overall frequency of male-induced recombination was 5% and the extract was shown to be heat-labile and ineffective when boiled. There was some clustering of the effect which might imply a premeiotic response but, unfortunately, the experiment has never been repeated.

A.2 Sex-limited translocations

A distinct, and very different, type of two-track heredity is found in species characterized by sex-linked and sex-limited systems of translocation heterozygosity (Table 7.2). This is quite different from the sex-independent systems of permanent translocation heterozygosity found in some plants, notably *Oenothera* (Cleland, 1972) and *Isotoma* (James, 1965, 1984). Here all, or a majority of individuals, regardless of sex, are structural heterozygotes in which meiotic pairing no longer leads exclusively to bivalent formation but involves the production of ring or chain multiples, of varying sizes, in which individual chromosomes occur in a constant linear order. At orientation, every other chromosome in the multiple is directed toward the same spindle pole. This, coupled with a distal localization of the chiasmata between successive members of the multiple, results in only two classes of gametes with respect to those chromosomes which comprise the multiple. Consequently, each of these classes, which carries one of the two alternate chromosome groups as defined by their position in the multiple, behaves genetically as a single linkage unit with respect to the interchange chromosomes.

In both of these plant genera, where interchange heterozygotes are obligate or near-obligate inbreeders, additional mechanisms have evolved to reinforce the system. These include balanced lethal alleles, which ensure that neither of the two alternate chromosome groups can produce viable homozygous combinations at fertilization or else gametic-lethals, which ensure that pollen can only fertilize eggs containing the opposite chromosome group. For this reason it has been argued that permanent heterozygosity of this type precludes the production of allelic homozygotes of the kind expected from inbreeding and so conserves heterozygosity (Darlington, 1956). Where this has been tested, using protein electrophoresis, there is certainly evidence to support a significantly higher level of allelic heterozygosity in some of the structurally heterozygous populations (James *et al.*, 1983). Exceptions to this have been accounted for in terms of immigrant structural heterozygotes which have been involved in occasional interpopulation hybridization.

Simple chromosome translocations, involving the X and an autosome of an X0 system, and leading to the production of a chain of three multiple (Fig. 7.2), are well known in animals (White, 1973). Here, too, they automatically lead to a state of permanent heterozygosity, though one which is necessarily confined to the heterogametic sex. Thus, an X0♂XX♀ sex chromosome mechanism may be converted by translocation into an X_1X_2Y♂$X_1X_1X_2X_2$♀ system. In a male of this type the component chromosomes of the chain of three-multiple must necessarily segregate preferentially to ensure the production of genetically balanced X_1X_2- or Y-gametes. By linking other autosomes into this system through further translocations, these autosomes too automatically become locked into the

Table 7.2. *Species characterized by large, heterozygous, male sex-limited translocation multiples*

Species	♂ 2n	Multiple size	Reference
Delena cancerides	43	CV, CIX	Rowell, 1985, 1987
Kalotermes approximatus	32	CXI, CXIII, CXIV, CXV, RXVI	Luykx & Syren, 1979, 1981a; Syren & Luykx, 1981
Incisitermes schwarzi	32	CXI, RXIV, RXVI, RXVIII	
Neotermes castaneus	38	RVI	
Otocryptops sexguttatus	15	CV, CIX	Ogawa, 1954
Otocryptops rubiginosus	25	CIII, CIV, CV	Ogawa, 1961
Viscum fischeri	23	CIX	Barlow & Wiens, 1976
Viscum hildebrandtii	28	RIV, RVI, RVIII	Wiens & Barlow, 1979
Viscum anceps, continuum and engleri	28	RIV, RVI	
Viscum combreticola	28	RVI, RVIII	

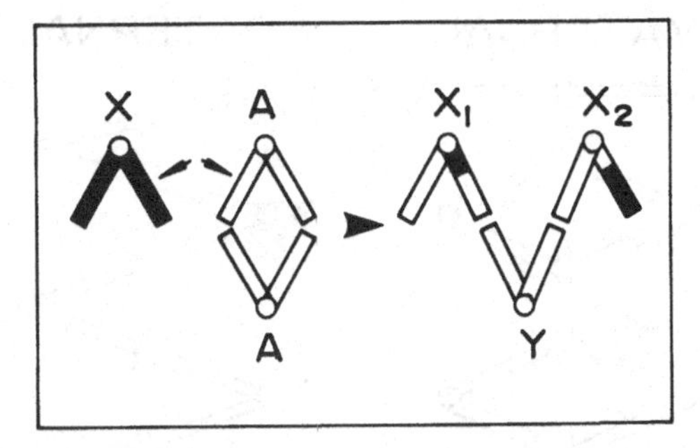

Fig. 7.2. The production of an X_1X_2Y sex chromosome multiple following a reciprocal translocation between the X-chromosome of an X0 male and one of a pair of homologous autosomes (A).

preferential segregation that governs the sex-determining mechanism. This is precisely what appears to have happened in the centipede *Otocryptops* where, in addition to basic XY-males, there are species with sex multiples of three, four, five and nine (Table 7.2).

The system is best illustrated in the spider *Delena cancerides* where the relationship to defined sex chromosomes is readily apparent although the historical details of the origin of the multiple involved are not known with certainty. Huntsman spiders usually have three X-chromosomes in the male and these segregate as a unit in a manner similar to that which we encountered in *Tegenaria* (Chapter 7B1.2). In *D. cancerides* itself, two cytologically distinct kinds of population exist in different geographical regions within Australia (Rowell, 1985). One of these has an all-telocentric chromosome complement with 2n = 43 = 20II + 3X which is equivalent to other huntsman spiders belonging to the genera *Heteropoda*, *Isopoda*, *Olios* and *Pediana* and which can be reasonably regarded as the basic situation. In the second kind of population, however, there are 21 metacentrics and one telocentric. Eight of the metacentrics and the one telocentric consistently give rise to a chain of nine-multiple at male meiosis-1 (2n = 22 = 6II + C.IX + X.I). One of the eight metacentrics in this chain represents an X.A-fusion product and the X-arm of this metacentric invariably forms one of the two terminal members of the chain. The two other X-chromosomes have fused with each other to produce a single metacentric X which is present as a univalent at first male meiosis and as a bivalent at first female meiosis. In the male this univalent shows a loose association with the terminal X-arm of the chain multiple at early prophase-1 but by diplotene this association has invariably lapsed. Nevertheless, the metacentric X-univalent segregates to the same pole as the X.A-fusion metacentric of the chain multiple so that the system of preferential X-chromosome segregation which is present in the progenitor 3X form is maintained.

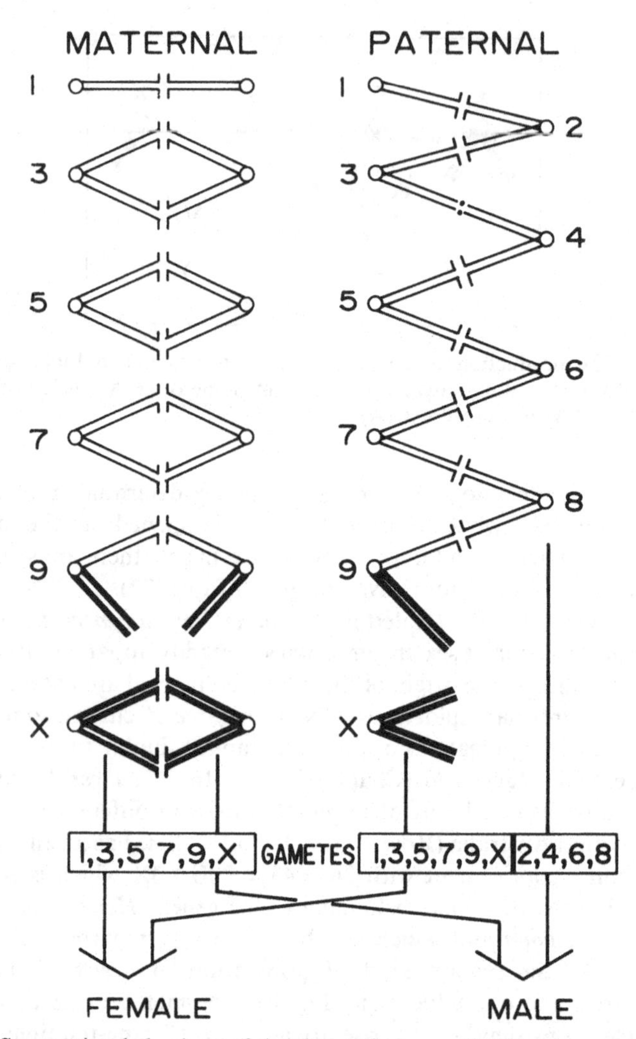

Fig. 7.3. Segregation behaviour of the chain of nine multiple in the male meiosis of the huntsman spider *Delena cancerides* compared with that of its homologous counterparts in the female (after Rowell, 1985).

At metaphase-1 the chain of nine adopts a predominantly alternate orientation. Consequently, females always receive the same subset of chromosomes from the multiple at fertilization (Fig. 7.3). If the chromosomes in the chain are numbered successively from the telocentric end, then numbers 1, 3, 5, 7 and 9 segregate with the free X-univalent, while numbers 2, 4, 6 and 8 are confined to the male line.

A similar pattern of behaviour characterizes a number of termites and mistletoes (Table 7.2). In both of these, however, the rearrangements involved in the production of multiples are more commonly interchanges

Fig. 7.4. Alternate orientation of the chain of nine multiple of the mistletoe *Viscum fischerii* at first metaphase in the PMC (2n = 23 = 7II + C.IX). Note the pronounced difference in the pattern of chiasma distribution in the seven bivalents compared with the members of the chain (photograph kindly supplied by Dr Brian Barlow).

rather than simple centric fusions of the kind present in *Delena*. Moreover, the sex chromosomes are not identifiable as discrete morphological entities in either of these two groups. Even so, laboratory hybrids between two different colonies of the termite *Incisitermes schwarzi*, with chains of 13 and 17, respectively, at male meiosis, indicate that both X- and Y-chromosomes, as defined in terms of their behaviour, have been involved in the evolution of the multiples in this species (Luykx & Syren, 1981a, b).

About 60% of the species belonging to the plant genus *Viscum* (mistletoes) are dioecious, and permanent sex-associated interchange heterozygotes are present in a majority of these. Here, 'male' plants consistently show multiple configurations at meiosis which range in size from four to 12. There are also high levels of floating interchange heterozygosity, though here the multiples produced are not sex-associated. Neither sex-limited- nor floating-interchange heterozygotes are present in monoecious species of *Viscum* (Barlow & Martin, 1984). The smallest and simplest sex-limited interchanges occur in dioecious African species where male plants have multiples of four or six, rarely of eight. Asian and European species are characterized by multiples of eight, nine, 10 or rarely 12. Male plants of *V. fischerii*, for example, consistently produce 7II + C.IX at meiosis and regular alternate orientation results in a 4:5 segregation from this chain at anaphase-1 (Fig. 7.4). This species is an obligate outcrosser and female plants form 11 bivalents. The sex ratio gives a close fit to 1♂:2♀ rather than the conventional 1:1 ratio. Barlow & Wiens (1976), who made these observations, were of the opinion that the sex-limited translocation system of *V. fischerii* serves to maintain the integrity of 'highly adaptive gene complexes'. Although they do not comment specifically on the point there are, as Figure 7.4 indicates, striking

Table 7.3. *Chiasma distribution in non-chain- and chain-forming populations of Delena cancerides*

Population type	Number of cells scored	Chiasma distribution P[a]	I[a]	D[a]
All telocentric	6	138	42	20
Metacentric	6	70	12	18
Chain populations	19			
CIX		32	51	68
6 Free bivalents		55	53	74

After Rowell, 1987.
[a] P = proximal, I = interstitial and D = distal.

differences in the chiasma pattern of the bivalents compared with the members of the chain multiple in male meiocytes, with a pronounced distal localization in the latter. This would certainly provide a rational basis for maintaining gene complexes within members of the chain but critical evidence in support of their argument is lacking.

From an analysis of chiasma distribution in *Delena cancerides* (Rowell, 1987), it is clear that, somewhat surprisingly, chiasmata are not restricted to distal locations within the chain multiple, though there is certainly a significant increase in the number of both interstitial and distal chiasmata in the chain of nine (Table 7.3). Since the translocations involved in the formation of this chain are all simple centric fusions, rather than interchanges as in the case of *Viscum*, the occurrence of interstitial chiasmata is not expected to lead to segregation problems and even the proximal chiasmata need not do so. What is clear, however, is that the presence of interstitial chiasmata can be expected to result in some recombination between the two segregating complexes within the chain.

A.3 Haplo-diploidy

Sexual reproduction involves two events, meiosis and fertilization, and the vast majority of eukaryotes reproduce by this means. A minority of them have, however, abandoned sexual reproduction, either partially or entirely, and have reverted to parthenogenesis; that is, to the development of eggs without fertilization.

The most extreme form of two-track heredity is in fact found in cases where parthenogenesis is mixed with sexuality. Here all males are haploid and arise from unfertilized eggs, while all females are diploid and are produced by normal sexual reproduction involving the sperm produced by the haploid males. Such haplo-diploidy is thus both a reproductive mechanism and a means of sex determination. It is also a mode of reproduction that readily facilitates the production of biased sex ratios with a preponderance of females.

Obligate haploid males, resulting from unfertilized eggs (*arrhenotoky*), are known in each of eight different groups of invertebrate animals (White, 1973). In some of these it is a rare occurrence and found only in isolated species. This is true of *Xyleborus xylosandrus* and *Micromalthus debilis*, the only known coleopterans to have evolved male haploidy. In other groups, such as monogodont rotifers (Birky & Gilbert, 1971) and acarines (Oliver, 1971), it occurs in many species; while in still others, it is the norm. Thus, with the exception of some all-female parthenogens, the remainder of the large order Hymenoptera are male haploid.

In mites and ticks, virgin females of *Ophionyssus natricis*, *Ornithonyssus bacoti* and *O. sylviatum* lay haploid eggs without mating whereas *Dermanyssus gallinae* will not oviposit without mating, though in all four species mated females produce either haploid or diploid eggs (Oliver, 1967). Cases where mating is a necessary prerequisite for the oviposition of haploid eggs are analogous to thelytokous gynogenesis (see Chapter 7B).

The males in haplo-diploids produce haploid sperm by a mode of cell division which is unconventional but may show clear signs of having been derived secondarily from meiosis (reviewed in White, 1973). In some wasps, for example, there is an abortive first division involving the production of an atypical spindle with no true metaphase plate and the elimination of a non-nucleated cytoplasmic bud. This is followed by a second equational division which gives rise to two haploid spermatids. Some bees, too, show an abortive first division but in others it is completely lacking. Moreover, in all bees the second division involves an unequal cytokinesis leading to the production of only one functional sperm. Biparental diploid males are sometimes produced in *Habrobracon* but their meiotic behaviour is identical with that of conventional haploid males (Torvik-Greb, 1935). That is, there is no synapsis despite the diploidy, which implies that the modified form of division is genetically regulated and is not simply a consequence of male haploidy.

In the beetle *Micromalthus debilis* a unipolar spindle is formed at the abortive first metaphase with all the chromosomes located at the diffuse end. The second division is again equational and gives rise to two sperms. Finally, in mites and iceryine coccids male meiosis is replaced by a single equational division, each primary spermatocyte giving rise to two sperms.

By comparison with female meiosis, which is normal in all haplo-diploids, the males invariably lack any recombination. There is either an abortive first division or no first division at all and the female is the sole determiner of the sex ratio.

In the Coccidae, a family of homopteran bugs, there is a unique diploid form of arrhenotoky where males begin life as haploid, having been produced parthenogenetically, but then become diploid through the fusion of the first two haploid cleavage nuclei. After a few divisions, however, one set of chromosomes becomes heterochromatinized, as is the convention in diploid sexual males. Meiosis in the diploid arrhenotokous males also follows that of sexual males (see Chapter 8). From one to all of the heterochromatic chromosomes are lost prior to prophase-1 or else, if they are retained, there is no pairing between them and their euchromatic homologues. In either event, the first division is equational so that, again, there is no male recombination (Nur, 1971, 1980).

In summary what all three categories of two-track heredity share in common is a difference between males and females in respect of their recombination characteristics. Precisely how these differences have come about, and what their consequences are for the organisms in question, remain as unresolved issues.

B ONE-TRACK HEREDITY – FEMALE PARTHENOGENESIS

In contrast to haplo-diploidy, female parthenogenesis, or *thelytoky*, has occurred repeatedly in animal evolution. Here organisms reproduce solely by eggs which develop without fertilization into females (Table 7.4). This mode of parthenogenesis is especially well represented in many arthropods and equally notable for its absence in some well studied animal groups (birds and mammals) and its rarity in others (Coleoptera, Lepidoptera, Hemiptera). Whereas arrhenotoky is invariably facultative, thelytoky may be either facultative or obligatory. Indeed, *tychoparthenogenesis*, that is, rare or accidental female parthenogenesis, is widespread in a number of insect groups.

The production of these all-female forms, which stems from the absence of fertilization, requires some modification – either in the process of meiosis itself or else in the products of this process – to ensure that the chromosome number remains constant over successive generations. Almost every conceivable way of avoiding reduction has been employed by one thelytokous form or another but there are two principal possibilities. Either meiosis is replaced by a single equational division where one of the two diploid products becomes the egg nucleus (*apomictic* or *ameiotic parthenogenesis*) or else there is a compensatory doubling of the chromosome number in conjunction with some form of meiosis (*automictic* or *meiotic parthenogenesis*). There is no known case in which a female may produce both apomictic and automictic eggs (Bell, 1982) though, when there is no associated meiotic recombination, the automixis of thelytokous forms is functionally equivalent to apomixis.

Apomictic parthenogenesis is the most common type in insects and is found in representatives of the Orthoptera, Diptera, Coleoptera, Homoptera and Hymenoptera (Suomalainen Saura & Lokki, 1976;

Table 7.4. *Interrelationship of meiosis and fertilization in reproductive systems*

Component		Fertilization	
		Present	Absent
Meiosis	Present	Sexual Reproduction	Meiotic ♀ parthenogenesis
	Absent	♂ parthenogenesis	Ameiotic ♀ parthenogenesis

White, 1970). It is also the simplest form from a genetical point of view since meiosis is totalling lacking; only one maturation division takes place in the egg and this is equational. Although the chromosomes divide equationally, as in mitosis, they are often strongly contracted, in a manner reminiscent of meiotic behaviour. Additionally in some cases, as in *Otiorhyncus sulcatus*, there is a rudimentary first division as well as an equational division.

Ameiotic systems are thus sub-sexual in character and not, as is often stated, asexual. While they may not always retain obvious remnants of meiosis, the products of their germ line divisions, which strictly speaking are not gametes, are nevertheless unreduced eggs. In apomictic forms all sibships are necessarily isogenic and retain the maternal genotype. This means that when that genotype is heterozygous its characteristic heterozygosity will be conserved. However, while apomixis is an efficient way of conserving heterozygosity it does so at the expense of limiting genetic diversity compared to amphimictic systems.

As far as the automictic, or meiotic, type is concerned, this system embraces three rather distinct mechanisms which result in different ways of either avoiding, or else compensating for, chromosome reduction (Table 7.5). In the first of these a pre-meiotic or peri-meiotic DNA doubling occurs resulting in a compensatory chromosome replication. The precise time of its occurrence is not always known with certainty since the evidence supporting it is often indirect and based simply on the presence of a diploid number of bivalents at metaphase-1. In theory, such DNA doubling may involve a failure of anaphase separation at the last pre-meiotic mitosis, an endoreduplication event during the pre-meiotic interphase or else during prophase-1.

Unless the individual is a hybrid combining different genomes (e.g., AB), the first three of these result in the production of an 8C, essentially autopolyploid state (AA→AAAA), a kind of internal fertilization; otherwise it produces an allopolyploid condition (AB→AABB). These polyploidization events are, of course, additional to any polyploidy that may occur initially. So far as is known this behaviour is always

Table 7.5. *Forms of automictic female parthenogenesis in invertebrate animals*

Category	Occurrence
(*1*) *DNA doubling*	
(a) Failure of last oogonial mitosis	*Eisenia*, *Eiseniella*, *Allolobophora* (Oligochaeta)
(b) Premeiotic DNA doubling	Presumed, though not proven, in *Warramba virgo* (Orthoptera)
(c) Pachytene DNA endo-replication	*Caraussius morosus* and *Sipyloidea sipylus* (Phasmatodea)
(d) Endo-replication at interkinesis	*Hybsibius dujardinii* (Tardigrada)
(*2*) *Fusion of meiotic products*	
(a) Central fusion of 1st division products	*Solenobia lichenella* and *Apterona helix* (Lepidoptera) *Lymnophyes biverticillatus* and *L. virgo* (Diptera, Orthocladiinae) *Lundstroemia parthenogenetica* (Diptera, Chironomidae)
(b) Terminal fusion	
(i) Fusion of 2nd polar nucleus with the female pronucleus	*Diprion polytomium* (Hymenoptera), *Lecanium hesperidium* and *L. hemisphaericum* (Homoptera)
(ii) Fusion of inner half of 1st polar nucleus with 2nd polar nucleus	*Solenobia triquetrella* (Lepidoptera), *Drosophila mangabeirai* and *Lonchoptera dubia* (Diptera)
(iii) Fusion of cleavage nuclei (= pronuclear duplication)	*Diplolepis rosae* (Hymenoptera)[a] and *Euhadenoecus insolitus* (Orthoptera)[b]

After Bell, 1982; Cuellar, 1987.
[a] Stille & Dävring, 1980.
[b] Lamb & Willey, 1987.

followed by the pairing of homogenic sister replication products leading to the formation of autobivalents. A pairing preference between identical, as opposed to homologous, chromosomes, equivalent to that which occurs in autobivalent formation, has been demonstrated in spontaneous tetraploid spermatocytes of *Euchorthippus* using a C-band marker polymorphism to distinguish the two chromosome types (Giraldez & Santos, 1981). Crossing-over in autobivalents has no genetical significance and the chiasmata that do form serve simply to hold homologues together until segregation is accomplished. Under these circumstances, while meiosis occurs its consequences are essentially mitotic. This system operates in the diploid parthenogenetic grasshopper *Warramaba* (formerly *Moraba*) *virgo* (White, Cheney & Key, 1963), a species which appears to have arisen by the hybridization of two extant diploids which are still taxonomically undescribed but have been given the provisional designations of P169 and P196 (Hewitt, 1975). No direct demonstration of premeiotic DNA doubling has been provided in the case of *W. virgo* but the presence of a diploid number of autobivalents leaves no doubt that some form of DNA doubling must occur either prior to, or during, first prophase.

W. virgo is also a fixed heterozygote for several chromosome differences which have, in the main, also resulted from its hybrid origin. No natural hybrids are known between P169 and P196 although these two species are sympatric over part of their range in Western Australia. In the laboratory they copulate regularly. The cross 196♀ × 169♂ gives rise only to mosaic haplo-diploid females which die at the first instar (White *et al.*, 1977; White, 1980). The reciprocal, 169♀ × 196♂, cross produces both male and female progeny. Some of the males die as embryos and the remainder die as nymphs within a few days of hatching because they are unable to feed. Many of the female embryos also die but a number survive to become imagines, some in the fifth instar others not until the sixth or seventh instar. These hybrid females are externally normal, but on dissection, most commonly show large paired oviducts but only vestigal ovaries. While these diploid female hybrids are largely sterile they do lay a few eggs and three embryos from such eggs when examined showed the same karyotype as the mother that produced them.

More direct evidence on DNA doubling is available in the stick insects *Sipyloidea sipylus* (Pijnacker & Ferwerda, 1978) and *Carausius morosus* (Koch, Pijnacker & Kreke, 1972; Pijnacker & Ferwerda, 1986). Here there is an additional replication of the pachytene chromosomes in the growing oocyte. In the latter species ($2n = 64 = 61 + XXX$), pachytene in the female is followed by a duplication of desynapsed chromosomes resulting in a tetrapachytene stage consisting of paired sister chromosomes (*autobivalents*). SCs are formed only during tetrapachytene and then only in the autosomal autobivalents. RN-like structures are also present at this time. No SC forms in the sex autobivalents.

As a result of autobivalent formation, meiosis proceeds normally, despite the presence of both structural heterozygosity and uneven polyploidy, and leads to the production of unreduced eggs. Bivalents decondense after tetrapachytene resulting in a diffuse stage during which the SCs completely disassemble. SC formation is thus not a prerequisite for the initial synapsis of homologues in the female parthenogenesis of *C. morosus* and, as in the diptera where somatic pairing persists into meiosis (see Chapter 4C.1), SCs are formed between already aligned sister chromatids and pairing takes place simultaneously over the entire length of each bivalent.

Fatherless males (2n = 63 = 61 + XX) and intersexes may arise in *C. morosus*, either spontaneously or else following X-ray or high temperature treatment. Here, two kinds of meiosis occur and, usually, both are found in the same individual. SCs are formed only once in both types, namely during pachytene in cases where chromosome doubling does not take place and during tetrapachytene where it does. Spermatocytes which go through an unreplicated meiosis may, however, contain up to 10% additional DNA which is synthesized during zygotene. No SCs form in the two X-chromosomes of rare males or the three X-chromosomes of intersexes where the Xs are characterized by thickened lateral elements.

Pre-meiotic doubling appears to be especially common in hermaphrodite taxa, as exemplified by parthenogenetic earthworms where, according to Omodeo (1951), the last oogonial mitosis fails so that cells enter meiosis with twice the number of chromosomes which in turn produce a diploid number of autobivalents. A distinctive form of extra replication has been claimed for the tardigrade *Hypsibius dijardini* (2n = 10) in which female parthenogenesis is obligatory (Ammermann, 1967). Five telocentric bivalents are present at metaphase-1. The precise details of prophase-1 in this organism have not been clarified but the metaphase-1 bivalents look achiasmate. Completion of the first division produces two groups each with five half bivalents. These fall apart to produce 10 single chromatids. Since 10 chromatids subsequently pass to each pole at second anaphase, it is clear that there must be an additional replication event during interkinesis in this species.

A second mode of automictic female parthenogenesis involves either the fusion of first division products, so called *central fusion*, resulting from the collapse of first anaphase, or else a *terminal fusion* event follows the second division. In these cases, the fusion of nuclei of different parentage of the kind involved in sexual fertilization is replaced by the fusion of two nuclei from the same individual. Central fusion has been documented in a few parthenogenetic moths where, because the female is heterogametic (ZW), fusion between non-sister nuclei is critical for preventing the formation of both ZZ-eggs, which would develop into males, and WW-eggs, which would be expected to be lethal (Narbel, 1946; Narbel-Hofstetter, 1950). It also occurs in dipterans (Scholl, 1956, 1960). The most common form

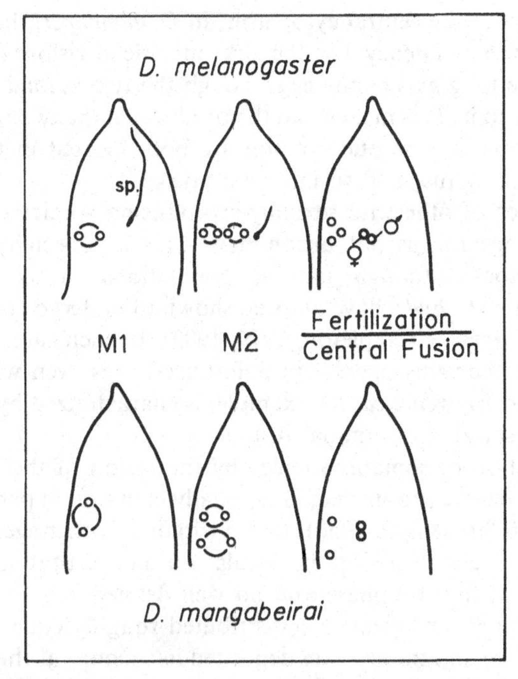

Fig. 7.5. Differential formation of first and second division egg spindles in the sexually reproducing *Drosophila melanogaster* and the parthenogenetically reproducing *Drosophila mangaberei* (after Templeton, 1982 and with the permission of Springer-Verlag).

of terminal fusion involves the second polar nucleus and the female pronucleus. Less commonly the two central nuclei, each derived from a different set of dyads, fuse.

Drosophila mangabeirai is the only species of *Drosophila* to have become an obligate natural parthenogen. This species consists of a single biotype invariably heterozygous for the same three inversions. Most individuals are diploid, though polyploidy frequently arises. In diploids, both meiotic divisions occur in the eggs and four haploid nuclei result. Autogamous fusion occurs between two of these but spindle orientations are such that, in 60 % of the cases, this takes place between the haploid products of different secondary oocytes, effectively reconstructing the original maternal condition. The remaining 40 % of eggs are produced by some other form of fusion and are non-viable (Templeton, 1982). This case thus involves the permanizing of structural heterozygosity by selective autogamy. Additionally, if no recombination occurs between an individual locus and the centromere then central fusion will also maintain heterozygosity at that locus. In most species of *Drosophila* the spindles at both meiotic divisions are oriented at right angles to the long axis of the egg so that the pro-nuclei align transversely in the egg (Fig. 7.5). Here the innermost usually becomes the egg nucleus so that all subsequent cleavage divisions occur in the central cytoplasm. In *D. mangaberi*, however, there is a pronounced tendency for the first meiotic division to take place parallel to the long axis of the egg, though the two second divisions are perpendicular to it. This means that the products of the two nuclei that are separated at the first mitotic division are both located in the inner egg region where they fuse and so initiate cleavage.

In a number of otherwise sexually-reproducing species of *Drosophila* tychoparthenogenesis is not uncommon. This is especially the case in laboratory stocks, though natural populations of *D. robusta*, *D. mercatorum* and *D. hydei* have all been shown to undergo tychoparthenogenetic development (Templeton, 1979, 1983). In such cases a mixture of automictic mechanisms operate in unfertilized eggs even within a single female. *D. parthenogenetica*, for example, is characterized by about equal numbers of central and terminal fusion.

The restitution of somatic number by the fusion of the second polar body and the female pro-nucleus is especially common in parthenogenetic enchytraeids (Christensen, 1960). In the triploid parthenogenetic form of *Lumbricellus lineatus* there is a spectacular form of restitution. Here, there is no pairing at first prophase and no well defined equatorial plate. At anaphase-1 the 39 univalents are distributed roughly equally at the two poles (Table 7.6) and the eggs are deposited in cocoons at this stage. Since it is extremely unlikely that all three genomes in this triploid share so little homology, or homoeology, that no pairing is possible between them, this species is presumably an allotriploid in which univalents are formed not for structural but for genotypic reasons. Once the eggs are laid, the spindle elongates and the two chromosome groups move further apart. The spindle then bends into a V-shape and anaphase movement is arrested. The two chromosome groups now form second division plates and there is a phase of second anaphase separation. The two chromosome clusters that move to the apex of the V-shaped spindle now fuse to form a small peripheral interphase nucleus. Subsequently, the two nuclei at the tip of the two arms of the spindle also move toward one another and fuse to form a large central interphase nucleus (Fig. 7.6). Thus, each of these two nuclei contain a complete triploid set regardless of the actual distribution of chromosomes at meiosis-1. The central nucleus initiates the first cleavage division and so provides for the development of a triploid parthenogenetic female.

A further variant, involving a unique form of central fusion, occurs in the thelytokous midge *Lundstroemia parthenogenetica* (2n = 3x = 9). Here there is an abortive first meiotic division, involving the formation of a restitution spindle, and this is followed by an equational division (Porter, 1971).

Yet a third form of automictic thelytoky involves a normal meiosis followed by a fusion of haploid cleavage nuclei in the early embryo. The

Table 7.6. *The distribution of chromosomes at anaphase-1 in oocytes of the triploid thelytokous enchytraeid Lumbricillus lineatus (2n = 3x = 39)*

Distribution	Number of anaphases
19–20	19
18–21	10
17–22	7
16–23	3
15–24	2
14–25	1
13–26	1
Total	43

After Christensen, 1960.

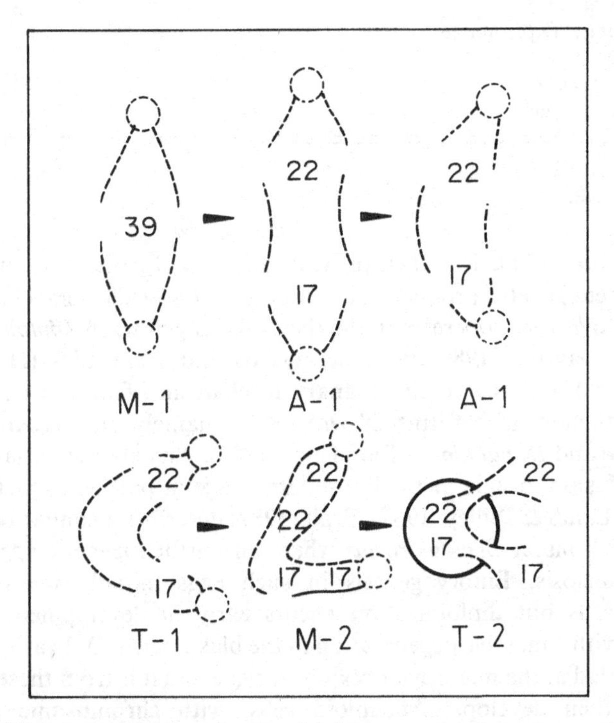

Fig. 7.6. Meiotic restitution in the triploid oocyte (2n = 3 × = 39) of the parthenogenetic euchytraeid worm *Lumbricellus lineatus* (after Christensen, 1960).

Table 7.7. *Ploidy levels of thelytokous animals*

	Automicts (Meiosis compensated)		Apomicts (Meiosis suppressed)	
Taxon	Diploid	Polyploid	Diploid	Polyploid
Nematoda, Heteroderidae				3x–5x
Oligochaeta, Enchytraeidae		4x– 6x		
Lumbricidae		3x–10x		
Tardigrada	+	3x– 4x		3x–4x
Insecta				
Orthoptera, *Saga peda*	+	4x		
Warramaba virgo	+			
Dictyoptera, *Pycnoscelus*		+		
Diptera, Chaemomyidae		3x		
Chironomidae			+	3x
Drosophilidae	+			
Coleoptera, Chrysomelidae				3x
Curculionidae				3x–6x
Hymenoptera, Diprionidae		4x		
Vertebrata				
Pisces, *Poecilia*			+	3x[a]
Reptilia, Teiidae	+	3x		

After Bell, 1982.
[a] Turner, 1982.

resulting diploid nucleus then provides the basis for the development of the parthenogenetic product. This process, of so-called *gamete* or *pronuclear duplication*, operates in the thelytokous gall wasp *Diplolepis rosae* (Stille & Dävring, 1980) in some coccids and white flies (Hemiptera, Homoptera) and is also the primary mechanism of automixis in three tychoparthenogenetic forms of *Drosophila*, namely *D. mercatorum*, *D. ananassae* and *D. pallidosa* (Templeton, 1983). An extreme variant of this form of fusion, occurs at the blastoderm stage in parthenogenetic camel crickets (Lamb & Willey, 1987). Scali (1982) describes a similar behaviour in the stick insect *Bacillus rosius* where all parthenogenetic eggs have a normal meiosis. Embryogenesis in such eggs initially involves only haploid cells but diploidization occurs early in development. This is coupled with a massive degeneration in the blastoderm. Only a few diploid cells, located at the micropylar region, survive and it is from these that the embryo then develops. Aneuploid cells, with chromosome numbers ranging from 28 to 39, are also found with hypoploid cells being more common than hyperploids. These decline in frequency during larval development.

Table 7.8. *Relationship between mode of reproduction (Sex. = sexual, Parth. = parthenogenetic) and the occurrence of polyploidy in the Oligochaeta*

	Family							
	Lumbricidae		Tubificidae		Enchytraeidae		Totals	
Ploidy	Sex.	Parth.	Sex.	Parth.	Sex.	Parth.	Sex.	Parth.
Diploid	26	1	9	0	58	1	93	2
Polyploid	5	14	0	11	31	17	36	42

After Christensen, 1980.

There is a strong correlation between parthenogenesis and polyploidy in several animal groups (Table 7.7). Thus of 64 races of parthenogenetic weevils found in 50 different species, only two are diploid. Of the remainder, 38 are triploid, 17 are tetraploid, five are pentaploid and two are hexaploid (Lokki & Saura, 1980). Similarly, about one-third of the 30 known parthenogenetic vertebrates are triploid. In lumbricid and enchytraeid worms, both of which are hermaphrodite, polyploid cytotypes are more common in parthenogenetic than in amphimictic species (Table 7.8). Additionally, 81 % of the polyploids are also achiasmate whereas in chiasmate forms only 25 % are polyploid. The correlation between polyploidy and parthenogenesis is paralleled by a high degree of allelic heterozygosity in polyploid apomicts in comparison with related diploid bisexuals (Table 7.9). Indeed, Stebbins (1980) has argued that polyploidy and heterozygosity are key factors favouring the occurrence, the establishment and the subsequent success of apomicts. The psychid moth *Solenobia triquetrella* is an exception. Here both diploid and tetraploid parthenogenetic races have the same levels of heterozygosity as the diploid amphimictic race. Since female moths are heterogametic (ZW), thelytokous lines can only be propagated if automixis conserves heterozygosity. In thelytokous females of *Solenobia triquetrella* this is achieved by central fusion between two non-sister nuclei resulting from the second meiotic division. This, coupled with the fact that female meiosis is achiasmate in moths, means that the heterozygosity of the parental form is precisely conserved. The central fusion practised in this case involves a modification of an existing pattern of behaviour in the bisexual race. Here, too, the two central polar nuclei fuse, but the product does not develop any further when the egg is fertilized. In the absence of fertilization, however, the fusion of these two nuclei provides the basis for the parthenogenetic development of the egg.

Table 7.9. *Heterozygosity per locus per individual, assayed in terms of electrophoretic enzyme variation, in seven species of thelytokous insects*

	Race				
		Apomictic			
Species	Bisexual 2x	2x	3x	4x	Reference
Coleoptera					
Bromius obscurus	0.18		0.34		Lokki, 1976a
Polydrosus mellis	0.14	0.36	0.37		Lokki, 1976b
Otiorhynchus scaber	0.31[a]		0.25	0.38	Suomalainen & Saura, 1973; Saura *et al.*, 1976a
O. salicis	0.12		0.24		Saura *et al.*, 1976b
O. singularis			0.37		Suomalainen & Saura, 1973
Strophosomus melanogrammus			0.30		Suomalainen & Saura, 1973
Lepidoptera					
Solenobia triquetrella	0.23	0.23		0.23	Lokki *et al.*, 1975

[a] For causes that remain to be identified, the level of heterozygosity in this diploid bisexual is exceptionally high.

The correlation between parthenogenesis and polyploidy can, however, be accounted for in two rather different ways:

(1) By parthenogenesis preceding polyploidy. The backcrossing of a diploid parthenogen to a genetically distinct sexual taxon would be expected to result in a triploid which, because of difficulties in consistently producing balanced gametes by conventional meiosis, could only survive by maintaining the parthenogenetic mode of reproduction. The increase in ploidy following such a hybridization event is then a result of the female parthenogen producing an unreduced egg, with the sexual male adding a haploid complement. Seiler & Schäffer (1960), for example, were of the opinion that in *Solenobia triquetrella*, a flightless moth with a diploid sexual and both diploid and tetraploid parthenogenetic races, parthenogenesis evolved first. Suomalainen *et al.* (1976) suggest that this also applies to most polyploid animal parthenogens. Triploidy, as we have seen, is the most common level of ploidy found in parthenogenetic weevils and is also common in parthenogenetic vertebrates. Thelytoky of non-hybrid origin is, of course, only possible given a capacity for tychoparthenogenesis; that is, for spontaneous diploidization resulting in the production of a diploid parthenogenetic female. As already noted, a number of sexual species of *Drosophila* have a latent capacity for parthenogenetic development and gene mutations resulting in intranuclear fusion have been identified in a number of organisms. The subsequent production of a polyploid parthenogen carries the additional rider that there is no absolute barrier to reproduction between the diploid parthenogenetic female so produced and related bisexual males.

(2) By polyploidy preceding parthenogenesis. Obligate parthenogens might be expected to arise in cases where conventional sexual reproduction experiences difficulties. In theory, these stem either from impediments to meiosis, which are common in hybrids or polyploids and especially uneven polyploids, or else from impediments to fertilization. In some hybrids the poor interaction of the divergent genomes combined in the hybrid can sometimes be circumvented by the addition of another chromosome set from one of the parental species. For example, in frogs, hybrids between ♀ *Rana catesbiana* and ♂ *R. clamitans* arrest at the exogastrula stage and die. The hybrid can, however, be rescued experimentally by suppression of the second polar body which then adds a second set of *catesbiana* chromosomes to the embryo, so producing a triploid hybrid. This triploid still expresses genes of both species but now develops normally. Triploid hybrids also arise spontaneously at a low frequency in this cross

(Elinson & Briedis, 1981). Thus, the occurrence of triploids in some parthenogenetic vertebrates might also be accounted for by the fact that the developmental inviability of hybrids can be circumvented by the addition of a second chromosome set from the female parent by some form of thelytokous automixis. Certainly, a number of parthenogenetic triploid lizards have been described which have the morphological, karyological and biochemical characteristics expected of hybrids (Cole, 1978). Indeed, the extremely high levels of heterozygosity which characterize many parthenogenetic populations are also to be expected if such populations have arisen through a hybrid ancestry.

An additional point worth noting is that no instance of meiotic reduplication has ever been demonstrated in any case of tycho-parthenogenesis. Such behaviour is known, however, from two groups of experimental hybrids. In a series of specific and sub-specific hybrids in the genus *Xenopus* (Amphibia: Anura) a considerable number of polyploid oocytes were observed and, in four of these cases, triploid females containing exclusively triploid oocytes were obtained. Ovaries of seven of the diploid hybrids contained a mixture of diploid and tetraploid oocytes. The latter invariably formed 36 bivalents. Consequently such hybrids gave rise to some diploid eggs. Likewise, in females obtained by the back-crossing of triploid female *X. laevis*/*X. gilli* hybrids to diploid *X. laevis* males, some of the ovaries were composed of a mixture of triploid and hexaploid oocytes and these latter again contained only bivalents (Müller, 1977).

In a second study, involving grasshoppers of the genus *Caledia*, up to 70% of the meiocytes present in males from the cross Daintree taxon♀ × Moreton taxon♂ were tetraploid. Here it was possible to show that the doubling involved endoreduplication during the pre-meiotic mitosis. These endoreduplicated cells then entered meiosis and formed only bivalents (Shaw & Wilkinson, 1978). In both these cases some of the F_1 hybrids produced proved capable of sub-sexual reproduction. In *Caledia* this was not associated with polyploidy (Shaw, pers comm.) but in *Xenopus* the reduplication process was retained in subsequent generations of triploid back-crosses. Thus, at face value, meiotic doubling can result from a perturbation of the genetic control of pre-meiotic events induced directly by hybridization and this, in turn, may lead to parthenogenetic development sometimes coupled with polyploidy.

That a causal relationship might exist between the interaction of the different genomes brought together by hybridization and the adoption of parthenogenesis is not unexpected. Thelytoky requires a modification in meiosis and such modifications commonly arise following hybrid production. Even so, the failure to resynthesize parthenogenetic forms by laboratory crosses of their presumed progenitors suggests that only very specific hybrid genotypes are able to generate parthenogens (Turner, 1982). Similarly, in cases where eggs have been artificially induced to develop parthenogenetically, as in the silkmoth *Bombyx mori*, it is clear that some strains have a stronger genetic predisposition towards parthenogenesis (Astaurov, 1940).

The origin of parthenogenetic vertebrates, in particular, has been the subject of some controversy. The best known of these occur in reptiles. Here all-female populations are known in the *Lacerta saxicola* group and in the genus *Cnemidophorus* where they occur in conjunction with triploidy and pre-meiotic doubling. The prevailing view is that these parthenogens have arisen through interspecific hybridization resulting in allodiploid parthenogenetic hybrids which by back-crossing to the ancestral amphimicts have given rise secondarily to allotriploid parthenogens. While accepting that triploidy in this case is likely to have arisen only through hybridization, Cuellar (1987) is strongly opposed to the argument that it has also been involved in the production of diploid parthenogens. He maintains that spontaneous genetic change within an amphimictic diploid has produced a parthenogenetic diploid which, by crossing to the same or another amphimict, then produces a parthenogenetic auto or allotriploid.

There is certainly compelling evidence to support the involvement of hybridization in the production of diploid all-female forms by both gynogenesis and hybridogenesis in teleost fish. *Gynogenesis* is a category of unisexual propagation in which an all-female population requires the donation of sperm from a related bisexual host species to activate the cleavage of unreduced eggs, whether diploid or polyploid. Gynogenetic offspring are thus genetically identical with their mothers and gynogenesis is considered by some to be a specialized form of female parthenogenesis. Hybridogenesis also involves the propagation of an all-female line but here the eggs are reduced and fertilization by sperm from a related bisexual species is required not only to trigger cleavage but also to support somatic development. Clearly this is not a form of parthenogenesis since both maternal and paternal genomes are expressed in the all-female progeny, though only the maternal genome is in fact transmitted by the reduced egg.

In the viviparous teleost fish *Poeciliopsis* seven bisexual species are known. Five of these have been involved in the hybrid origin of three allodiploid and three allotriploid unisexual forms and one of them, *P. monacha*, has contributed a genome to all six of the unisexual biotypes (Schultz, 1980, 1982). The best characterized of the all-female diploids is that produced from the hybridization of ♀*P. monacha* and ♂*P. lucida*. While this hybrid develops as a diploid, the paternal *lucida* genome is eliminated just prior to meiosis in a cell division characterized by a unipolar spindle on which only the *monacha* chromosomes align. The *lucida* chromosomes remain scattered in the cytoplasm and are not included in the reconstituted haploid nucleus. A single equational meiotic division follows so that only the maternal *monacha* chromosome set is transmitted by the haploid eggs. This diploid, all-female hybrid, *P. monacha–lucida*, lives with *P. lucida* and uses its sperm both to trigger, and to ensure, normal somatic development. Mating thus compensates for the *lucida* genome that is eliminated from the germ line. Stocks of female *P. monacha–lucida* have been maintained in the laboratory for over 25 generations, by mating with males of *P. lucida*, without ever producing male offspring (Schultz, 1980). The ability to experimentally synthesize this unisexual biotype confirms that hybridization can provide a route to unisexuality, though to unisexuality of a rather distinctive kind.

A second all-female hybrid biotype, *P. monacha–occidentalis*, has arisen following the hybridization of ♀*P. monacha* with ♂*P. occidentalis* and this hybrid too has been successfully synthesized in the laboratory. The third hybridogenic unisexual diploid, *P. monacha–latideus*, on the other hand, has not been so synthesized since all the products of experimental hybridization in this case have turned out to be sterile males. Since all three diploid hybridogenic forms of *Poeciliopsis* contain a *monacha* genome it is clear that when placed in a heterospecific background this genome undergoes a modification of its normal meiotic activity. Moreover, when *P. monacha–lucida* females are experimentally mated to males of *P. monacha* all the *lucida* chromosomes are lost and *P. monacha* offspring of both sexes result. Hybridogenesis may thus revert to bisexuality.

In nature, the diploid unisexual *P. monacha–lucida* has given rise, by further hybridization, to three triploid unisexual biotypes. Two of these are referred to in terms of their mode of origin as *P. monacha–2 lucida* and *P. 2 monacha–lucida*. The third has yet to be characterized. All three again produce only female offspring, though this time through gynogenesis and not hybridogenesis, using sperm from either *P. lucida* or *P. monacha* to trigger egg development. In these triploids there is a pre-meiotic endoreduplication ($3x \rightarrow 6x$) with autobivalent formation and two subsequent meiotic divisions.

A second, related, genus, *Poecilia*, shows a similar scenario. Here hybrids between the diploids *P. mexicana* and *P. latipinna* have given rise to a gynogenetic diploid unisexual hybrid known as *P. formosa*, the Amazon molly (= *P. mexicana–latipinna*). This, in turn, has led to the production of two gynogenetic triploid unisexuals, *P. mexicana–2 latipinna* and *P. 2 mexicana–latipinna* (Turner, 1982). It was formerly assumed that diploid eggs were produced by *P. formosa* through a pre-meiotic endoreduplication but DNA measurements have shown that this is not so. Oogenesis in triploid *Poecilia* is known to be ameiotic and the same may be true for *P. formosa* (Turner, 1982). Gynogenetic diploid *P. formosa* may reconvert to bisexuality when mated to the distantly related *P. vittata*. By contrast, gynogenetic strains of *P. formosa* rarely produce competent males. This is true, too, of obligate thelytokous parthenogens, though rare, non-functional, males have been reported in some of them. In such cases it seems likely that the loss of male function would have been coincident with the adoption of obligate thelytoky.

In sum, whereas hybridization in *Poeciliopsis* has given rise to diploid hybridogenic and triploid gynogenetic unisexual biotypes, in *Poecilia* both diploid and triploid unisexuals are gynogenetic. Electrophoresis indicates that the unisexual populations of both genera are made up of multiple clones and that clonal diversity of the diploid unisexuals is greater than that of the triploids, which are also less abundant in most localities.

Several all-female salamander populations, derived by hybridization between species of *Ambystoma*, occur in the region of the Great Lakes in eastern North America. Hybrids between *A. laterale* and *A. jeffersonianum* are triploid and have been assigned species status with *A. tremblayi* combining two *laterale* genomes with one from *jeffersonianum* (LLJ). Diploid *A. laterale* hybrids and triploid *A. texanum* hybrids occur in northwest Ohio and on islands in Lake Erie while *A. nothagenes* is a trihybrid combining the genomes of *A. texamum*, *A. laterale* and *A. tigrinum*. Tetraploids, assumed to represent combinations of *A. platineum* and *A. texamum*, have been found in eastern central Illinois. It has been claimed that these various all-female hybrids are perpetuated by gynogenesis in which males from the progenitor bisexual species provide sperm to initiate egg development. However, since the frequency of suitable males is often very low, parthenogenesis has also been assumed to occur (see discussion in White, 1970; Bogart, 1980). In triploid females of the *A. jeffersonianum* complex, for example, pre-meiotic doubling has been inferred from the presence of 42 bivalents at prophase-1 of meiosis.

Bogart & Licht (1986) have challenged both these assumptions. Their analysis of 283 individuals of *Ambystoma*, sampled from Pelée Island on Lake Erie, indicated that only 14 were male. Despite their infrequency, there was a high incidence of sperm-positive females, that is females in which viable sperm were recovered after flushing the cloaca. Three different levels of genome combinations of *A. texanum* and *A. laterale* were identified in their sample, namely TT, LL and LT (diploids), TTT, TLL and TTL (triploids) and TTLL and TTTL (tetraploids). Offspring produced by either gynogenesis or by strict female parthenogenesis are expected to be identical in ploidy level with the female from which they arose. 20 of the 269 females, however, produced some larvae that differed from their mother in ploidy and 14 of these involved progeny of mixed ploidies. Thus diploid females produced both diploid and triploid larvae while triploid females produced both triploid and tetraploid larvae as too

Table 7.10. *Modes of agamospermy (seed development without fertilization) in angiosperm plants*

Category	Description
Gametophytic apomixis	An unreduced female gametophyte (embryo sac) is formed.
Diplospory	The female gametophyte arises directly from a diploid embryo sac mother cell without reduction; meiosis is circumvented (Compositae).
Apospory	The female gametophyte arises from somatic cells of the nucellus or from the chalaza of the ovule (Rosaceae).
Adventitious embryony	There is no gametophytic stage; both meiosis and fertilization are omitted., Embryos arise directly from sporophytic (somatic) cells of the ovule. There are no embryo sacs or egg cells (Liliaceae and Orchidaceae).

After Asker, 1984.

did tetraploid females. None of the sperm-negative females produced any progeny and the presence in progeny larvae of electromorphs which were not represented in the mother confirmed that sperm are required for successful development in Pelée Island females. That is, these females are capable of producing both reduced and unreduced eggs but only those that are fertilized give rise to larvae.

Thus, as Bogart & Licht emphasize, there is no evidence from their data for either parthenogenesis or gynogenesis. Sperm is required for the successful development of all eggs, whether reduced or unreduced, produced by all three ploidy classes of female (2x, 3x and 4x).

It is also worth turning to the situation in plants where there can be no doubt about the hybrid origin of all-female forms. Here, apomixis is synonymous with agamospermy; that is, to seed development in the absence of fertilization (Asker, 1984). Three different systems are involved (Table 7.10). That which equates most directly with animal parthenogenesis is diplospory.

Allium tuberosum, Chinese chives, is a diplosporous autotetraploid ($2n = 4x = 32$). At male meiosis, quadrivalents (85%), trivalents (1%), bivalents (13%) and univalents (1%) are present at first metaphase. In female meiosis, on the other hand, there are invariably 32 bivalents because the chromosomes are doubled pre-meiotically and only autobivalents form. Consequently, all seeds are derived from unreduced eggs and parthenogenesis is regular (Gohil, & Kaul, 1981). A second species, *A. odorum*, a close relative of *A. tuberosum*, shows an equivalent endoreduplicational female meiosis (Håkansson & Levan, 1957). Although apospory and apogamy have no real equivalents in animals, the mechanisms involved in these systems still show interesting parallels with the situation in animals. Obligate apogamy, for example, is well known in homosporous ferns, which are monoecious, and again tends to be associated with polyploidy (Manton, 1950). Roughly three-quarters of apogamous ferns are polyploids and most are triploids of the type ABC. Sporocytic doubling, the result of a failure in the last pre-meiotic mitosis, leads to the formation of a restitution nucleus with double the chromosome number and hence to meiocytes with the constitution AABBCC. Meiosis is thus regularized and only autobivalents form (Wagner & Wagner, 1980). In other cases the first division of meiosis is either aberrant, reduced, arrested or else completely replaced by a single equational division (reviewed in Golubovskaya, 1979).

While diploid apomicts are certainly known in plants, there is a close association between agamospermy, hybridization and polyploidy. Most agamospermous flowering plants are species hybrids or hybrid derivatives, for agamospermy is one of the most successful ways of allowing hybrid genotypes to breed true by seed. It is true that hybrids which are semi- or completely sterile sexually can sometimes be maintained by vegetative propagation. Seed production, however, has the additional advantages of dormancy, resistance to adverse conditions and, especially, the capacity for dispersal (Grant, 1981).

The observed high frequency of both odd polyploids (*anisopolyploids*) and high polyploids, including secondary aneuploids, among agamospermous angiosperms indicates not only that agamospermy can succeed where amphiploidy fails but, additionally, that the recombination potential of meiosis is still able to function if the opportunity is provided for occasional out-crossing (Asker, 1984). Episodes of sexual reproduction are known to intervene in several agamospermous lineages, giving rise to new hybrid genotypes. These may then reproduce themselves by a further cycle of agamospermy. The net result of this is the production of an agamic complex. Such complexes include some of the most successful known plant genera including *Rubus*, *Hieracium*, *Poa* and *Bouteloua*. The main evolutionary significance of these agamic complexes is that they have permitted hybridization to occur on a far greater scale than in polyploid sexual complexes.

The *Bouteloua curtipendula* complex, the gamma grasses, for example, rests on a sexual base which includes both diploid and tetraploid forms. Superimposed on this, however, is an agamic complex ranging from 5 to 10 x. These euploid agamospermous forms have, in turn, given rise to an extensive and nearly continuous series of secondary aneuploids which results from a combination of natural hybridization, irregular meiosis and agamospermous reproduction (Fig. 7.7). The plants with high chromosome numbers are often characterized by a very unequal segregation at

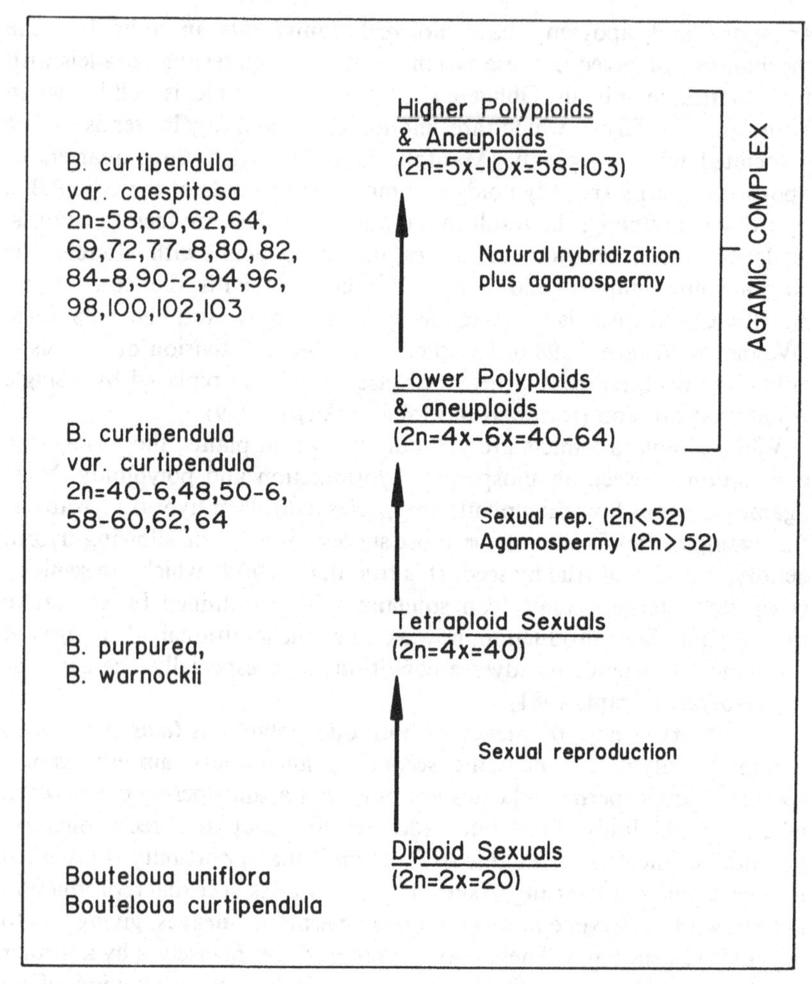

Fig. 7.7. The origin and organization of the agamic complex in the plant *Bouteloua curtipendula* (after Grant, 1981).

anaphase-1 with a majority of chromosomes passing to one pole. Meiotic products with these high chromosome numbers yield viable pollen grains which, in turn, produce aneuploid progeny via occasional sexuality. Such progeny once formed are then perpetuated by agamospermy. Some polyploid animal apomicts can also tolerate aneuploidy to a limited extent, e.g. the Curculionidae.

Stebbins (1980) draws attention to an important difference in the reproductive cycle of animals and plants which has a bearing on the adoption of agamospermy. In animals, where meiosis is followed immediately by fertilization, both of the requirements for parthenogenesis affect the oocyte directly. Thus, on the one hand, female meiosis must either be circumvented or eliminated while, on the other hand, the oocyte must ultimately be capable of developing without the stimulus provided by fertilization. In angiosperms, on the other hand, not only does a gametophytic stage intervene between meiosis and fertilization but a nutritive endosperm is required to support the development of the fertilized egg. In some plant apomicts both embryo and endosperm formation take place autonomously so that no pollination is required and, often, pollen does not form. Most apomictic plants, however, still produce functional pollen which is required for the second fertilization that occurs in angiosperms and which leads to the formation of the endosperm. This pseudogamy is reminiscent of the gynogenetic mechanism found in some animals, though it serves a quite different purpose. The endosperm is a unique tissue which is neither gametophytic nor sporophytic. It has a distinctive chromosome complement, most commonly triploid, with two sets of female chromosomes and one male set. The formation of the endosperm thus results in an additional factor being involved in the regulation of egg development, namely a workable proportional relationship between the chromosome numbers of the embryo sac, the embryo and the endosperm. In a normal diploid this relationship takes the form x (embryo sac):2 x (embryo):3 x (endosperm). Not all combinations of chromosome numbers are, however, functional. In diploid/tetraploid crosses, for example, where the relationship is x:3 x:4 x, the endosperm breaks down. In reciprocal tetraploid/diploid crosses it becomes x:3 x:5 x and here development is often much retarded so that viable seed is rarely formed. A successful apomict can, therefore, only evolve given a workable relationship between the three components and in diploid apomicts this takes the form 2 x:2 x:4 x.

The switch from bisexuality to parthenogenesis has been held to have demographic as well as genetic effects. It has been claimed, for example, that because of their increased egg-laying capacity, thelytokous animals can be expected to show an inherent increase in reproduction rate which will be doubled relative to that of related sexual populations, other things being equal (Maynard-Smith, 1978). But as White (1970) points out, other things are not necessarily, or indeed often, equal. Individuals of the arctic parthenogenetic simuliid *Gymnopais sp.* lay only 20 eggs by comparison with the several hundred produced by temperate bisexual black flies (Downes, 1964). Moreover, the actual reproductive capacity of a parthenogen is most objectively gauged not simply by the number of eggs produced but by the number of viable offspring. In *Drosophila mangaberei*, for example, only 59 % of the eggs produced eventually hatch (Templeton, 1982). In the Acrididae, Tetrigidae and Phasmatodea where facultative thelytoky is present it is invariably automictic, usually involves post-meiotic restitution and is associated with the production of a substantial number of inviable eggs leading to high levels of embryonic mortality.

Equivalent situations have been noted for other parthenogens by Lamb & Willey (1979) and by Bell (1982). The large percentage of inviable progeny reported by these authors presumably reflects imperfections in the meiotic mechanisms involved.

Parthenogenetic forms have sometimes been regarded as evolutionary blind alleys with only limited potential. There are, however, three facts that militate against such a simplistic interpretation (Lynch & Gabriel, 1983; Lynch, 1984):

(1) Many parthenogens have a broader geographical range, or less restrictive habitat requirements, than do related sexuals. In some cases, at least, this has been shown to be a consequence of the joint distribution of a number of distinct clones. Electrophoretic surveys of *Solenobia triquetrella*, for example, indicate that thelytokous alpine populations consist of multiple clones which are highly variable (Lokki *et al.*, 1975). Parthenogenesis is in fact rare among lepidopterans with mobile adults and all seven known obligate parthenogens, including *S. triquetrella*, have apterous and immobile females, which clearly favours clonal diversity.

(2) The prevalence of non-segregative automixis and of apomixis among parthenogens leads to many of them exhibiting enhanced levels of heterozygosity compared to their sexual relatives. Where meiosis is achiasmate, as in female lepidopterans, then, regardless of the precise means of avoiding reduction, the genetic consequences are indistinguishable from those of mitosis. Heterozygosity is completely conserved. This is true too of those chiasmate forms where reduction is circumvented by some form of DNA doubling either prior to meiosis or else at early prophase-1. Here, while crossing-over does occur it involves genetically identical DNA strands. Where reduction is avoided either by the suppression of one or other of the two meiotic divisions, or else by post-meiotic compensatory fusion, the genetic composition of the products will vary with the pattern of segregation which depends on chiasma frequency and chiasma distribution relative to the centromere (Table 7.11). Apomicts are expected to have, and to conserve, relatively high levels of heterozygosity since heterozygotes will not be lost through segregation. Additionally, not only will heterozygosity be greater in apomicts compared to amphimicts, but it will be even greater in apomicts of higher, compared with those of lower, ploidy.

(3) A number of well studied obligate parthenogens are only imperfectly reproductively isolated from their sexual progenitors with which they show occasional crossing. Alternatively, there may be one or more

Table 7.11. *The consequences of thelytoky on pre-existing heterozygosity*

Mechanism	Consequence
Ameiotic – omission of meiosis	Heterozygosity preserved unaltered
Meiotic	
Pre-meiotic doubling	If synapsis is restricted to exact molecular copies, heterozygosity is preserved unaltered.
Failure of M-1 or A-1	Loss of fraction 'r' of pre-existing heterozygosity where r = recombination fraction between a locus and the centromere.
Fusion of second division sister nuclei	Loss of heterozygosity at a rate of '1–2r'.
Fusion of second division non-sister nuclei	Loss of heterozygosity at a rate of 'r'.
Post-meiotic fusion of mitotic products	Complete and immediate homozygosity for all loci.

After Crozier, 1975.

unisexual generations within a life cycle interrupted by a single sexual generation, as is the case in most rotifers, digenetic trematodes, cladocerans and aphids, as well as in many plant apomicts.

8

Evolutionary aspects of meiosis

I shall take it for granted not merely that selection tends to favour more fit at the expense of less fit phenotypes, but that the phenotypes which are actually present are more fit than any alternative phenotypes...The force of this axiom is to compel the belief that genetic systems are highly precise adaptations. Otherwise we must believe that they are makeshifts, that several of them may perform the same function equally well and that a particular system has been adopted only by chance.

Graham Bell

No attempt to deal with the facts of meiosis would be complete without some consideration of the evolution of the various modes and mechanisms to which the greater part of this monograph has been devoted.

The data summarized in Chapter 6 indicate that all the component parts of the meiotic mechanism are under genetic control and show heritable variation in the form of mutations. Since selection will operate on any set of entities with the properties of heredity, variation and a capacity for multiplication, it has commonly been assumed, as the quote which introduces this chapter also initially assumes, that changes in the character of meiosis stemming from mutation will necessarily reflect adjustments relating to selection for enhanced fitness. Inherent in this assumption is the belief that different meiotic systems have different consequences which allow them to respond to different selection pressures.

It is certainly possible to demonstrate a response to artificial selection for a number of aspects of meiotic behaviour. Shaw (1971b), for example, was able to alter the mean cell chiasma frequency of male meiocytes by selecting grasshoppers with high and low mean cell chiasma frequencies and using these as parents in experimental crosses. In these organisms, it is possible to obtain a testis biopsy, which can be used to score chiasma frequency, without affecting male fertility. The male so sampled can then be used for mating. Using this technique on a base stock of *Schistocerca gregaria*, six males with the highest and five males with the lowest chiasma frequency were each mated in single pair crosses with one of their sisters (Shaw, 1972). These sampling and mating procedures were then repeated over four generations of disruptive selection. At the end of the selection program the base stock had an unaltered mean cell chiasma frequency

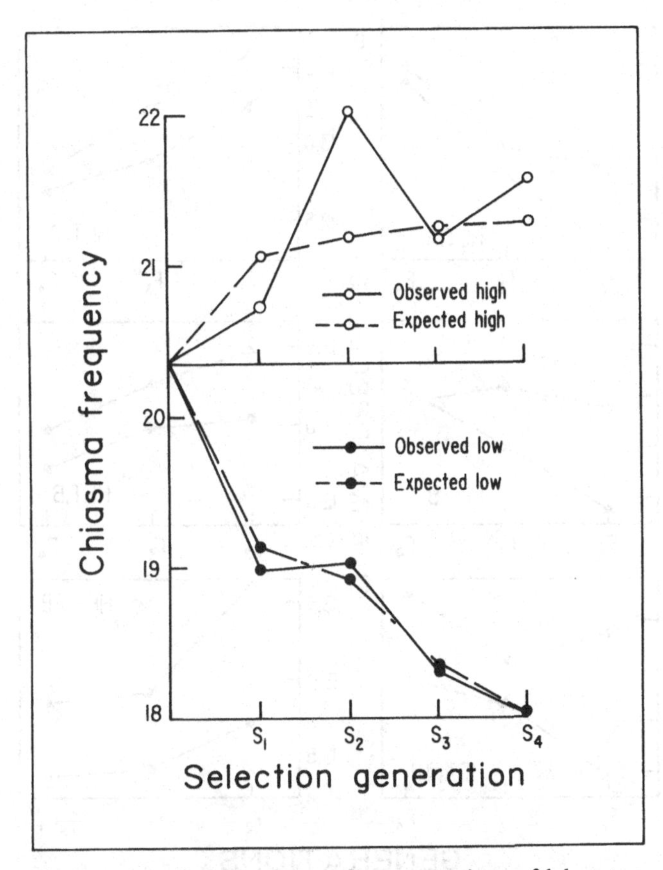

Fig. 8.1. The asymmetrical response to four generations of laboratory selection for high and low mean cell chiasma frequency in the desert locus *Schistocerca gregaria* (after Shaw, 1972 and with the permission of Springer-Verlag).

(initial 20.35 versus final 20.2) whereas the values in the high and low selection lines were 21.6 (+1.2) and 18.0 (−2.4), respectively (Fig. 8.1). The observed response was thus asymmetrical and was both more rapid and more consistent in the low lines. Indeed the difference was only significant in this line.

Similarly, selection for high seed set in interchange heterozygotes of rye has been shown to be accompanied by, and presumably achieved through, an increase in the frequency of alternate orientation (Lawrence, 1958) conditioned, in turn, by a reduction in chiasma frequency (Fig. 8.2). Thus, with a reduction from five to four chiasmata per single interchange multiple there was an overall increase of 26% in alternate orientation per multiple (Rees & Sun, 1965). In a colchicine-derived autotetraploid of *Lolium perenne*, fertility as assessed by seed setting was increased from 67 to 74% by seven generations of selection. Crowley & Rees (1968) were

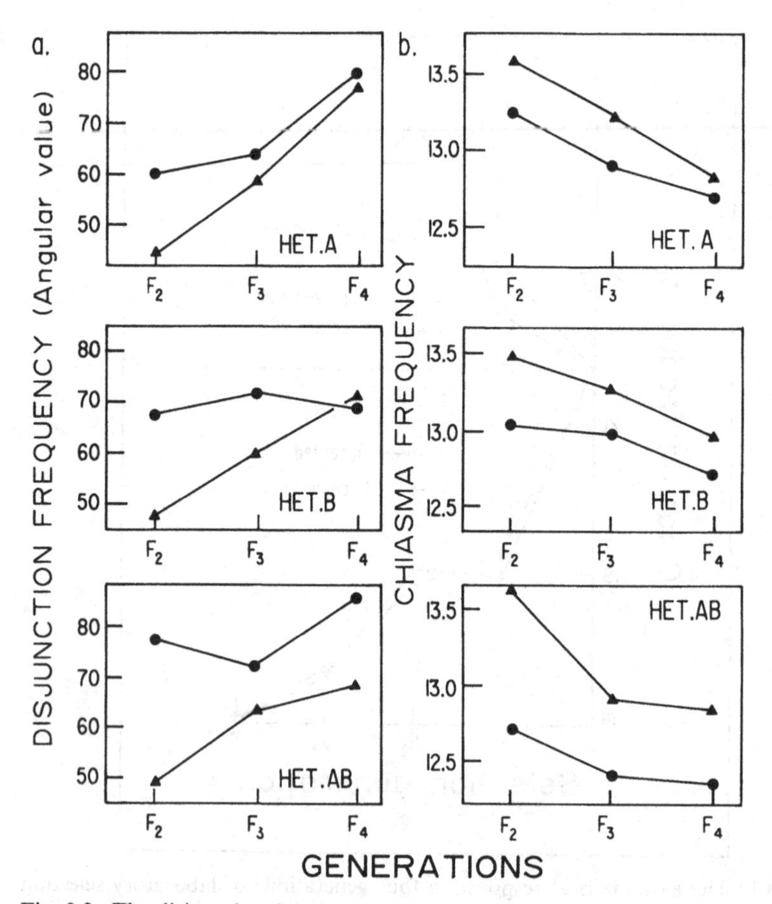

Fig. 8.2. The disjunction frequencies (*a*) in high (circles) and low (triangles) lines of the plant *Secale cereale* over three generations of selection (F_2–F_4) in three classes of interchange heterozygotes (Het.A, Het.B and Het.AB) compared with the mean cell chiasma frequencies (*b*) of these same lines and generations (after Sun & Rees, 1967 and with the permission of *Heredity*).

able to establish that this increase was associated with an increased frequency of quadrivalents resulting from a redistribution of, rather than a frequency change in, chiasmata.

The key question, however, is not whether meiotic characteristics can be induced to respond to selection under experimental conditions but the extent to which they have done so under natural conditions. The meiotic mutations that have been identified almost always impair the efficiency of the process (Chapter 6A.1) and so are not sound candidates for producing evolutionary adjustments which might serve to enhance fitness. The achiasmate female meiosis induced by the *c(3)G* mutation in *Drosophila melanogaster*, for example, is only a poor imitation of the achiasmate sequence that characterizes normal males of this species. Thus, some crossing-over may still occur in mutant homozygotes and, additionally, the mutation is associated with enhanced non disjunction of both the X-chromosome and autosome-4 (see Chapter 6A.2). Golubovskaya (1979) describes an induced recessive mutation, *afd-W23*, in maize where prophase-1 is entirely lacking in microsporocytes. Consequently homologues do not pair and 20 univalents are present at first metaphase. These, however, are all auto-oriented on the spindle and give a consistent 20:20 distribution at first anaphase. Golubovskaya suggests that a mutation with a similar effect in the female might be a promising candidate for the development of apomictic reproduction. However, the tetrads produced by mutant PMCs are, of course, all abnormal since at meiosis-2 the single chromatids move at random to the two poles. Female plants homozygous for *afd* are also sterile though their cytology has not been defined.

Marshall & Brown (1981), on the other hand, point out that apomictic mutants appear to be rare in plant populations. They attribute this to the fact that agamospermy requires simultaneous mutation at two or more loci because an apomictic cycle involves genetic changes in both meiosis and fertilization. Since interspecific hybrids are either partially or completely sterile, the occurrence of female sterility resulting from such hybridity requires only a change in meiosis in order to lead to the establishment of apomixis. It is in these terms that they explain the close association between apomixis, polyploidy and interspecific hybridization.

Artificial selection has also been shown to lead to a 60-fold increase in the level of diploid thelytoky in laboratory strains of *Drosophila mercatorum* (Carson, 1967) though even this apparently substantial response only lifts the frequency from $< 0.1\%$ in wild strains to 6% in laboratory strains. A similar improvement has been achieved in *D. parthenogenetica* (Stalker, 1956) though here a proportion of the parthenogens proved to be triploid. Nevertheless, in species which regularly produce a small proportion of unreduced eggs it is evident that this proportion can be increased by artificial selection.

There seems little doubt that, in nature, circumstances may force females into adopting a parthenogenetic mode of reproduction. The forcing agent may be geographical or ecological isolation (environmental) or, alternatively, hybridization (genetic). In the former event, organisms with a capacity for tychoparthenogenesis would be favoured. This is precisely what appears to have happened in the stick insect *Sipyloidea sipylus*. In its original habitat, in southwest Asia and the Pacific, this species reproduces sexually. Following its accidental introduction into Madagascar, however, it has become an obligate thelytoke there (Bergerard, 1962).

What may well prove to be a very much more common and realistic situation has been described by Evans & Aung (1985). They screened 30 diploid populations of *Lolium multiflorum* for genes capable of modifying meiosis in species hybrids using a standard genotype of *L. temulentum* as a tester stock. Only two plants from these 30 populations gave modified pairing in the diploid hybrids. In one of these, the incidence of univalents was shown to be due to heterozygosity for genes that suppress chiasmate associations between homoeologous chromosomes. This rarity of diploidizing genes within diploid populations is not surprising and may well stem from the fact that the mutations concerned are genetically neutral in such populations. However, should mutations of this kind become incorporated by chance into an interspecific, or even an intergeneric, polyploid hybrid they would be expected to lead to an immediate advantage. This results, according to Evans & Aung, not only from the higher fertility of the hybrids themselves but also from the conservation of hybridity in the progeny of such hybrids.

Darlington (1939) was the original proponent of the idea that modifications of meiosis and fertilization were to be interpreted not only in terms of the control of recombination but, additionally, in the pursuit of hybridity. His hypothesis has been strongly supported by James (1984) from his studies in the plant *Isotoma petraea*. Here two types of natural populations can be distinguished. In the one, a majority of individuals form seven bivalents at meiosis but with about 10% carrying a small interchange multiple of four, less commonly of six. In the other, every plant is heterozygous for multiple interchanges which, in different isolates, include two multiples of six, one of 12 or one of 10 plus one of four. The distal chiasma localization which characterizes *Isotoma* undoubtedly facilitates the adoption of permanent interchange heterozygosity, which itself involves the suppression of independent assortment, and both these characters can be argued to function in the pursuit of hybridity. In line with this argument, allozyme variability in the interchange populations is present as fixed heterozygosity some 12 times more frequently than in the bivalent-forming populations and are also said to be superior to the bivalent-formers in several important, though unspecified, biological attributes (James, 1984).

These examples remind us that while there is no precise definition of fitness (Stearns, 1982) it is generally assumed to have two principal components. One rests directly on development and metabolism and determines the viability of the individual. The other refers to reproduction and determines the fertility and fecundity of the individual (Lewis, 1966). Viability is a physiological property of the individual organism which relates directly to the environment in which that individual lives. The fertility and fecundity of an individual, however, are not concerned with the adaptation of that individual to its environment. Rather, they relate to the ability of the individual to produce offspring. Consequently, viability, on the one hand, and fertility and fecundity, on the other hand, are quite different fitness factors and there are very clear instances where they are in blatant opposition. Thus, while the fertility of individuals of *Chironomus nuditarsis*, heterozygous for four different spontaneous translocations, was invariably impaired, there was no effect on the viability of the individuals concerned (Fischer, 1978). Likewise hybridity, for example, may lead to vigour (increased viability) but to sexual sterility (reduced fertility). Changes that affect the ability of an organism to survive need not affect the number of germ cells it produces or the competence of these germ cells to fertilize, to be fertilized or to subsequently develop following fertilization. Meiosis is one of the key components for regulating fertility. How then do changes in meiosis evolve given that such changes cannot influence the viability of the individual in which they originate, though they may certainly influence the fertility of that individual as well as the viability of the progeny resulting from these changes. A solution to this problem depends, in part, on whether a comparison of present day patterns of meiosis can provide sensible clues to the steps involved in their past evolution.

Homopteran bugs show a greater range of meiotic mechanisms than probably any other single group of eukaryotes. For this reason they offer a unique opportunity to analyse, in retrospect, the ways in which different forms of meiosis may have evolved. There are two groups of homopterans to consider – the aphids and the coccids.

We have already touched on some of the mechanisms which operate in aphids (see Chapter 2C), the group in which animal parthenogenesis was first discovered. Here the thelytoky is apomictic with a single equational division giving rise to an egg that develops parthenogenetically, though a partial and temporary pairing of homologues may occur in early prophase (Blackman, 1980). Apomixis in aphids is, however, often cyclical and alternates with a single sexual generation. This change is triggered by environmental factors. In temperate climates, aphids typically reproduce through the summer months by thelytokous parthenogenesis. They produce males and sexual females only in autumn in response to reduced photoperiod and temperature. These environmental triggers act via the neuroendocrine system and through the activity of juvenile hormones. Apomictic females, like sexual females, are XX in constitution. The reversion to sexuality thus necessitates a reactivation of the sex-determining mechanism, leading to the appearance of males. This involves the loss of one of the X-chromosomes of the apomictic females and depends on a modification of female meiosis. The details are best known in *Tetraneura ulni* (Orlando, 1974). Here it is achieved by the formation of an X-bivalent involving an end-to-end pairing of the two X-homologues in the otherwise asynaptic nucleus. This bivalent itself undergoes desynapsis at a late stage in the first division of meiosis. One of the X-homologues then moves undivided into the first polar nucleus while the other divides equationally, in conjunction with the autosomes, giving rise to an X0-egg which develops as a male. That is, reduction takes place in the sex chromosomes but not in the autosomes. Blackman & Hales (1986) have described a variant form of behaviour in *Amphorophora tuberculata*. In pre-growth phase oocytes the two X-chromosomes are associated end-to-end. This terminal connection, which is governed by heterochromatic association, is retained as oocytes enter the growth phase whereas the two autosomes appear as separate entities. In oocytes destined to develop into males the X-association swells into a large spherical body which transforms into a strongly contracted X-bivalent towards the end of the growth phase, at which time the two autosomes are still unpaired. At anaphase, three of the X-chromatids pass into the polar body following a 3:1 separation of the four chromatids in the X-bivalent.

The aphid example is instructive in two respects. First, it indicates that the initial change which led to the adoption of thelytoky is reversible. Second, males in fact rarely appear in the first generation following the development of conditions optimal for a return to sexuality. After an initial batch of all-female progeny there is a pause and then a complete switch to the production of males. Once effected this switch remains irreversible for the remainder of the reproductive life of the mother in question. Taken together these two features indicate that the modifications of meiosis which led to this pattern of cyclical parthenogenesis involved a novel form of gene regulation rather than any fundamental change in the genetic constitution of the organism itself. This case thus supports the epigenetic paradigm of Waddington who argued that biological organisms have a strong tendency to adapt epigenetically to the environments they meet. It also emphasizes the fact that meiosis is an epigenetic event within the germ line. Like equivalent epigenetic events in the somatic lineage of an organism, it is liable to change through a modification of the regulatory systems which govern the gene pathways involved. The reversion to meiosis in unicellular protozoans and yeasts in response to an unsatisfactory environment is a further example of the operation of this paradigm. Both cases show that environmental factors, able to switch existing gene control systems, have sometimes played a key role in both the induction of meiosis and in the production of a changed mode of meiosis. In about one in every 30 of the recognized species of aphids the sexual phase has been completely lost so that the organism is completely parthenogenetic. In some cases this has resulted from the fact that the reversion is permanently inhibited by the environment itself. In other cases, however, it is the result of a genetic change which has led to an inability of the organism to respond to the environmental stimulus that normally leads to a return of sexuality. Here, presumably, the genetic change involved is itself secondary, stemming from a primary environmentally-induced effect.

In the coccids, the evolutionary changes in meiosis have been more numerous, involving as they do both modified sexual systems and a wide range of different categories of parthenogenesis. The coccids are the most highly specialized of the homopterans which, in addition to the aphids, includes the cicadas and the leaf hoppers. Like the aphids, male coccids are characterized by holocentric chromosomes and an inverted meiosis. While we know nothing about the ultimate mode of origin of these two features, they appear, as Nur (1980) has emphasized, to have played key roles in pre-adapting the group to several of the unusual modes of male meiosis which they show. As in the male, oogenesis in sexual females is also inverted (Chandra, 1962), though otherwise appears orthodox.

The simplest category of coccids is characterized by an XX♀ X0♂ sex-determining system, of the kind which is also present in aphids, and is chiasmate. This is assumed to represent the progenitor system from which all others have evolved (Fig. 8.3). The most striking of the changes in this basic sexual system has been the evolution of parahaploid males which, while they arise from fertilized eggs, do not transmit the paternal set of chromosomes (Brown, 1977). No differentiated sex chromosomes are present in this parahaploid lineage and what is presumed to represent the first step in its evolution has been referred to as the *lecanoid* system. Here there is a heterochromatinization of the paternal set of chromosomes in early male embryogeny. This persists during the development of the male germ line and consequently into meiosis. As a result of this, homologues do not pair at prophase-1 and the first division is equational for both the heterochromatic (H) and euchromatic (E) chromosomes. At second division a double metaphase plate forms with the H-chromosomes on one side and the E-chromosomes on the other. This partitioning appears to result from the initial development of a rigid half spindle, in association with the H-set, which is replaced gradually by a bipolar spindle (Nelson-Rees, 1963). The opposing orientation of the H- and E-chromosomes results in their segregation at second anaphase. As in most other sexual coccids the four products of male meiosis form a single, quadrinucleate, spermatid. Only the two E-nuclei form functional sperm; the H-nuclei degenerate. Consequently, only chromosomes of maternal origin are transmitted by males.

In the second step, exemplified by the Comstockiella system (Brown, 1963), there is a destruction of from one to all of the H-chromosomes just prior to first male prophase. The first meiotic division remains equational but the course of the second division depends on the number of H-chromosomes which are destroyed prior to the onset of meiosis. If all of them are destroyed there is no second division. Where one or more H-chromosomes persist they segregate from the E-set at second division as in the conventional lecanoid system. The number of H-chromosomes which are destroyed may, however, vary from cyst to cyst within a given individual, as well as from species to species. Moreover, in *Nicholiella*

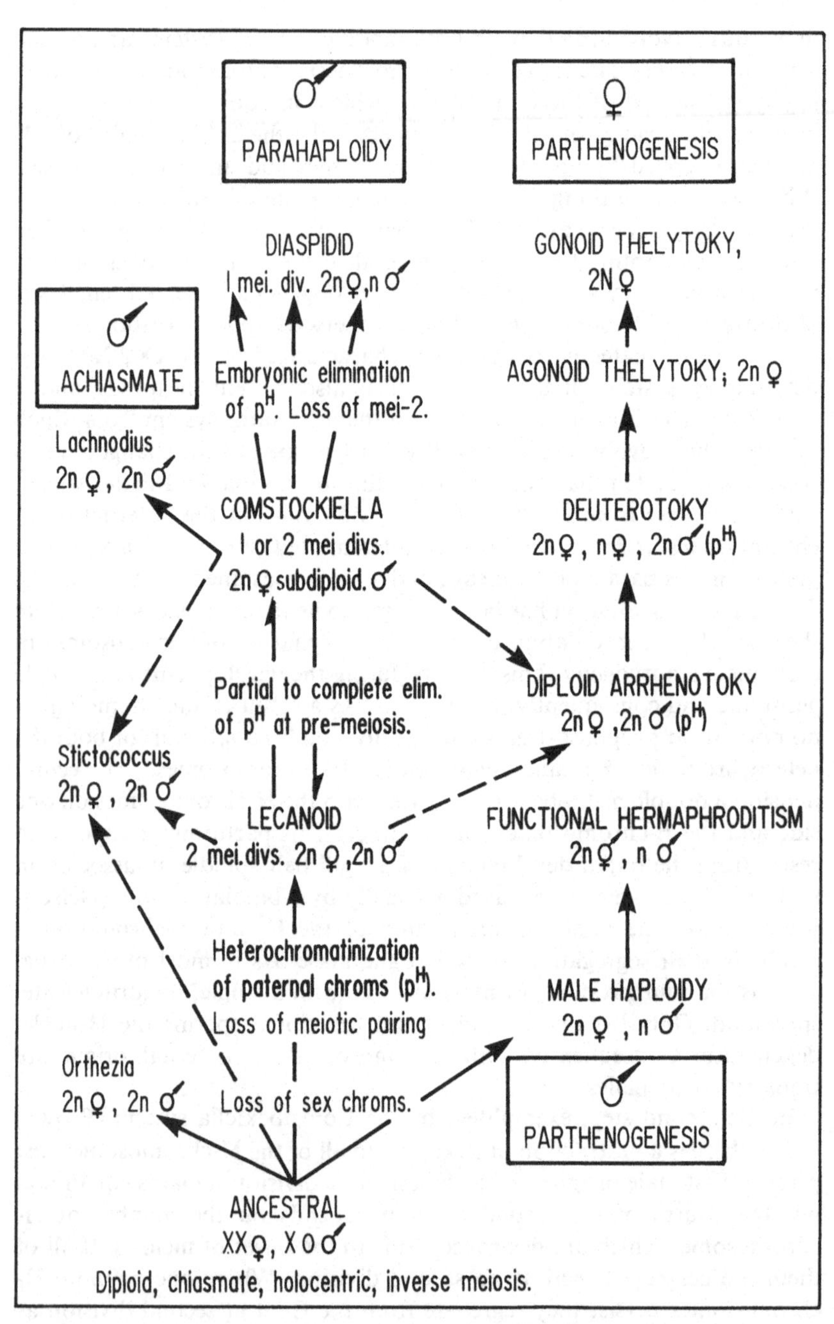

Fig. 8.3. Stages in the evolution of meiosis in coccid bugs (after Brown, 1977; Nur, 1980).

bumeliae, typical lecanoid and Comstockiella systems occur in different sections of the same testis or else in different testes of the same individual. Thus, there are two quite different systems, both based on heterochromatinization, in male coccids. In the one, all the heterochromatic chromosomes are retained throughout both meiotic divisions and degenerate only after they have separated from the euchromatic set at anaphase-2 (lecanoid type). In the other, from one to all of the heterochromatic chromosomes undergo intranuclear destruction (Comstockiella type). Here, those chromosomes that are not destroyed before meiosis are then eliminated after meiosis in typical lecanoid fashion.

In the third and final step, known as the *diaspidid system*, the paternal chromosome set is eliminated not at meiosis but by anaphase lagging of the paternal set at a cleavage division prior to that at which heterochromatinization takes place in equivalent lecanoid–Comstockiella systems. If elimination fails at this time, the paternal set becomes heterochromatinized and, if not then eliminated at the next cleavage division, the entire nucleus degenerates. Consequently the male is effectively haploid and male meiosis involves a single equational haploid division and the production of a binucleate, rather than a quadrinculeate spermatid.

The lecanoid–diaspidid evolutionary sequence involves modifications in male meiosis only (Fig. 8.4); the female retains an orthodox behaviour. Consequently, this constitutes yet another example of two-track heredity. The existence of an inverse meiosis may well have been pre-adaptive to the adoption of the lecanoid system since the suppression of homologue pairing, which follows the heterochromatinization of the paternal chromosome set, would have been easily accommodated in a system of inverse meiosis. The male meiotic changes involved in this lineage thus again appear to be essentially regulatory and to stem from the initial heterochromatinization of the paternal set in early development. The coexistence of both Comstockiella and lecanoid systems in the same individual, or even the same testis, testifies to a capacity to revert to a lecanoid state and is again indicative of a regulatory type change. So too, as pointed out by Brown (1977), is the change from the Comstockiella to the diaspidid system.

Two other categories of change from the basic chiasmate bisexual system have been documented in coccids. First, the adoption of an achiasmate male meiosis coupled with the loss of differentiated sex chromosomes. So far as is known this has occurred only in the genus *Orthezia* (Brown, 1958), though equivalent achiasmate male meioses have been independently derived from the lecanoid–Comstockiella lineage on two occasions (*Stictococcus* and *Lachnodius*). Second, the adoption of male haploidy and, via this route, of hermaphroditism. In the functional secondary hermaphroditism that occurs in the iceryine coccids, and which

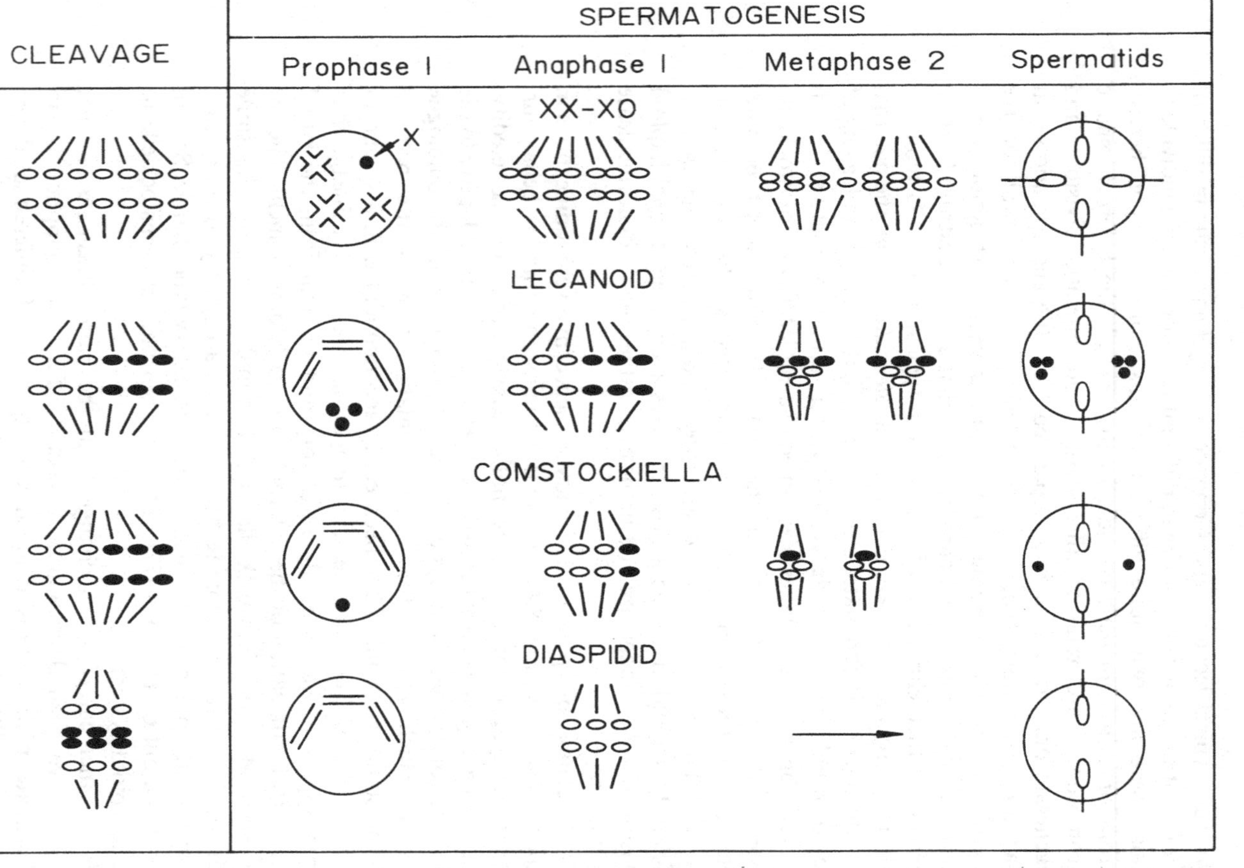

Fig. 8.4. Diagrammatic representation of chromosome behaviour at early cleavage and during spermatogenesis in the four holokinetic chromosome systems with inverted meiosis involved in the evolution of male paraphaploidy in coccid bugs. Euchromatic chromosomes are shown in outline while heterochromatic chromosomes are shown solid. The lagging heterochromatic chromosomes of the diaspidid system are eliminated during cleavage (after Nur, 1980 and with

has evolved independently on at least three occasions, the gonad is a mosaic ovotestis and self-fertilization takes place. The ovotestis develops in what corresponds in body form, habit and mating instinct to true females in related species. The core of the ovotestis becomes haploid following the degeneration of one haploid set of chromosomes during early gonadal development. Spermatogenesis is then accomplished, as in true haploid males, by a single equational division with both products developing into sperm. The cortex of the ovotestis is ovarian and diploid, as too are all the somatic cells. These hermaphrodites are thus essentially arrhenotokous rather than thelytokous. While the bulk of the population arise from fertilized eggs and are diploid and hermaphrodite, rare haploid males may originate from reduced eggs of the hermaphrodite that escape fertilization and develop parthenogenetically. Such males, when successful in fertilization, introduce recombinational variability into an otherwise closed system of self-fertilization.

These systems apart, at least four major categories of parthenogenesis appear to have developed from the parahaploid lineage. One of these involves a very distinctive type of facultative parthenogenesis which is a variant form of haplo-diploidy. Here, although males begin to develop as haploids they become diploid through the fusion of the first two haploid cleavage nuclei. This diploid phase is, however, short-lived and is soon followed by the heterochromatinization of one set of chromosomes in a manner reminiscent of the lecanoid system. This is known as *diploid arrhenotoky* and has been described in four species of the Coccidae, namely *Lecanium cerasifex*, *L. putmani*, *Pulvinaria innumeralis* and *P. floccifera*. In only one of these, *L. putmani*, has meiosis been studied in detail and here it is basically of the lecanoid type. The same family also exhibits a form of parthenogenesis which is closely related to diploid arrhenotoky but differs in that some of the unfertilized eggs develop into females. This has been termed *deuterotoky* and has been recorded in two of the four species which show diploid arrhenotoky, namely *L. putmani* and *P. innumeralis*. In the two other deuterotokous species, *P. mesembryanthemi* and *P. hydrangae*, the haploid males are nonfunctional. For this reason Nur (1980) suggests that these species are evolving to thelytoky but that suppression of heterochromatinization has not yet stabilized. Thus, embryos mosaic for tissues both with and without an H-set have been observed in *P. mesembryanthemi* (Nur, 1980). Whether fertilized eggs develop into females in deuterotokous forms is not known but would appear likely. Deuterotoky is also known in one species of acarine, *Histiophorus numerousus* (Oliver, 1971).

The final form of parthenogenesis present in coccids is thelytoky. Here two categories can be distinguished. In the one, heterochromatinization of one of the two haploid sets does not take place in some of the eggs and a haploid number of bivalents is then present at first metaphase. In this

Table 8.1. *Parthenogenetic systems in coccid bugs (Hemiptera, Homoptera)*

Type	Fate of eggs: Fertilized	Fate of eggs: Unfertilized	Occurrence	Mode of restoration of 2n number in unfertilized eggs
Haploid arrhenotoky	♀	♂	*Icerya littoralis, I. purchasi*	
			Steattococcus tuberculatus,	
Diploid arrhenotoky	♀	♂	*Lecanium cerasifex, Pulvinaria floccifera*	
Deuterotoky	♀(?)	♂ and ♀	*Lecanium putmani, Pulvinaria innumerabilis*	Fusion of cleavage nuclei
			P. mesembraynthemi[a] and *P. hydrangae*[a]	
Agonoid thelytoky	—	♀	*Pulvinaria hydrangae*	
		♀	*Coccus hesperidium, Saissetia coffeae*	Fusion of pronucleus with Polar nucleus II
Gonoid thelytoky	—	♀	*Coccus hesperidium, Saissetia coffeae*	
		♀	*Coccus hesperidium, Aspidiotus hederae*	No reduction

After Nur, 1971, 1980 and with the permission of Blackwell Scientific Publishers Limited.
[a] The males in these two species are non-functional.

event the maternal chromosome number is restored either by fusion of the female pro-nucleus with the second polar nucleus or, more commonly, by the fusion of the first two cleavage nuclei. Nur (1980) refers to this as *agonoid thelytoky* and assumes that it has evolved from diploid arrhenotoky via deuterotoky. In the second category, referred to as *gonoid thelytoky*, two modes of meiosis are possible. In the one, there is only a single maturation division in which the chromosomes divide equationally. In the other, although the chromosomes do not pair, there are two maturation divisions and the reduced chromosome number is then restored to diploidy by the fusion of the female pro-nucleus and the second polar nucleus (Nur, 1979). This provides grounds for suggesting that these categories of gonoid thelytoky evolved from the comparable agonoid forms through the suppression of homologue pairing. The characteristics of these different modes of parthenogenesis found in coccids are summarized in Table 8.1.

Unlike female coccids, which feed throughout their life, male coccids do not feed after the third pre-pupal stage. They are, therefore, much more susceptible to death by dessication. They are also short-lived as adults compared to uninseminated females. Thelytoky and hermaphroditism present themselves as a means of dispensing with such fickle males. Additionally, both male haploidy and the lecanoid system offer the potential for a much more flexible sex ratio. In the case of male haploidy, a shortage of males in one generation leads inevitably to an increase in the frequency of uninseminated females in that generation. These, in turn, produce only males and so provide automatically for an increased frequency of males in the next generation. A parallel ability to adjust the sex ratio is also inherent in the lecanoid system and, presumably also, in the derivative Comstockiella and diaspidid systems. Here, the longer the time that elapses before an adult female is inseminated, the higher the proportion of males in her progeny (Nelson-Rees, 1960). This, again, is most probably achieved by an epigenetic change in the mechanism that regulates the imprinting of the egg and so determines whether that egg will develop as male or female. The evolution of diploid arrhenotoky, in which males develop parthenogenetically, may represent a further step in coping with local shortages of males.

While the rationale for many of the changes that have occurred in coccid meiosis thus seem reasonably clear, the precise mechanisms by which they were introduced are far less clear. Nur (1980) inclines to accepting natural selection as the principal causative factor but recognizes that some of the changes might have resulted from what Brown (1963, 1964, 1977), on theoretical grounds, referred to as *automatic frequency response*. Such a response, he predicted, would occur if the mutations involved in given changes were to increase in frequency simply because of the efficiency with which they were transmitted, with no influence from

Table 8.2. *Permanent, sexually-reproducing, uneven polyploid species with complementary gametic elimination*

Species	Ploidy	Genome structure	Gametes		Reference
			♀	♂	
Andropogon ternatus (Gramineae)	3x = 30	SST	S(10)	ST(20)	Norrmann & Quarin, 1987
Rosa canina (Rosaceae)	5x = 35	AABCD	ABCD(28)	A(7)	Klášterská & Natarajan, 1974a
Leucopogon juniperinum (Styphelae)	3x = 12	AAB	AB(8)	A(4)	Smith-White, 1948, 1955
Enchytraeus lacteus (Enchytraeidae)	9x + 8 = 170	9(A) + 8	4	9(A) + 4	Christensen & Jensen, 1964

either natural selection or from random fluctuations. On this basis, Brown argued that selection has been largely neutral in the evolution of both male haploidy and parahaploidy in coccids and developed models to support his argument (Brown, 1964; Hartl & Brown, 1970). While the lecanoid–diaspidid series would undoubtedly have had considerable evolutionary potential once established, there is no reason to believe that any obvious selective advantage would have accrued from the meiotic innovations involved at their inception. It was for this very reason that Brown postulated automatic frequency response and autoselection. Indeed, he was of the opinion that, initially, the Comstockiella system would have been somewhat crude and, consequently, could have been expected to have yielded defective sperm more often than the progenitor lecanoid system (Brown, 1963). Presumably, such autoselected changes would, sooner or later, have become exposed to natural selection so that their ultimate fate may well have been determined by a combination of autoselection and natural selection. Bell (1982) suggests that both automixis and apomixis could have been autoselected relative to amphimixis and it is not difficult in principle to accept that secondary hermaphroditism, like parthenogenesis, could, once established, have had a selective value in species where there are extreme hazards to mating.

The coccid system does, however, clarify two points. First, it would appear that in this group of organisms different meiotic systems do appear to perform the same function equally well. Second, many of the modifications to meiosis have involved a radical alteration, most commonly a simplification, of the genetic control systems operative in normal meiosis. In the extreme there has been a complete suppression of the genes involved in pairing and recombination coupled with a switch to a single equational division. The difficulty is to appreciate how at their inception such systems would have been able to compete with those already in existence unless they were pre-adapted to work efficiently from the outset.

The most difficult meiotic changes to accommodate from an evolutionary point of view are unquestionably those involving complementary gametic elimination. There are four well documented examples of this behaviour (Table 8.2) all of which allow numerically uneven polyploids to breed true via a conventional sexual cycle. Since the details vary in each species they merit brief individual comment.

(1) *Andropogon ternatus*. This triploid grass from South America has the constitution SST with $2n = 3\times = 30$. 10 (S) bivalents and 10 (T) univalents are present at first meiosis in both PMCs and EMCs. In the PMCs the univalents lag at first anaphase and then move late to one pole where they form a separate nucleus. The products of the first division are thus, respectively, uninucleate (10S) and binucleate (10S, 10T). After the second meiotic division, two types of microspore are

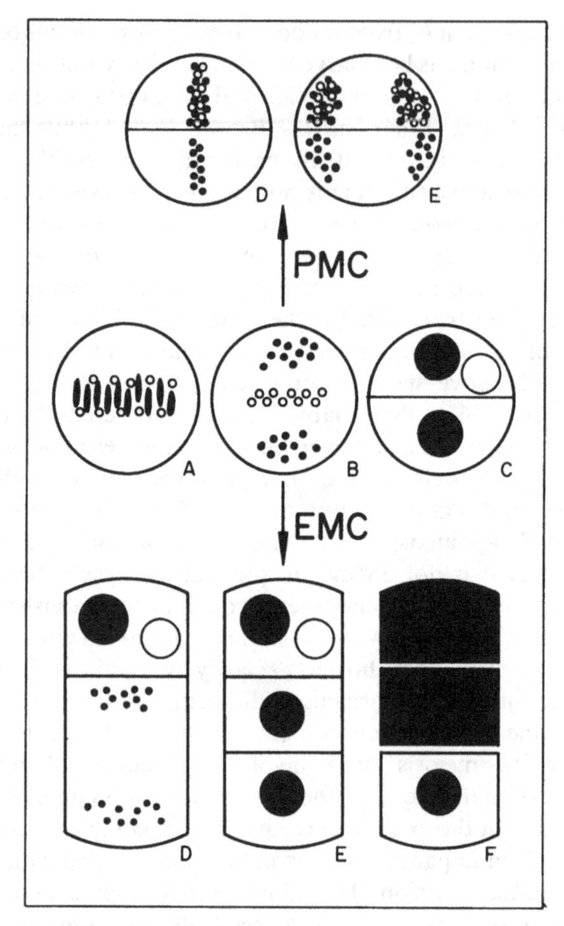

Fig. 8.5. Complementary gametic elimination in the triploid grass (2n = 3x = 30) *Andropogon ternatus* (after Norrmann & Quarin, 1987).

formed with either 10 or 20 chromosomes. Only the latter form functional microspores. Meiosis proceeds in an identical fashion in the EMCs. Here, however, the 10 univalents are always included in the micropylar cell, again as a separate nucleus. The binucleate micropylar cell then degenerates while the chalazal cell divides equationally. Only the terminal chalazal product becomes a functional megaspore and this has 10(S) chromosomes (Fig. 8.5).

(2) *Rosa canina*. This is a pentaploid (2n = 5x = 35) in which seven bivalents (AA) and 21 univalents (BCD) are present in both PMCs and EMCs. In the former case the univalents lag and divide at second division where their delay in moving on the spindle is even greater.

Table 8.3. *Chromosome distribution in 586 PMCs and 7 EMCs of Leucopogon juniperinum*

	Univalent distribution at metaphase-1				
Cell type	0–4	1–3	2–2	Incomplete	χ^2_{n-2}
PMC					
Observed	242	136	56	152	750.02
Expected	54.25	217	162.75	–	$P \ll 0.001$
EMC	Segregation micropylar/chalazal				
	0/4	1/3	2/2	2/1	4/0
Observed	1	—	1	3	2

After Smith-White, 1948, 1955.

Consequently, many fail to be incorporated into the tetrad nuclei. Pollen grains with varying chromosome numbers form but over 10% of these have a haploid (7A) content. These, and only these, form functional microspores. At first anaphase in the EMC all 21 univalents move undivided to the spindle pole located nearer to the micropyle. They then divide equationally, together with the seven half bivalents, at second anaphase. As a result, the linear tetrad consists of two similar nuclei with 28(ABCD) chromosomes situated at the micropylar end and two with only 7(A) chromosomes at the chalazal end. On selfing, therefore, these roses breed true. Thus, two kinds of chromosomes are present in the somatic set of these hybrids. Those which form bivalents are distributed in a conventional manner. Those which form univalents are distributed solely through the mother plant and are eliminated in PMCs.

(3) *Leucopogon juniperinum*. This is a triploid of the type AAB (2n = 3x = 12) and four A bivalents together with four B univalents regularly form in both PMCs and EMCs. In both, too, the behaviour of the univalents is random. Pollen development in *Leucopogon* is of an unusual type which characterizes all other genera of the family (Styphelae) to which it belongs. Three of the four microspore nuclei cluster at one end of the cell where they form small nuclei which subsequently degenerate. Only the one remaining, large, microspore nucleus continues development. The most common arrangement at first metaphase in the PMC is with all four univalents lying at one of the poles (Table 8.3) and, in this event, all four are excluded from the one nucleus (A) that continues to develop after meiosis, the other three

degenerating. In over 25% of the EMCs there is also a full polarization of the four univalents to the micropylar end of the linear tetrad whereas on a random basis one expects only 5%. Since, here, the functional megaspore develops from the micropylar end only the A+B product is functional. Thus, in *Leucopogon*, chromosome behaviour on the female meiotic spindle is again polarized though not to the same extent as in *Rosa*. However, there is an additional and complementary polarized disjunction in the PMCs too. Consequently a substantial number of zygotes carry a balanced triploid complement.

In *Leucopogon*, as in *Rosa*, the univalents are carried in an essentially apomictic manner since they are transferred only through the maternal line. In the case of *Leucopogon* two critical conditions pre-adapt this system for success. In both the PMC and the EMC there is an intracellular gradient controlling univalent behaviour. In the PMC the unusual type of monad pollen development, coupled with cytoplasmic polarity, leads to the elimination of the univalents. Within the EMC, a tissue gradient governs the development of the micropylar megaspore whereas in most angiosperms this gradient favours the chalazal megaspore.

In all three cases, therefore, the polarity of the EMC plays a fundamental role in determining the composition of the functional megaspore by regulating the behaviour of the univalent chromosomes that form. Notice, however, that this polarity is opposite in effect in *Rosa* and in *Andropogon*, while in *Leucopogon* an opposite polarity occurs in PMCs and EMCs of the same species. These polarity systems all reflect cytoplasmic gradients operative within the meiocytes which are pre-adaptive to the system.

(4) An equivalent mechanism is involved in the oligochaete worm *Enchytraeus lacteus* which includes five different cytotypes. Four of these, representing di-($2n = 2x = 36$), tetra-($2n = 4x = 72$), hexa-($2n = 6x = 108$) and octoploid ($2n = 8x = 144$) forms, have a normal meiosis. The fifth, however, has $2n = 170$ and includes 162 small chromosomes ($\equiv$ nonoploid, $2n = 9x$) plus eight large chromosomes of unknown origin. At female meiosis only the eight large chromosomes form bivalents and these separate into half bivalents in a conventional manner. The 162 univalents remain at the spindle equator at first anaphase. One of the groups of four half bivalents becomes the first polar body. The other rejoins the univalents at the spindle equator and divides with them in a second meiotic division to give a second polar body and a female pro-nucleus. The first meiotic division in the male is a repeat of that in the female but this time only the small nuclei produced in association with the four large bivalents enter a second meiotic division. The large central nucleus containing

the 162 small chromosomes does not divide. The functional sperm thus carry 4L chromosomes and complement the egg with its 4L + 162S chromosomes (Christensen & Jensen, 1964).

How complementary changes of these several kinds could have developed in the meiotic programmes of the two sexes within each species is a problem that has never been seriously addressed. Whether it involves an imprinting mechanism of the kind which operates in coccids is unknown but would appear probable. Such behaviour highlights again the importance of pre-adaptation in facilitating a change in the meiotic mechanism, this time in respect of the differential behaviour of chromosomes with a distinct origin on a common meiotic spindle.

Postlogue

There's no difficulty in recognizing the obvious. One should be slower to believe it.
P. D. James

Every aspect of meiosis that we have considered is evidently beset with both controversial and unresolved issues. Far from having solved long-standing problems, the new information now available to us has simply served to put them into a different context. It is for this reason that we still lack a complete theory of meiosis.

Over and above individual issues there are two rather more fundamental features of meiosis which are also in need of a solution. The first of these is a developmental matter. The meiocyte is simply a specialized cell with a distinctive pattern of cytodifferentiation though, unlike most other categories of cytodifferentiation, that of the meiocyte principally involves changes in the chromosomes themselves rather than chromosome-induced changes in the cell cytoplasm (Stern & Hotta, 1984). In *S. pombe*, cell-type determination is controlled by a class of master regulatory, *mat*, genes, whereas cell differentiation is regulated by protein kinases (see Section 4A). These two systems are associated through the *mei-3*$^{+}$ gene which is transcriptionally regulated by the *mat* genes and whose product, in turn, inhibits the *ran-1*$^{+}$ protein kinase (McLeod & Beach, 1988). Precisely how the differentiated state of the meiocyte is initiated in other eukaryotes remains unsolved, though the molecular events associated with it are now tolerably well known. Some of them involve the activity of genes, enzymes and proteins available to, and used by, mitotic cells. This includes those genes responsible for such ubiquitous cell products as actins, tubulins and kinetochores, as well as some of the enzymes involved in the replication and recombination of DNA. Meiocyte differentiation also involves novel gene products. These proteins, while not represented in non-germ line tissues, have somatic analogues with distinctive biochemical properties. For example, the optimal temperature at which m-rec protein functions differs from that of s-rec protein (Stern, 1986). Some of these novelties are likely to represent divergent consequences of gene duplication. This may well be true of the beta-3 tubulin of *Aspergillus nidulans* (Weatherbee *et al.*, 1985) and the variant forms of meiotic histones. Still others, like the proteins of the synaptonemal complex and the recombination nodule, as well as certain of the recombination enzymes, are meiosis-specific.

This implies that some of the DNA sequences present in eukaryotic genomes function specifically in relation to particular meiotic needs (Stern & Hotta, 1984, 1985). One such sequence category provides transcriptive sites for the production of meiosis-specific proteins. Others regulate the structural organization of meiotic chromosomes during first prophase either by providing sites for homologue synapsis (Zyg-DNA) or else for recombination (Psn-DNA). As we saw earlier (Chapter 4D.2.2), complete pairing of homologues is necessary for the stable production of both Psn-RNA and of R-protein prior to nicking and pachytene DNA repair synthesis. Normal lily meiocytes have a much higher level of Psn-DNA sites, and of nick repair activity, than do microsporocytes of the diploid achiasmatic hybrid Black Beauty. While limited incubation of such microsporocytes with meiotic endonuclease or DNase II results in very low levels of DNA nicking, the accessibility of nuclease to Psn-DNA regions is significantly enhanced by the addition of relatively high concentrations of Psn-RNA (Stern & Hotta, 1983). Taken at face value this suggests that pachytene DNA metabolism facilitates the interaction of Psn-RNA, and presumably Psn-protein, with Psn-DNA and that this interaction is necessary before endonuclease nicking can occur.

The occurrence of variant, or even novel, proteins, specialized for particular functions, is a commonplace event in cell differentiation. To this extent, the meiocyte is no different from other differentiated cell types. The one advantage that the meiocyte offers, by comparison with most forms of somatic cell differentiation, is that the genetic dissection of the sequence of events involved has now progressed to the point in *Drosophila* and yeast where it is possible to provide a temporal ordering of the genes that control that sequence. What is now lacking is a precise characterization of the products of these genes and the way in which their products interact with one another and with specific sites on the meiotic chromosomes. It is precisely for this reason that we are, as yet, unable to relate the biochemical information now available on meiotic events to the gene systems which underlie those events. While we can anticipate that the application of the techniques of molecular biology will ultimately resolve this impasse it is not likely to be a simple or straightforward task. A number of the key enzymes and proteins are rapidly synthesized, short-lived and act only at limited chromosomal sites.

The second problem is of a very different kind. It relates to the fact that meiosis must have originated as a component part of the process of sexual reproduction. Relative to this mode of origin, it is the principal event in the life cycle of eukaryotes which includes a form of genetic recombination capable of influencing the subsequent development of the products of sexual heredity. A detailed resolution of the biochemistry of recombination cannot be expected to resolve the role of recombination in sexual reproduction. Neither will a description of the salient biochemical features of the variant forms of meiosis provide an explanation for how, and why, these variants evolved. This is not the place to enter into a detailed discussion relating to the question of 'what use is sex' (Maynard-Smith, 1978) or 'what role does recombination play in evolution' (Williams, 1975) though, clearly, these matters are pertinent to the origin and evolution of meiosis.

In classical terms 'recombination' was defined as the production of new allelic combinations – whether by crossing-over, and the subsequent segregation of the products of this process, or by the random assortment of chromosomes during meiosis. In molecular terms, as Hastings (1988) has emphasized, this definition can be extended to cover not only the assortment of different DNA molecules but, additionally, any form of interaction between homologous DNA molecules that generates a sequence that has been derived from more than one source. This includes interactions between sister chromatids (SCEs) or even between identical sequences, neither of which lead to actual changes in genotype. Recombination leading to chiasma formation certainly plays a critical role in the meiotic mechanism of many organisms since it is required for the regularized disjunction of homologues. However, it plays an important general role in the life of all cells since it provides one of the major mechanisms involved in DNA repair. Recombinationless mutants are thus extremely sensitive to DNA damaging agents (Hastings, 1988).

Recombination, in all its various forms, seems to depend on the breakage and reunion of DNA molecules and at least some of the enzymes involved are similar to those used to repair damaged DNA. This has led to the suggestion that genetic recombination might have evolved as a by-product of selection for DNA repair systems. Certainly it seems highly unlikely that the enzymes required for recombination could have evolved because of any selective advantage conferred by sexual reproduction since this advantage would not have existed until the mechanism itself was operative. The existence of prokaryotic systems which provide for limited recombination, coupled with a capacity for conjugation, has led some to conclude that sex preceded meiosis. It is, however, inconceivable that a sexual cycle of the kind which characterizes eukaryotes could have come into being without the prior presence of linear chromosomes and the capacity to organize a division spindle, neither of which are known in a presumptive form in prokaryotes.

Darlington (1939, 1958) argued that it was impossible to imagine the origin of meiosis and sexual reproduction to have resulted from the gradual accumulation of changes, each of which had a value as an adaptation. He was also firmly of the opinion that the capacity for sexual reproduction would not have conferred any advantage on the generation in which it first appeared and, equally, that no improvement in meiosis could have benefited the individual in which it first arose. How efficient any alteration in the meiotic mechanism would have to be at its inception to persist, depends on how it would have affected the viability or the fertility of the individual in which it first arose. If fertility were to be only partially restored to an otherwise sterile hybrid by a modification of its meiotic behaviour then, provided that the hybrid in question was superior in viability to its parents, its relative reproductive inefficiency would surely have been of secondary importance. In this instance it is the hybridity of the individual that is crucial to its competitive efficiency but without an altered meiosis this superior viability would have had no opportunity for expression. What we lack in most other cases is a detailed understanding of precisely what effects alterations in the meiotic mechanism have, or have had, on these two fitness characters. Neither do we have any clear indication of the extent to which environmentally-induced epigenetic changes leading to altered gene regulation, chromosome imprinting mechanisms, or automatic frequency response systems, have played a role in the evolution of meiosis.

The truth is that, while we can now provide answers to a number of previously undefined or ill-defined problems of meiosis, answers are not solutions and what we still lack is a solution to these problems.

References

Get your facts right first, then you can distort them as much as you please
Mark Twain

Abbott, A. G. & Gerbi, S. A. (1981). Spermatogenesis in *Sciara coprophila* II. Precocious chromosome orientation in meiosis-2. *Chromosoma*, **83**, 19–27.

Abirached-Darmency, M., Zickler, D. & Cauderon, Y. (1983). Synaptonemal complex and recombination in rye (*Secale cereale*). *Chromosoma*, **88**, 299–306.

Akstein, E. (1962). The chromosomes of *Aedes aegypti*, and of some other species of mosquitoes. *Bulletin of the Research Council of Israel, Section B, Zoology*, **11**, 146–55.

Albertson, D. G. & Thomson, J. N. (1982). The kinetochores of *Caenorhabditis elegans*. *Chromosoma*, **86**, 409–28.

Albini, S. M. & Jones, G. H. (1984). Synaptonemal complex-associated centromeres and recombination nodules in plant meiocytes prepared by an improved surface-spreading technique. *Experimental Cell Research*, **155**, 588–92.

Allen, J. W. (1979). BrdU-dye characterisation of late replication and meiotic recombination in Armenian hamster germ cells. *Chromosoma*, **74**, 189–207.

Ammermann, D. (1967). Die Cytologie der Parthenogenese bei dem Tardiagraden *Hypsibius dujardini*. *Chromosoma*, **23**, 203–13.

Anderson, E. G. (1929). Studies on a case of high non-disjunction in *Drosophila melanogaster*. *Zeitschrift für induktive Abstammungs- und Vererbungslehre*, **51**, 397–44.

Anderson, L. K., Stack, S. M. Sherman, J. D. (1988). Spreading synaptonemal complexes from *Zea mays*, 1. No synaptic adjustment of inversion loops during pachytene. *Chromosoma*, **96**, 295–305.

Arrighi, F., Hsu, T. C., Pathak, S. & Sawada, H. (1974). The sex chromosomes of the Chinese hamster: constitutive heterochromatin deficient in repetitive DNA sequences. *Cytogenetics and Cell Genetics*, **13**, 268–74.

Ashley, T. (1984a). A re-examination of the case for homology between the X and Y chromosomes of mouse and man. *Human Genetics*, **67**, 372–7.

Ashley, T. (1984b). Application of the spreading techniques to structural heterozygotes. *Chromosomes Today*, **8**, 80–9.

Ashley, T. (1985). Is crossover between the X and Y a regular feature of meiosis in mouse and man. *Genetica*, **66**, 161–7.

Ashley, T. (1987). Meiotic behaviour of sex chromosomes: what is normal? *Chromosomes Today*, **9**, 184–95.

Ashley, T. & Moses, M. J. (1980). End association and segregation of the achiasmatic X and Y chromosomes of the sand rat *Psammomys obesus*. *Chromosoma*, **78**, 203–10.

Asker, S. (1984). Apomixis and biosystematics. In *Plant Biosystematics*, ed. W. F. Grant, pp. 237–48. Toronto: Academic Press.

Astaurov, B. L. (1940). Artificial parthenogenesis in the silkworm (*Bombyx mori L.*). *Academy of Sciences USSR: Moscow*, 240 pp.

Ault, J. G. (1984). Unipolar orientation stability of the sex univalent in the grasshopper (*Melanoplus sanguinipes*). *Chromosoma*, **89**, 201–5.

Ault, J. G. (1986). Stable versus unstable orientations of sex chromosomes in two grasshopper species. *Chromosoma*, **93**, 298–304.

Ault, J. G., Lin, H.-P. P. & Church, K. (1982). Meiosis in *Drosophila melanogaster* IV. The conjunctive mechanism of the XY bivalent. *Chromosoma*, **86**, 309–17.

Avivi, L. & Feldman, M. (1980). Arrangement of chromosomes in the interphase nucleus of plants. *Human Genetics*, **55**, 281–95.

Ayonoadu, U. & Rees, H. (1968). The influence of B chromosomes on chiasma frequencies in Black Mexican sweet corn. *Genetica*, **39**, 75–81.

Bajer, A. (1968). Behaviour and fine structure of spindle fibers during mitosis in endosperm. *Chromosoma*, **25**, 249–81.

Baker, B. S., Carpenter, A. T. C., Esposito, M., Esposito, R. E. & Sandler, L. (1976). The genetic control of meiosis. *Annual Review of Genetics*, **10**, 53–134.

Baker, B. S. & Hall, J. C. (1976). Meiotic mutants: genetic control of meiotic recombination and chromosome segregation. In *The Genetics and Biology of Drosophila*, vol. 1c, ed. M. Ashburner & E. Novitski, pp. 351–433. London: Academic Press.

Barlow, B. A. & Martin, N. J. (1984). Chromosome evolution and adaptation in mistletoes. In *Plant Biosystematics*, ed. W. F. Grant, pp. 117–40. Toronto: Academic Press.

Barlow, B. A. & Wiens, D. (1976). Translocation heterozygosity and sex ratio in *Viscum fischeri*. *Heredity*, **37**, 27–40.

Barlow, P. W. & Vosa, C. G. (1970). The effect of supernumerary chromosomes on meiosis in *Puschkinia libanotica* (Liliaceae). *Chromosoma*, **30**, 344–55.

Barry, E. G. (1969). The diffuse diplotene stage of meiotic prophase in *Neurospora*. *Chromosoma*, **26**, 119–29.

Bastmeyer, M., Stetten, W. & Fuge, H. (1986). Immunostaining of spindle components in tipulid spermatocytes using a serum against pericentriolar material. *European Journal of Cell Biology*, **42**, 305–10.

Beadle, G. (1932). A possible influence of the spindle fiber on crossing over in *Drosophila melanogaster*. *Proceedings of the National Academy of Sciences*, **18**, 160–5.

Becker, H. J. (1976). Mitotic recombination. In *The Genetics and Biology of Drosophila*, vol. 1c, ed. M. Ashburner & E. Novitski, pp. 1019–87. London: Academic Press.

Bedo, D. G. (1987). Specific recognition and differential affinity of meiotic X-Y pairing sites in *Lucilia cuprina* males (Diptera: Caliphoridae). *Chromosoma*, **95**, 126-35.

Beermann, S. (1977). The diminution of heterochromatic chromosomal segments in *Cyclops* (Crustacea, Copepoda). *Chromosoma*, **60**, 297–344.

Bell, G. (1982). *The Masterpiece of Nature: The Evolution and Genetics of Sexuality*. London: Croom Helm.

Benavente, R. (1982). Holocentric chromosomes of arachnids: presence of kinetochore plates during meiotic divisions. *Genetica*, **59**, 23–7.

Benavente, R. & Wettstein, R. (1977). An ultrastructural cytogenetic study on the evolution of sex chromosomes during the spermatogenesis of *Lycosa malitiosa* (Arachnida). *Chromosoma*, **64**, 255–77.

Benavente, R. & Wettstein, R. (1980). Ultrastructural characterisation of the sex chromosomes during spermatogenesis of spiders having holocentric chromosomes and a long diffuse stage. *Chromosoma*, **77**, 69–81.

Benirschke, K. (1968). The chromosome complement and meiosis of the North American porcupine. *Journal of Heredity*, **59**, 71–6.

Bennett, J. H., Hayman, D. L. & Hope, R. M. (1986). Novel sex differences in linkage values and meiotic chromosome behaviour in a marsupial. *Nature*, **323**, 59–60.

Bennett, M. D. (1973). The duration of meiosis. In *The Cell Cycle in Development and Differentiation*, ed. M. Balls & F. S. Billett, pp. 111–31.

Bennett, M. D. (1976). The cell in sporogenesis and spore development. In *Cell Division in Higher Plants*, ed. M. M. Yeoman, pp. 161–98. London: Academic Press.

Bennett, M. D. (1982). Nucleotypic basis of the spatial ordering of chromosomes in eukaryotes and the implications of the order for genome evolution and phenotypic variation. In *Genome Evolution*, ed. G. A. Dover & R. B. Flavell, pp. 239–61. London: Academic Press.

Bennett, M. D. (1984). Premeiotic events and meiotic chromosome pairing. In *Controlling Events in Meiosis*, ed. C. W. Evans & H. G. Dickinson, *Symposia for the Society of Experimental Biology*, **38**, 87–121.

Bennett, M. D., Chapman, V. & Riley, R. (1971). The duration of meiosis in pollen mother cells of wheat, rye and *Triticale*. *Proceedings of the Royal Society of London*, **B178**, 259–75.

Bennett, M. D., Finch, R. A., Smith, J. B. & Rao, M. K. (1973a). The time and duration of female meiosis in wheat, rye and barley. *Proceedings of the Royal Society of London*, **B183**, 301–19.

Bennett, M. D., Rao, M. K., Smith, J. B. & Bayliss, M. W. (1973b). Cell development in the anther, the ovule and the young seed of *Triticum aestivum* L. cv. Chinese Spring. *Philosophical Transactions of the Royal Society of London*, **B266**, 39–81.

Bennett, M. D. & Rees, H. (1970). Induced variation in chiasma frequency in rye in response to phosphate treatments. *Genetical Research*, **16**, 325–31.

Bennett, M. D., Smith, J. B., Simpson, S. & Webb, B. (1979). Intranuclear fibrillar material in cereal pollen cells. *Chromosoma*, **71**, 289–332.

Bennett, M. D. & Stern, H. (1975). The time and duration of preleptotene chromosome condensation stage in *Lilium* hybrid c.v. Black Beauty. *Proceedings of the Royal Society of London*, **B188**, 477–93.

Bennett, M. D., Toledo, L. A. & Stern, H. (1979). Effect of colchicine on meiosis in *Lilium speciosum* c.v. 'Rosemede'. *Chromosoma*, **72**, 175–89.

Bergerard, J. (1962). Parthenogenesis in the Phasmidae. *Endeavour*, **21**, 137–43.

Bernelot-Moens, C. & Moens, P. B. (1986). Recombination nodules and chiasma localization in two Orthoptera. *Chromosoma*, **93**, 220–6.

Bier, K., Kunz, W. & Ribbert, D. (1969). Insect oogenesis with and without lampbrush chromosomes. *Chromosomes Today*, **2**, 107–15.

Birky, C. W. & Gilbert, J. J. (1971). Parthenogenesis in rotifers: the control of sexual and asexual reproduction. *American Zoologist*, **11**, 245–66.

Bishop, C., Weissenbach, J., Casanova, M., Bernheim, A. & Fellows, M. (1985). DNA sequences and analysis of the human Y chromosome. In *The Y Chromosome, Part A: Basic Characteristics of the Y Chromosome*, ed. A. A. Sandberg, pp. 141–76. New York: Alan R. Liss.

Blackman, R. L. (1976). Cytogenetics of two species of *Euceraphis* (Homoptera, Aphididae). *Chromosoma*, **56**, 393–408.

Blackman, R. L. (1980). Chromosomes and parthenogenesis in aphids. In *Insect Cytogenetics*, ed. R. L. Blackman, G. M. Hewitt & M. Ashburner, pp. 133–48. Oxford: Blackwell Scientific.

Blackman, R. L. (1985). Spermatogenesis in the aphid *Amphorophoa tuberculata* (Homoptera, Aphididae). *Chromosoma*, **92**, 357–62.

Blackman, R. L. & Hales, D. F. (1986). Behaviour of the X chromosomes during growth and maturation of parthenogenetic eggs of *Amphorophora tuberculata* (Homoptera, Aphididae), in relation to sex determination. *Chromosoma*, **94**, 59–64.

Bogart, J. P. (1980). Evolutionary implications of polyploidy in amphibians and reptiles. In *Polyploidy: Biological Relevance*, ed. W. H. Lewis, pp. 341–78. New York: Plenum Press.

Bogart, J. P. & Licht, L. E. (1986). Reproduction and the origin of polyploids in hybrid salamanders of the genus *Ambystoma*. *Canadian Journal of Genetics and Cytology*, **28**, 605–17.

Bogdanov, Yu. F. (1977). Formation of cytoplasmic synaptonemal-like complexes at leptotene and normal synaptonemal complexes at zygotene in *Ascaris suum* male meiosis. *Chromosoma*, **61**, 1–21.

Bogdanov, Yu. F., Kolomiets, O. L., Lyapunova, E. A., Yanina, I. Yu. & Mazurova, T. R. (1986). Synaptonemal complexes and chromosome chains in the rodent *Ellobius talpinus* heterozygous for ten Robertsonian translocations. *Chromosoma*, **94**, 94–102.

Bogdanov, Yu. F., Liapunova, N. A., Sherudilo, A. I. & Antropova, E. N. (1968). Uncoupling of DNA and histone synthesis prior to prophase-1 of meiosis in the cricket *Gryllus (Acheta) domesticus* L. *Experimental Cell Research*, **52** 59–70.

Bogdanov, Yu. F., Strokov, A. A. & Reznickova, S. A. (1973). Histone synthesis during meiotic prophase in *Lilium*. *Chromosoma*, **43**, 237–45.

Bojko, M. (1983). Human meiosis VIII. Chromosome pairing and formation of the synaptonemal complex in oocytes. *Carlsberg Research Communications*, **48**, 457–83.

Bojko, M. (1985). Human meiosis IX. Crossing over and chiasma formation in oocytes. *Carlsberg Research Communications*, **50**, 43–72.

Bokhari, F. S. & Godward, M. B. E. (1980). The ultrastructure of the diffuse kinetochore in *Luzula nivea*. *Chromosoma*, **79**, 125–36.

Borisy, G. G. & Gould, R. R. (1977). Microtubule-organizing centers of the mitotic spindle. In *Mitosis: Facts and Questions*, ed. M. Little, N. Paweletz, C. Petzelt, H. Ponstingl, D. Schroeter & H.-P. Zimmermann, pp. 78–87. Berlin: Springer-Verlag.

Borts, R. H., Lichten, M., Hearn, M., Davidow, L. S. & Haber, J. E. (1984). Physical monitoring of meiotic recombination in *Saccharomyces cerevisiae*. *Cold Spring Harbor Symposia in Quantitative Biology*, **49**, 67–76.

Boswell, R. E. & Mahowald, A. P. (1985). Tudor, a gene required for assembly of the germ plasm in *Drosophila melanogaster*. *Cell*, **43**, 97–104.

Bottke, W. (1973). Lampenbürstenchromosomen und Amphinukleolen in Oocytenkernen der Schnecke *Bithynia tentaculata* L. *Chromosoma*, **42**, 175–90.

Bouchard, R. A. & Stern, H. (1980). DNA synthesised at pachytene in *Lilium*: a non-divergent subclass of moderately repetitive sequences. *Chromosoma*, **81**, 349–63.

Bownes, M. (1983). Interactions between germ cells and somatic cells during insect oogenesis. In *Current Problems in Germ Cell Diffentiation*, ed. A. McLaren & C. C. Wylie, pp. 41–69. Cambridge: Cambridge University Press.

Braselton, J. P. (1971). The ultrastructure of the non-localized kinetochores of *Luzula* and *Cyperus*. *Chromosoma*, **36**, 89–99.

Braselton, J. P. (1981). The ultrastructure of meiotic kinetochores of *Luzula*. *Chromosoma*, **82**, 143–51.
Brenner, S. L. & Brinkley, B. R. (1982). Tubulin assembly sites and the organization of microtubule arrays in mammalian cells. *Cold Spring Harbor Symposia in Quantitative Biology*, **46**, 241–54.
Brinkley, B. R. (1985). Microtubule organizing centers. *Annual Review of Cell Biology*, **1**, 145–72.
Brinkley, B. R., Brenner, S. L., Hall, J. M., Tousson, A., Balczon, R. D. & Valdiva, M. M. (1986). Arrangements of kinetochores in mouse cells during meiosis and spermatogenes. *Chromosoma*, **94**, 309–17.
Brooker, T. R. & Lehman, I. R. (1971). Branched DNA molecules: intermediates in T4 recombination. *Journal of Molecular Biology*, **60**, 131–49.
Brown, L. M. & Jones, R. N. (1976). B chromosome effects at meiosis in *Crepis capillaris*. *Cytologia*, **41**, 493–506.
Brown, S. W. (1958). The chromosomes of *Orthezia* species (Coccoidea-Homoptera). *Cytologia*, **23**, 429–34.
Brown, S. W. (1963). The *Comstockiella* system of chromosome behaviour in the armoured scale insects (Coccoidea: Diaspididae). *Chromosoma*, **14**, 360–406.
Brown, S. W. (1964). Automatic frequency response in the evolution of male haploidy and other coccid chromosome systems. *Genetics*, **49**, 797–817.
Brown, S. W. (1977). Adaptive status and genetic regulation in major evolutionary changes of coccid chromosome systems. *Nucleus*, **20**, 145–57.
Brown, S. W. & Cleveland, C. (1968). Meiosis in the male of *Puto albicans* (Coccoidea-Homoptera). *Chromosoma*, **24**, 210–32.
Buck, R. C. (1967). Mitosis and meiosis in *Rhodnium prolixus*: the fine structure of the spindle and diffuse kinetochore. *Journal of Ultrastructure Research*, **18**, 489–501.
Buckle, V., Mondello, C., Darling, S., Craig, I. W. & Goodfellow, P. W. (1985). Homologous expressed genes in the human sex chromosomes pairing regions. *Nature*, **317**, 739–41.
Burns, J. A. (1972). Preleptotene chromosome contraction in *Nicotiana* species. *Journal of Heredity*, **63**, 175–8.
Buss, M. E. & Henderson, S. A. (1971). The induction of orientational instability and bivalent interlocking at meiosis. *Chromosoma*, **35**, 153–83.
Byskov, A. G. & Saxén, L. (1976). Induction of meiosis in fetal mouse testis in vitro. *Developmental Biology*, **52**, 193–200.
Calarco-Gillam, P. D., Siebert, M. C., Hubble, R., Mitchison, T. & Kirschner, M. (1983). Centrosome development in early mouse embryos as defined by autoantibody against pericentriolar material. *Cell*, **35**, 621–9.
Callan, H. G. (1941). The sex-determining mechanism of the earwig, *Forficula auricularia*. *Journal of Genetics*, **41**, 349–74.
Callan, H. G. (1957). The lampbrush chromosomes of *Sepia officinalis* L., *Anilocra physodes* L. and *Scyllium catulus* Cuv. and their structural relationship to the lampbrush chromosomes of Amphibia. *Pubblicazioni della Stazione Zoologica di Napoli*, **29**, 329–46.
Callan, H. G. (1972). Replication of DNA in the chromosomes of eukaryotes. *Proceedings of the Royal Society of London*, **B181**, 19–41.
Callan, H. G. (1973). Replication of DNA in eukaryotic chromosomes. *British Medical Bulletin*, **29**, 192–5.
Callan, H. G. & Perry, P. E. (1977). Recombination in male and female meiocytes contrasted. *Philosophical Transactions of the Royal Society of London*, **B277**, 227–33.

Callan, H. G. & Taylor, J. H. (1968). A radioautographic study of the time course of male meiosis in the newt *Triturus vulgaris*. *Journal of Cell Science*, **3**, 615–26.

Callow, R. S. (1985). Comments on Bennett's model of somatic chromosome disposition. *Heredity*, **54**, 171–7.

Camenzind, R. & Fux, T. (1977). Dynamics and ultrastructure of monocentric chromosome movement. *Caryologia*, **30**, 127–50.

Camenzind, R. & Nicklas, R. B. (1968). The non-random chromosome segregation in spermatocytes of *Gryllotalpa hexadactyla*. A micromanipulation study. *Chromosoma*, **24**, 324–35.

Cameron, F. M. & Rees, H. (1967). The influence of B chromosomes on meiosis in *Lolium*. *Heredity*, **22**, 446–50.

Carlenius, C., Ryttman, H., Tegelström, H. & Jansson, H. (1981). R-, G- and C-banded chromosomes in the domestic fowl *(Gallus domesticus)*. *Hereditas*, **94**, 61–6.

Carothers, E. E. (1917). The segregation and recombination of homologous chromosomes as found in two genera of Acrididae (Orthoptera). *Journal of Morphology*, **28**, 445–521.

Carpenter, A. T. C. (1973). A meiotic mutant defective in distributive disjunction in *Drosophila melanogaster*. *Genetics*, **73**, 393–428.

Carpenter, A. T. C. (1975a). Electron microscopy of meiosis in *Drosophila melanogaster* females I. Structure, arrangement and temporal change of the synaptonemal complex in wild type. *Chromosoma*, **51**, 157–82.

Carpenter, A. T. C. (1975b). Electron microscopy of meiosis in *Drosophila melanogaster* females II. The recombination nodule – a recombination associated structure at pachytene. *Proceedings of the National Academy of Sciences*, **72**, 3186–9.

Carpenter, A. T. C. (1979a). Synaptonemal complex and recombination nodules in wild type *Drosophila melanogaster* females. *Genetics*, **92**, 511–41.

Carpenter, A. T. C. (1979b). Recombination nodules and synaptonemal complex in recombination-defective females of *Drosophila melanogaster*. *Chromosoma*, **75**, 259–92.

Carpenter, A. T. C. (1981). E.M. autoradiographic evidence that DNA synthesis occurs at recombination nodules during meiosis in *Drosophila melanogaster* females. *Chromosoma*, **83**, 59–80.

Carpenter, A. T. C. (1982). Mismatch repair, gene conversion and crossing over in two recombination defective mutants of *Drosophila melanogaster*. *Proceedings of the National Academy of Sciences*, **79**, 5961–5.

Carpenter, A. T. C. (1984a). Genic control of meiosis. *Chromosomes Today*, **8**, 70–9.

Carpenter, A. T. C. (1984b). Recombination nodules and the mechanism of crossing-over in *Drosophila*. In *Controlling Events of Meiosis*, ed. C. W. Evans & H. G. Dickinson. *Symposia of the Society of Experimental Biology*, **38**, 233–43.

Carpenter, A. T. C. (1984c). Meiotic roles of crossing over and of gene conversion. *Cold Spring Harbor Symposia in Quantitative Biology*, **49**, 23–9.

Carpenter, A. T. C. (1987). Gene conversion, recombination nodules, and the initiation of meiotic synapsis. *BioEssays*, **6**, 232–6.

Carpenter, A. T. C. & Baker, B. S. (1974). Genic control of meiosis and some observations on the synaptonemal complex in *Drosophila melanogaster*. In *Mechanisms of Recombination*, ed. R. F. Grell, pp. 365–75. New York: Plenum Press.

Carson, H. L. (1967). Selection for parthenogenesis in *Drosophila mercatorum*. *Genetics*, **55**, 157–71.

Catcheside, D. G. (1977). *The Genetics of Recombination*. London: Edward Arnold.
Cawood, A. H. & Jones, J. K. (1980). Chromosome behaviour during meiotic prophase in the Solanaceae. *Chromosoma*, **80**, 57–68.
Chaganti, R. S. K., Schonberg, S. & German, J. (1974). A manyfold increase in sister chromatid exchanges in Bloom's syndrome lymphocytes. *Proceedings of the National Academy of Sciences*, **71**, 4508–12.
Chaly, N., Bladon, T., Setterfield, G., Little, J. E., Kaplan, J. G. & Brown, D. L. (1984). Changes in distribution of nuclear matrix antigens during the mitotic cell cycle. *The Journal of Cell Biology*, **99**, 661–71.
Chandley, A. C. (1982). A pachytene analysis of two male-fertile paracentric inversions in chromosome 1 of the mouse and in the male-sterile double heterozygote. *Chromosoma*, **85**, 127–35.
Chandley, A. C. (1986). A model for effective pairing and recombination at meiosis based on early replicating sites. (R-bands) along chromosomes. *Human Genetics*, **72**, 50–7.
Chandley, A. C., Goetz, P., Hargreave, T. B., Joseph, A. M. & Speed, R. M. (1984). On the nature and extent of XY pairing at meiotic prophase in man. *Cytogenetics and Cell Genetics*, **38**, 241–7.
Chandra, H. S. (1962). Inverse meiosis in triploid females of the mealybug *Planococcus citri*. *Genetics*, **47**, 1441–54.
Chauan, K. P. S. & Abel, W. O. (1968). Evidence for the association of homologous chromosomes during premeiotic stages in *Impatiens and Salvia*. *Chromosoma*, **25**, 297–302.
Chinnappa, C. C. (1980). Bivalent forming race of *Mesocyclops edax* (Copepoda, Crustacea). *Canadian Journal of Genetics and Cytology*, **22**, 427–31.
Chinnappa, C. C. & Victor, R. (1979). Achiasmatic meiosis and complex heterozygosity in female cyclopoid copepods (Copepoda, Crustacea). *Chromosoma*, **71**, 227–36.
Christensen, B. (1960). A comparative cytological investigation of the reproductive cycle of an amphimictic diploid and a parthenogenetic triploid form of *Lumbricillus lineatus* (O.F.M.) (Oligochaeta, Enchytraeidae). *Chromosoma*, **11**, 365–79.
Christensen, B. (1961). Studies on cyto-taxonomy and reproduction in the Enchytraeidae. *Hereditas*, **47**, 387–450.
Christensen, B. (1980). Annelida. In *Animal Cytogenetics*, vol. 2, ed. B. John, pp. 1–81. Stuttgart: Gebrüder Borntraeger.
Christensen, B. & Jensen, J. (1964). Sub-amphimictic reproduction in a polyploid cytotype of *Enchytraeus lacteus* Nielsen and Christensen (Oligochaeta, Enchytraeidae). *Hereditas*, **52**, 106–18.
Church, K. (1972). Meiosis in the grasshopper II. The preleptotene spiral stage during oogenesis and spermatogenesis in *Melanoplus femur-rubrum*. *Canadian Journal of Genetics and Cytology*, **14**, 397–401.
Church, K. & Lin, H.-P. P. (1982). Meiosis in *Drosophila melanogaster* II. The prometaphase-1 kinetochore microtubule bundle and kinetochore orientation in males. *The Journal of Cell Biology*, **93**, 365–73.
Church, K. & Lin, H.-P. P. (1985). Kinetochore microtubules and chromosome movement during prometaphase in *Drosophila melanogaster* spermatocytes studied in life and with the electron microscope. *Chromosoma*, **92**, 273–82.
Church, K. & Moens, P. B. (1976). Centromere behaviour during interphase and meiotic prophase in *Allium fistulosum* from 3-D, E.M. reconstruction. *Chromosoma*, **56**, 249–63.
Church, K., Nicklas, R. B. & Lin, H.-P. P. (1986). Micromanipulated bivalents can

trigger minispindle formation in *Drosophila melanogaster* spermatocyte cytoplasm. *The Journal of Cell Biology*, **103**, 2765–73.
Church, K. & Wimber, D. E. (1969). Meiotic timing and segregation of H^3-thymidine labelled chromosomes. *Canadian Journal of Genetics and Cytology*, **11**, 573–81.
Clarke, L. & Carbon, J. (1985). The structure and function of yeast centromeres. *Annual Review of Genetics*, **19**, 29–56.
Cleland, R. E. (1972). *Oenothera – Cytogenetics and Evolution*. New York: Academic Press.
Cole, C. J. (1978). Parthenogenetic lizards. *Science*, **201**, 1154–5.
Comings, D. E. & Okada, T. A. (1970). Whole mount electron microscopy of meiotic chromosomes and the synaptonemal complex. *Chromosome*, **30**, 269–86.
Comings, D. E. & Okada, T. A. (1971a). Triple chromosome pairing in triploid chickens. *Nature*, **231**, 119–21.
Comings, D. E. & Okada, T. A. (1971b). Fine structure of the synaptonemal complex. *Experimental Cell Research*, **65**, 104–16.
Comings, D. E. & Okada, T. A. (1972a). Holocentric chromosomes in *Oncopeltus*: kinetochore plates are present in mitosis but absent in meiosis. *Chromosoma*, **37**, 177–92.
Comings, D. E. & Okada, T. A. (1972b). The architecture of meiotic cells and mechanisms of chromosome pairing. *Advances in Cell and Molecular Biology*, **2**, 310–84.
Cooke, H. J., Brown, W. R. A. & Rappold, G. A. (1985). Hypervariable telomeric sequences from the human sex chromosomes are pseudoautosomal. *Nature*, **317**, 687–92.
Cooper, K. W. (1944a). Analysis of meiotic pairing in *Olfersia* and consideration of the reciprocal chiasmata hypothesis of sex chromosome conjunction in male *Drosophila*. *Genetics*, **29**, 537–67.
Cooper, K. W. (1944b). Invalidation of the cytological evidence for reciprocal chiasmata in the sex chromosome bivalent of male *Drosophila*. *Proceedings of the National Academy of Sciences*, **30**, 50–4.
Cooper, K. W. (1949). The cytogenetics of meiosis in *Drosophila*: mitotic and meiotic autosomal chiasmata without crossing over in the male. *Journal of Morphology*, **84**, 81–121.
Cooper, K. W. (1964). Meiotic conjunctive elements not involving chiasmata. *Proceedings of the National Academy of Sciences*, **52**, 1248–55.
Counce, S. J. & Meyer, G. F. (1973). Differentiation of the synaptonemal complex and the kinetochore in *Locusta* spermatocytes studied by whole mount electron microscopy. *Chromosoma*, **44**, 231–53.
Craig-Cameron, T. & Jones, G. H. (1970). The analysis of exchanges in tritium-labelled meiotic chromosomes. *Heredity*, **25**, 223–32.
Craig-Cameron, T. A., Southern, D. I. & Pell, P. E. (1973). Chiasmata and the synaptonemal complex in male meiosis of *Glossina*. *Cytobios*, **8**, 199–207.
Crowley, J. G. & Rees, H. (1968). Fertility and selection in tetraploid *Lolium*. *Chromosoma*, **24**, 300–8.
Crozier, R. H. (1975). Hymenoptera. In *Animal Cytogenetics*, vol. 3(7), ed. B. John, pp. 1–95. Stuttgart: Gebrüder Borntraeger.
Cuellar, O. (1987). The evolution of parthenogenesis: a historical perspective. In *Meiosis*, ed. P. B. Moens, pp. 43–104. Orlando: Academic Press.
Czaban, B. B. & Forer, A. (1985). The kinetic polarities of spindle microtubules in *vivo*, in crane fly spermatocytes I. Kinetochore microtubules that re-form after treatment with colcemid. *Journal of Cell Science*, **79**, 1–37.
Darlington, C. D. (1931). Meiosis. *Biological Reviews*, **6**, 221–64.

Darlington, C. D. (1932). *Recent Advances in Cytology*. London: Churchill.
Darlington, C. D. (1934). Crossing over of sex chromosomes in *Drosophila*. *American Naturalist*, **68**, 374–7.
Darlington, C. DF. (1939, 1958). *The Evolution of Genetic Systems*. Cambridge: Cambridge University Press; Edinburgh: Oliver and Boyd.
Darlington, C. D. (1956). Natural populations and the breakdown of classical genetics. *Proceedings of the Royal Society*, **B145**, 350–64.
Darlington, C. D. (1973). The place of the chromosomes in the genetic system. *Chromosomes Today*, **4**, 1–13.
Darlington, C. D. (1978). A diagram of evolution. *Nature*, **276**, 447–52.
David, C. N. (1983). Stem cell proliferation and differentiation in *Hydra*. In *Stem Cells*, ed. C. S. Potten, pp. 12–27. Edinburgh: Churchill-Livingstone.
Davidow, L. S. & Byers, B. (1984). Enhanced gene conversion and postmeiotic segregation in pachytene-arrested *Saccharomyces cerevisiae*. *Genetics*, **106**, 165–83.
Davies, E. D. G. & Jones, G. H. (1974). Chiasma variation and control in pollen mother cells and embryo sac mother cells of rye. *Genetical Research*, **23**, 185–90.
Davies, E. D. G. & Southern, D. I. (1977). Female meiosis in two species of Tetse fly (genus *Glossina*). *Genetica*, **47**, 173–5.
Davis, B. K. (1971). Genetic analysis of a meiotic mutant resulting in precocious sister-centromere separation in *Drosophila melanogaster*. *Molecular and General Genetics*, **113**, 251–72.
Dävring, L. & Sunner, M. (1973). Female meiosis and embryonic mitosis in *Drosophila melanogaster* I. Meiosis and fertilization. *Hereditas*, **73**, 51–64.
Dävring, L. & Sunner, M. (1976). Early prophase in female meiosis of *Drosophila melanogaster*. Further studies. *Hereditas*, **82**, 129–31.
Dävring, L. & Sunner, M. (1977). Late prophase and first metaphase of female meiosis in *Drosophila melanogaster*. *Hereditas*, **85**, 25–32.
Dawes, I. W. (1983). Genetic control and gene expression during meiosis and sporulation in *Saccharomyces cerevisiae*. In *Yeast Genetics: Fundamental and Applied Aspects*, ed. J. F. T. Spencer, D. M. Spencer & A. R. W. Smith, pp. 29–64. New York: Springer-Verlag.
Day, J. W. & Grell, R. F. (1976). Synaptonemal complexes during premeiotic DNA synthesis in oocytes in *Drosophila*. *Genetics*, **83**, 67–79.
De Brabander, M. (1982). A model for the microtubule organizing activity of the centrosomes and kinetochores in mammalian cells. *Cell Biology International Reports*, **6**, 901–15.
De Martino, C., Capanna, E., Nicotra, M. R. & Natali, P. G. (1980). Immunolocalization of contractile proteins in mammalian meiotic chromosomes. *Cell and Tissue Research*, **213**, 159–78.
Del Fosse, F. E. & Church, K. (1981). Presynaptic chromosome behavior in *Lilium* I. Centromere orientation and movement during premeiotic interphase in *Lilium speciosum* c.v. Rosemede. *Chromosoma*, **81**, 701–16.
Di Domenico, B., Kowalisyn, J., Frackman, S., Jensen, L., Easton-Esposito, R. E. & Elder, R. (1984). The *SPO 11* gene encodes a developmentally regulated specific transcript. *12th Conference on Yeast Genetics and Molecular Biology*, p. 267.
Dietz, R. (1954). Multiple Geschlechstchromosomen bei dem Ostracoden *Notodromas monacha*. *Chromosoma*, **6**, 397–418.
Dietz, R. (1955). Zahl und Verhalten der Chromosomen einiger Ostracoden. *Zeitschrift für Naturforschung*, **10b**, 92–5.
Dietz, R. (1958). Multiple Geschlechstchromosomen bei den cypriden Ostracoden, ihre Evolution und ihre Teilungsverhalten. *Chromosoma*, **9**, 359–440.

Dietz, R. (1959). Centrosomenfreie Spindelpole in Tipuliden–Spermatocyten. *Zeitschrift für Naturforschung*, **14b**, 749–52.
Dietz, R. (1966). The dispensability of centrioles in the spermatocyte division of *Pales ferruginea* (Nematocera). *Chromosomes Today*, **1**, 161–6.
Dietz, R. (1969). Bau und Funktion der Spindelapparats. *Naturwissenschaften*, **56**, 237–48.
Dollin, A. E. & Murray, J. D. (1984). Triple chromosome pairing in an aneuploid bull spermatocyte. *Canadian Journal of Genetics and Cytology*, **26**, 782–3.
Dover, G. A. & Riley, R. (1972). Prevention of pairing of homoeologous meiotic chromosomes of wheat by an activity of supernumerary chromosomes of *Aegilops*. *Nature*, **240**, 159–61.
Dover, G. A. & Riley, R. (1977). Inferences from genetical evidence on the course of meiotic pairing in plants. *Philosophical Transactions of the Royal Society of London*, **B277**, 313–26.
Dowling, E., Maloney, D. H. & Fogel, S. (1985). Meiotic recombination and sporulation in repair-deficient strains of yeast. *Genetics*, **109**, 283–302.
Downes, A. (1964). Arctic insects and their environment. *Canadian Entomologist*, **96**, 279–307.
Dresser, M. E. (1987). The synaptonemal complex and meiosis: an immunocytochemical approach. In *Meiosis*, ed. P. B. Moens, pp. 245–74. Orlando: Academic Press.
Drets, M. E., Corbella, E., Panzera, F. & Folle, G. A. (1983). C banding and non – homologous association II. The 'parachute' Xy_p sex bivalent and the behaviour of heterochromatic segments. *Chromosoma*, **88**, 249–55.
Drets, M. E. & Stoll, M. (1974). C-banding and non-homologous associations in *Gryllus argentinus*. *Chromosoma*, **48**, 367–90.
Dubey, D. D. & Raman, R. (1987). Factors influencing replicon organization in tissues having different S-phase durations in the mole rat, *Bandicota bengalensis*. *Chromosoma*, **95**, 285-9.
Dyer, A. F. (1964). Heterochromatin in American and Japanese species of *Trillium* III. Chiasma frequency and distribution and the effect on it of heterochromatin. *Cytologia*, **29**, 263–79.
Earnshaw, W. C. (1988). Mitotic chromosome structure. *BioEssays*, **9**, 147–50.
Earnshaw, W. C. & Heck, M. M. S. (1985). Localization of topoisomerase II in mitotic chromosomes. *The Journal of Cell Biology*, **100**, 1716–25.
Eddy, E. M. (1975). Germ plasm and the differentiation of the germ cell line. *International Review of Cytology*, **43**, 229–80.
Eddy, E. M. & Hahnel, A. C. (1983). Establishment of the germ cell line in mammals. In *Current Problems in Germ Cell Differentiation*, ed. A. McLaren & C. C. Wylie, pp. 41–69. Cambridge: Cambridge University Press.
Egel-Mitani, M., Olson, L. W. & Egel, R. (1982). Meiosis in *Aspergillus nidulans*: another example for lack of synaptonemal complexes in the absence of crossing over interference. *Hereditas*, **97**, 179–87.
Eichenlaub-Ritter, U. & Ruthmann, A. (1982). Holokinetic composite chromosomes with 'diffuse' kinetochores in the micronuclear mitosis of a heterotrichous ciliate. *Chromosoma*, **84**, 701–16.
Elder, F. F. B. & Pathak, S. (1980). Light microscopic observations on the behaviour of silver-stained trivalents in pachytene cells of *Sigmodon fulviventer* (Rodential, Muridae) heterozygous for centric fusion. *Cytogenetics and Cell Genetics*, **27**, 31–8.
Elinson, R. P. & Briedis, A. (1981). Triploidy permits survival of an inviable amphibian hybrid. *Developmental Genetics*, **2**, 257–67.
Esposito, M. S. (1978). Evidence that spontaneous mitotic recombination occurs

at the two-strand stage. *Proceedings of the National Academy of Sciences*, **75**, 4436–40.
Esposito, R. E. & Klapholz, S. (1981). Meiosis and ascospore development. In *The Molecular Biology of the Yeast Saccharomyces: Life Cycle and Inheritance*, ed. J. N. Strathern, E. W. Jones & J. R. Broach, pp. 211–87. New York: Cold Spring Harbor Laboratory.
Esposito, M. S. & Wagstaff, J. E. (1981). Mechanisms of mitotic recombination. In *The Molecular Biology of the Yeast Saccharomyces: Life Cycle and Inheritance*, ed. J. N. Strathern, E. W. Jones & J. R. Broach, pp. 341–70. New York: Cold Spring Harbor Laboratory.
Euteneuer, V. & McIntosh, J. R. (1981). Structural polarity of kinetochore microtubules in PtK_1 cells. *The Journal of Cell Biology*, **89**, 338–45.
Evans, E. P. & Burtenshaw, M. D. (1982). Meiotic crossing over between the X and Y chromosomes of male mice carrying the sex reversing (Sxr) factor. *Nature*, **300**, 443–5.
Evans, G. M. & Aung, T. (1985). Identification of a diploidising genotype of *Lolium multiflorum*. *Canadian Journal of Genetics and Cytology*, **27**, 498–505.
Evans, G. M. & Taylor, I. B. (1976). Genetic control of homoeologous chromosome pairing in *Lolium* hybrids. *Kew Chromosome Conference*, **1**, 57–66.
Evans, L., Mitchison, T. & Kirschner, M. (1985). Influence of the centrosome on the structure of nucleated microtubules. *The Journal of Cell Biology*, **100**, 1185–91.
Fawcett, D. W. (1956). The fine structure of chromosomes in the meiotic prophase of vertebrate spermatocytes. *The Journal of Biophysical and Biochemical Cytology*, **2**, 403–6.
Fincham, J. R. S. (1983). *Genetics*. Boston : Jones & Bartlett.
Finnerty, V. (1976). Gene conversion in *Drosophila*. In *the Genetics and Biology of Drosophila*, vol. 1a, ed. M. Ashburner & Novitski, pp. 331–49. London: Academic Press.
Fischer, J. (1978). Zum problem der chromosome-evolution durch translokationen bei *Chironomus* (Diptera). *Archiv für Genetik*, **51**, 73–98.
Fletcher, H. L. (1978). Localised chiasmata due to partial pairing: a 3D reconstruction of synaptonemal complexes in male *Stethophyma grossum*. *Chromosoma*, **65**, 247–69.
Fletcher, H. L. & Hewitt, G. M. (1980a). Effect of a 'B' chromosome on chiasma localisation and frequency in male *Euthystira brachyptera*. *Heredity*, **44**, 341–7.
Fletcher, H. L. & Hewitt, G. M. (1980b). A comparison of chiasma frequency and distribution between sexes in three species of grasshopper. *Chromosoma*, **77**, 129–44.
Flory, S. S., Tsang, J., Muniyappa, K., Bianchi, M., Gonda, D., Kahn, R., Azhderian, E., Egner, C., Shamer, S. & Radding, C. M. (1984). Intermediates in homologous pairing promoted by Rec A protein and correlations of recombination in vitro and in vivo. *Cold Spring Harbor Symposia in Quantitative Biology*, **49**, 513–23.
Fogel, S., Mortimer, R. K. & Lusnak, K. (1981). Mechanisms of gene conversion or 'wanderings on a foreign strand'. In *Molecular Biology of the Yeast Saccharomyces: Life Cycle and Inheritance*, ed. J. N. Strathern, E. W. Jones, & J. R. Broach, pp. 289–339. New York: Cold Spring Harbor Laboratory.
Fogel, S., Mortimer, R. K. & Lusnak, K. (1983). Meiotic gene conversion in yeast: molecular and experimental perspectives. In *Yeast Genetics: Fundamental and Applied Aspects*, ed. J. F. T. Spencer, D. M. Spencer & A. R. W. Smith, pp. 65–107. New York: Springer-Verlag.
Fogel, S., Mortimer, R. K., Lusnak, K. & Tavares, F. (1979). Meiotic gene

conversion: a signal of the basic recombination event in yeast. *Cold Spring Harbor Symposia in Quantitative Biology*, **43**, 1325–41.

Fogwill, M. (1958). Differences in crossing over and chromosome size in the sex cells of *Lilium* and *Fritillaria*. *Chromosoma*, **9**, 493–504.

Forer, A. (1966). Characterization of the mitotic traction system, and evidence that birefringent spindle fibers neither produce nor transmit force for chromosome movement. *Chromosoma*, **19**, 44–98.

Forer, A. (1974). Possible roles of microtubules and actin-like filaments during cell division. In *Cell Cycle Controls*, ed. G. M. Padilla, I. L. Cameron & A. Zimmerman, pp. 319–36. New York: Academic Press.

Forer, A. & Koch, C. (1973). Influence of autosome movements and of sex-chromosome movements in sex-chromosome segregation in crane fly spermatocytes. *Chromosoma*, **40**, 417–42.

Fox, D. P. (1973). The control of chiasma distribution in the locust, *Schistocerca gregaria* (Forskål). *Chromosoma*, **43**, 289–328.

Frankel, O. H. (1940). The causal sequence of meiosis I. Chiasma formation and the order of pairing in *Fritillaria*. *Journal of Genetics*, **41**, 9–34.

Fredga, K. & Santesson, B. (1964). Male meiosis in the Syrian, Chinese and European hamsters. *Hereditas*, **52**, 36–48.

Friedländer, M. & Hauschteck-Jungen, E. (1982). The regular divisions of the spermatocytes as related to a meiotic lysine-rich protein fraction. *Chromosoma*, **85**, 109–18.

Friedländer, M. & Hauschteck-Jungen, E. (1986). Regular and irregular divisions of Lepidoptera spermatocytes as related to the speed of meiotic prophase. *Chromosoma*, **93**, 227–30.

Friedman, C., Bouchard, R. A & Stern, H. (1982). DNA sequences repaired at pachytene exhibit strong homology among distantly related higher plants. *Chromosoma*, **87**, 409–24.

Fuge, H. (1972). Morphological studies on the structure of univalent sex-chromosomes during anaphase movement in spermatocytes of the crane fly *Pales ferruginea*. *Chromosoma*, **39**, 403–17.

Fuge, H. (1974). The arrangement of microtubules and the attachment of chromosomes to the spindle during anaphase in tipulid spermatocytes. *Chromosoma*, **45**, 245–60.

Fuge, H. (1977). Ultrastructure of mitotic cells. In *Mitosis: Facts and Questions*, ed. M. Little, N. Paweletz, C. Petzelt, H. Ponstingl, D. Schroeter & H.-P. Zimmermann, pp. 51–68. Berlin: Springer-Verlag.

Fuge, H. (1979). Synapsis, desynapsis and formation of polycomplex-like aggregates in male meiosis of *Pales ferruginea* (Diptera, Tipulidae). *Chromosoma*, **70**, 353–73.

Fuge, H. (1985). The three-dimensional architecture of chromosome fibres in the crane fly II. Amphitelic sex univalents in meiotic anaphase I. *Chromosoma*, **91**, 322–8.

Fuge, H., Bastmeyer, M. & Steffen, W. (1985). A model for chromosome movement based on lateral interaction of spindle microtubules. *Journal of Theoretical Biology*, **115**, 391–9.

Gassner, G. (1967). Synaptinemal complexes: recent findings. *Journal of Cell Biology*, **35**, 166A–7A.

Gassner, G. (1969). Synaptinemal complexes in the achiasmatic spermatogenesis of *Bolbe nigra* Giglio-Tos (Mantoidea). *Chromosoma*, **26**, 22–34.

Gatti, M. (1982). Sister chromatid exchanges in *Drosophila*. In *Sister Chromatid Exchange*, ed. S. Wolff, pp. 267–96. New York: John Wiley & Sons.

Gatti, M., Pimpinelli, S. & Baker, B. S. (1980). Relationships among chromatid interchanges, sister chromatid exchanges and meiotic recombination in *Drosophila melanogaster*. *Proceedings of the National Academy of Sciences*, **77**, 1575–9.

Gillies, C. B. (1972). Reconstruction of the *Neurospora crassa* pachytene karyotype from serial sections of synaptonemal complexes. *Chromosoma*, **36**, 119–30.

Gillies, C. B. (1975). An ultrastructural analysis of chromosome pairing in maize. *Comptes Rendus des Travaux du Laboratoire Carlsberg*, **40**, 135–61.

Gillies, C. B. (1979). The relationship between synaptinemal complexes, recombination nodules and crossing over in *Neurospora crassa* bivalents and translocation quadrivalents. *Genetics*, **91**, 1–17.

Gillies, C. B. (1984). The synaptonemal complex in higher plants. *CRC Critical Reviews in Plant Sciences*, **2**, 81–116.

Gillies, C. B. (1985). An electron microscopic study of synaptonemal complex formation at zygotene in rye. *Chromosoma*, **92**, 165–75.

Gillies, C. B., Kuspira, J.& Bhambhani, R. N. (1987). Genetic and cytogenetic analyses of the A genome of *Triticum monococcum* IV. Synaptonemal complex formation in autotetraploids. *Genome*, **29**, 309–18.

Gillies, C. B., Rasmussen, S. W. & von Wettstein, D. (1974). The synaptonemal complex in homologous pairing of chromosomes. *Cold Spring Harbor Symposia in Quantitative Biology*, **38**, 117–22.

Giraldez, R. & Santos, J. L. (1981). Cytological evidence for preference of identical over homologous but non-identical meiotic pairing. *Chromosoma*, **82**, 447–51.

Goday, C., Ciofi-Luzzatto, A. & Pimpinelli, S. (1985). Centromere ultrastructure in germ-line chromosomes of *Parascaris*. *Chromosoma*, **91**, 121–5.

Golhil, R. N. & Kaul, R. (1980). Studies on male and female meiosis in Indian *Allium* I. Four diploid species. *Chromosoma*, **77**, 123–7.

Gohil, R. N. & Kaul, R. (1981). Studies on male and female meiosis in Indian *Allium* II. Autotetraploid *Allium tuberosum*. *Chromosoma*, **82**, 735–9.

Goldstein, L. S. B. (1980). Mechanisms of chromosome orientation revealed by two meiotic mutants in *Drosophila melanogaster*. *Chromosoma*, **78**, 79–111.

Goldstein, L. S. B. (1981). Kinetochore structure and its role in chromosome orientation during the first meiotic division in male *Drosophila melanogaster*. *Cell*, **25**, 591–602.

Goldstein, P. (1984). Triplo-X hermaphrodite of *Caenorhabditis elegans*: pachytene analysis, synaptonemal complexes and pairing mechanisms. *Canadian Journal of Genetics and Cytology*, **26**, 13–7.

Goldstein, P. (1985). The synaptonemal complexes of *Caenorhabditis elegans* :a pachytene analysis of the Dp 1 mutant and disjunction regulator regions. *Chromosoma*, **93**, 177–82.

Goldstein, P. & Triantaphyllou, A. C. (1980). Karyotype analysis of the plant parasitic nematode *Heterodera glycines* by electron microscopy II. The tetraploid and an aneuploid hybrid. *Journal of Cell Science*, **43**, 225–37.

Goldstein, P. & Triantaphyllou, A. C. (1981). Pachytene karyotype analysis of tetraploid *Meloidogyne hapla* females by electron microscopy. *Chromosoma*, **82**, 405–12.

Golubovskaya, I. N. (1979). Genetic control of meiosis. *International Review of Cytology*, **58**, 247–90.

Gorbsky, G. J., Sammak, P. J. & Borisy, G. G. (1987). Chromosomes move poleward in anaphase along stationary microtubules that coordinately disassemble from their kinetochore ends. *The Journal of Cell Biology*, **104**, 9–18.

Gould, R. R. & Borisy, G. G. (1977). The pericentriolar material in Chinese

hamster ovary cells nucleates microtubule formation. *The Journal of Cell Biology*, **73**, 601–15.
Grant, V. (1981). *Plant Speciation*, 2nd edn. New York: Columbia University Press.
Greenbaum, I. F., Hale, D. W. & Fuxa, K. P. (1986). The mechanism of autosomal synapsis and the substaging of zygonema and pachynema from deer mouse spermatocytes. *Chromosoma*, **93**, 203–12.
Greenbaum, I. F. & Reed, M. J. (1984). Evidence for heterosynaptic pairing of the inverted segment in pericentric inversion heterozygotes of the deer mouse (*Peromyscus maniculatus*). *Cytogenetics and Cell Genetics*, **38**, 106–11.
Gregory, W. C. (1940). Experimental studies on the cultivation of excised anthers in nutrient solution. *American Journal of Botany*, **27**, 687–92.
Grell, R. F. (1976). Distributive pairing. In *The Genetics and Biology of Drosophila*, vol. 1a, ed. M. Ashburner & E. Novitski, pp. 435–86. New York: Academic Press.
Grell, R. F. (1985). The meiotic process and aneuploidy. In *Aneuploidy: Etiology and Mechanisms*, ed. V. L. Dellarco, P. E. Voytek & A. Hollaender, pp. 317–33. New York: Plenum Press.
Grell, R. F., Bank, H. & Gassner, G. (1972). Meiotic exchange without the synaptinemal complex. *Nature New Biology*, **240**, 155–7.
Gupta, M. L. (1966). A preliminary account of the meiotic mechanism in nineteen species of the Indian mantids. *Research Bulletin (N.S.) of the Panjab University*, **17**, 421–2.
Gutz, H. (1971). Site specific induction of gene conversion in *Schizosaccharomyces pombe*. *Genetics*, **69**, 317–37.
Haga, T. (1953). Meiosis in *Paris* II. Spontaneous breakage and fusion of chromosomes. *Cytologia*, **18**, 50–66.
Håkansson, A. & Levan, A. (1957). Endo-duplicational meiosis in *Allium odorum*. *Hereditas*, **43**, 179–200.
Hale, D. W. (1986). Heterosynapsis and suppression of chiasmata within heterozygous pericentric inversions of the Sitka deer mouse. *Chromosoma*, **94**, 425–32.
Hale, D. W. & Greenbaum, I. F. (1986). The behaviour and morphology of the X and Y chromosomes during prophase I in the Sitka deer mouse (*Peromyscus sitkensis*). *Chromosoma*, **94**, 235–42.
Hall, J. C. (1972). Chromosome segregation influenced by two alleles of the meiotic mutant *c(3)G* in *Drosophila melanogaster*. *Genetics*, **71**, 367–400.
Hartl, D. L. & Brown, S. W. (1970). The origin of male haploid genetic systems and their expected sex ratio. *Theoretical Population Biology*, **1**, 165–90.
Hartwell, L. H., Culotti, J., Pringle, J. R., & Reid, B. J. (1974). Genetic control of cell division cycle in yeast. *Science*, **183**, 46–51.
Hasenkampf, C. A. (1984). Synaptonemal complex formation in pollen mother cells of *Tradescantia*. *Chromosoma*, **90**, 275–84.
Hastings, P. J. (1988). Recombination in the eukaryotic nucleus. *BioEssays*, **9**, 61–4.
Hawley, R. S. (1980). Chromosomal sites necessary for normal levels of meiotic recombination in *Drosophila melanogaster* I. Evidence for and mapping of the sites. *Genetics*, **94**, 625–46.
Hays, T. S., Wise, D. & Salmon, E. D. (1982). Traction force on a kinetochore at metaphase acts as a linear function of kinetochore fibre length. *The Journal of Cell Biology*, **93**, 374–82.
Heath, I. B. (1974). Mitosis in the fungus *Thraustotheca clavata*. *The Journal of Cell Biology*, **60**, 204–20.

Heath, I. B. (1980). Variant mitoses in lower eukaryotes: indicators of the evolution of mitosis. *International Review of Cytology*, **64**, 1–80.
Hegner, R. W. (1914). *The Germ Cell Cycle in Animals*. New York: The Macmillan Company.
Henderson, S. A. (1961). The chromosomes of the British Tetrigidae (Orthoptera). *Chromosoma*, **12**, 553–72.
Henderson, S. A. (1963). Chiasma distribution at diplotene in a locust *Schistocerca gregaria*. *Heredity*, **18**, 173–90.
Henderson, S. A. (1964). RNA synthesis during male meiosis and spermiogenesis. *Chromosoma*, **15**, 345–66.
Henderson, S. A. (1969). Chiasma localisation and incomplete pairing. *Chromosomes Today*, **2**, 56–60.
Henderson, S. A. (1970a). Sex chromosomal polymorphism in the earwig *Forficula*. *Chromosoma*, **31**, 139–64.
Henderson, S. A. (1970b). The time and place of meiotic crossing over. *Annual Review of Genetics*, **4**, 295–324.
Hepler, P. K. & Wolniak, S. M. (1984). Membranes in the mitotic apparatus: their structure and function. *International Review of Cytology*, **90**, 169–238.
Heslop-Harrison, J. (1957). The experimental modification of sex expression in flowering plants. *Biological Reviews*, **32**, 38–90.
Hewitt, G. M. (1975). A new hypothesis for the origin of the parthenogenetic grasshopper *Moraba virgo*. *Heredity*, **34**, 117–23.
Hewitt, G. M. (1976). Meiotic drive for B-chromosomes in the primary oocytes of *Myrmeleotettix maculatus* (Orthoptera: Acrididae). *Chromosoma*, **56**, 381–91.
Hewitt, G. M. & John, B. (1965). The influence of numerical and structural chromosome mutations on chiasma conditions. *Heredity*, **20**, 123–35.
Heyting, C., Dettmers, R. J., Dietrich, A. J. J., Redeker, J. W. & Vink, A. C. G. (1988). Two major components of synaptonemal complexes are specific for meiotic prophase nuclei. *Chromosoma*, **96**, 325–32.
Hilliker, A. J. & Chovnick, A. (1981). Further observations on intragenic recombination in *Drosophila melanogaster*. *Genetic Research*, **38**, 281–96.
Hilliker, A. J., Clark, S. H. & Chovnick, A. (1988). Genetic analysis of intragenic recombination in *Drosophila*. In *The Recombination of Genetic Material*, ed. K. B. Low, pp. 73–90. San Diego: Academic Press.
Hinton, C. W. (1966). Enhancement of recombination associated with the *c(3)G mutant* of *Drosophila melanogaster*. *Genetics*, **53**, 157–64.
Hobolth, P. (1981). Chromosome pairing in allohexaploid wheat var. Chinese Spring. Transformation of multivalents into bivalents, a mechanism for exclusive bivalent formation. *Carlsberg Research Communications*, **46**, 129–73.
Høgset, A. & Øyen, T. (1984). Correlation between suppressed meiotic recombination and the lack of DNA strand breaks in the rDNA gene of *Saccharomyces cerevisiae*. *Nucleic Acids Research*, **12**, 7199–213.
Holliday, R. (1964). A mechanism for gene conversion in fungi. *Genetical Research*, **5**, 282–304.
Holliday, R. (1974). Molecular aspects of genetic exchange and gene conversion. *Genetics*, **78**, 273–87.
Holliday, R. (1977). Recombination and meiosis. *Philosophical Transactions of the Royal Society of London*, **B277**, 359–70.
Holliday, R., Raylor, S. Y., Kmiec, E. B. & Hollman, W. K. (1984). Biochemical characterisation of rec 1 mutants and the genetic control of recombination in *Ustilago maydis*. *Cold Spring Harbor Symposia in Quantitative Biology*, **49**, 669–73.
Holm, P. B. (1977a). Three dimensional reconstruction of chromosome pairing

during the zygotene stage of meiosis in *Lilium longiflorum* (Thunb.). *Carlsberg Research Communications*, **42**, 103–51.

Holm, P. B. (1977b). The premeiotic DNA replication of euchromatin and heterochromatin in *Lilium longiflorum* (Thunb.). *Carlsberg Research Communications*, **42**, 249–81.

Holm, P. B. (1986). Chromosome pairing and chiasma formation in allohexaploid wheat, *Triticum aestivum*, analyzed by spreading of meiotic nuclei. *Carlsberg Research Communications*, **51**, 239–94.

Holm, P. B. & Rasmussen, S. W. (1980). Chromosome pairing, recombination nodules and chiasma formation in diploid *Bombyx* males. *Carlsberg Research Communications*, **45**, 483–548.

Holm, P. B. & Rasmussen, S. W. (1983a). Human meiosis VI. Crossing over in human spermatocytes. *Carlsberg Research Communications*, **48**, 385–413.

Holm, P. B. & Rasmussen, S. W. (1983b). Human meiosis VII. Chiasma formation in human spermatocytes. *Carlsberg Research Communications*, **48**, 415–56.

Holm, P. B. & Rasmussen, S. W. (1984). The synaptonemal complex in chromosome pairing and disjunction. *Chromosomes Today*, **8**, 104–16.

Holm, P. B., Rasmussen, S. W., Zickler, D., Lu, B. C. & Sage, J. (1981). Chromosome pairing, recombination nodules and chiasma formation in the basidiomycete *Coprinus cinereus*. *Carlsberg Research Communications*, **46**, 305–46.

Holm, P. B. & Wang, X. (1988). The effect of chromosomes 5B on synapsis and chiasma formation in wheat, *Triticum aestivum* c.v. Chinese Spring. *Carlsberg Research Communications*, **53**, 191–208.

Horesh, O., Simchen, G. & Friedmann, A. (1979). Morphogenesis of the synapton during yeast meiosis. *Chromosoma*, **75**, 101–15.

Hotta, Y., Bennett, M. D., Toledo, L. A. & Stern, H. (1979a). Regulation of R-protein and endonuclease activities in meiocytes by homologous chromosome pairing. *Chromosoma*, **72**, 191–201.

Hotta, Y., Chandley, A. C. & Stern, H. (1977). Biochemical analysis of meiosis in the male mouse. *Chromosoma*, **62**, 255–68.

Hotta, Y., Chandley, A. C., Stern, H., Searle, A. G. & Beechey, C. V. (1979b). A disruption of pachytene DNA metabolism in male mice with chromosomally-derived sterility. *Chromosoma*, **73**, 287–300.

Hotta, Y, de la Pěna, A., Tabata, S. & Stern, H. (1985a). Control of enzyme accessibility to specific DNA sequences during meiotic prophase by alterations in chromatin structure. *Cytologia*, **50**, 611–20.

Hotta, Y., Ito, M. & Stern, H. (1966). Synthesis of DNA during meiosis. *Proceedings of the National Academy of Sciences*, **56**, 1184–91.

Hotta, Y. & Shepard, J. (1973). Biochemical aspects of colchicine action in meiotic cells. *Molecular and General Genetics*, **122**, 243–60.

Hotta, Y. & Stern, H. (1971). Analysis of DNA synthesis during meiotic prophase in *Lilium*. *Journal of Molecular Biology*, **55**, 337–55.

Hotta, Y. & Stern, H. (1974). DNA scission and repair during pachytene in *Lilium*. *Chromosoma*, **46**, 279–96.

Hotta, Y. & Stern, H. (1976). Persistent discontinuities in late replicating DNA during meiosis in *Lilium*. *Chromosoma*, **55**, 171–82.

Hotta, Y. & Stern, H. (1978a). DNA unwinding protein from meiotic cells of *Lilium*. *Biochemistry*, **17**, 1872–80.

Hotta, Y. & Stern, H. (1978b). Absence of satellite DNA synthesis during meiotic prophase in mouse and human spermatocytes. *Chromosoma*, **69**, 323–30.

Hotta, Y. & Stern, H. (1984). The organization of DNA segments undergoing repair synthesis during pachytene. *Chromosoma*, **89**, 127–37.

Hotta, Y., Tabata, S., Bouchard, R. A., Piñon, R. & Stern, H. (1985b). General recombination mechanisms in extracts of meiotic cells. *Chromosoma*, **93**, 140–51.

Hotta, Y., Tabata, S. & Stern, H. (1984). Replication and nicking of zygotene DNA sequences: control of a meiosis-specific protein. *Chromosoma*, **90**, 243–53.

Hotta, Y., Tabata, S., Stubbs, L. & Stern, H. (1985c). Meiosis-specific transcripts of a DNA component replicated during chromosome pairing: homology across the phylogenetic spectrum. *Cell*, **40**, 785–93.

Hughes-Schrader, S. (1924). Reproduction in *Acroschismus wheeleri* Pierce. *Journal of Morphology and Physiology*, **39**, 157–205.

Hughes-Schrader, S. (1943). Polarization, kinetochore movements and bivalent structure in the meiosis of male mantids. *Biological Bulletin Woods Hole*, **85**, 265–300.

Hughes-Schrader, S. (1947). The 'pre-metaphase stretch' and kinetochore orientation in phasmids. *Chromosoma*, **3**, 1–21.

Hughes-Schrader, S. (1948). Cytology of coccids (Coccoidea-Homoptera). *Advances in Genetics*, **2**, 127–203.

Hughes-Schrader, S. (1969). Distance segregation and compound sex chromosomes in mantispids (Neuroptera: Mantispidae). *Chromosoma*, **27**, 109–29.

Hughes-Schrader, S. & Schrader, F. (1961). The kinetochore of the Hemiptera. *Chromosoma*, **12**, 327–50.

Hurst, D. D., Fogel, S. & Mortimer, R. K. (1972). Conversion associated recombination in yeast. *Proceedings of the National Academy of Sciences*, **69**, 101–5.

Ito, M. & Takegami, M. H. (1982). Commitment of mitotic cells to meiosis during the G_2 phase of premeiosis. *Plant and Cell Physiology*, **23**, 943–52.

Ito, M., Hotta, Y. & Stern, H. (1967). Studies on meiosis in vitro II. Effect of inhibiting DNA synthesis during meiotic prophase on chromosome structure and behaviour. *Developmental Biology*, **16**, 54–77.

Ito, M., Takegami, M. H. & Noda, S. (1983). Achiasmate meiosis in the *Fritillaria japonica* group II. Formation of synaptinemal complexes in microsporocytes. *The Japanese Journal of Genetics*, **58**, 377–81.

Ito, S. (1969). The lamellar systems of cytoplasmic membranes in dividing spermatogenic cells of *Drosophila virilis*. *Journal of Biophysical and Biochemical Cytology*, **7**, 433–42.

James, S. H. (1965). Complex hybridity in *Isotoma petraea* I. The occurrence of interchange heterozygosity, autogamy and a balanced lethal system. *Heredity*, **20**, 341–53.

James, S. H. (1984). The pursuit of hybridity and population divergence in *Isotoma petraea*. In *Plant Biosystematics*, ed. W. F. Grant, pp. 169–77. Toronto: Academic Press.

James, S. H., Wylie, A. P., Johnson, M. S., Carstairs, S. A. & Simpson, G. A. (1983). Complex hybridity in *Isotoma petraea* V. Allozyme variation and the pursuit of hybridity. *Heredity*, **51**, 653–63.

Janssens, F. A. (1909). La théorie de la chiasmatypie. Nouvelle interpretation des cinèses de maturation. *La Cellule*, **25**, 387–411.

Janssens, F. A. (1924). La chiasmatypie dans les insectes. *La Cellule*, **34**, 135–359.

Jenkins, G. (1985a). Synaptomemal complex formation in hybrids of *Lolium temulentum* × *L. perenne* (L.) I. High chiasma frequency diploid. *Chromosoma*, **92**, 81–8.

Jenkins, G. (1985b). Synaptonemal complex formation in hybrids of *Lolium temulentum* × *L. perenne* (L.) II. Triploid. *Chromosoma*, **92**, 387–90.

Jenkins, G. (1986). Synaptonemal complex formation in hybrids of *Lolium temulentum* × *L. perenne* (L.) III. Tetraploid. *Chromosoma*, **93**, 413–19.

Jenkins, G. & Rees, H. (1983). Synaptonemal complex formation in a *Festuca* hybrid. *Kew Chromosome Conference*, **2**, 233–42.
Jensen, C. & Bajer, A. (1973). Spindle dynamics and arrangement of chromosomes. *Chromosoma*, **44**, 73–89.
John, B. (1957). XY segregation in the crane fly *Tipula maxima* (Diptera: Tipulidae). *Heredity*, **11**, 209–15.
John, B. (1973). The cytogenetic systems of grasshoppers and locusts II. The origin and evolution of supernumerary segments. *Chromosoma*, **44**, 123–46.
John, B. (1976). Myths and mechanisms of meiosis. *Chromosoma*, **54**, 295–325.
John, B. (1987). The orientation behaviour of multiple chromosome configurations in acridid grasshoppers. *Genome*, **29**, 292–308.
John, B. (1988). The biology of heterochromatin. In *Molecular and Structural Aspects of Heterochromatin*, ed. R. S. Verma, pp. 1–147. New York: Cambridge University Press.
John, B. & Claridge, M. F. (1974). Chromosome variation in British populations of *Oncopsis* (Hemiptera: Cicadellidae). *Chromosoma*, **46**, 77–89.
John, B. & Freeman, M. (1975). Causes and consequences of Robertsonian exchange. *Chromosoma*, **52**, 123–36.
John, B. & Freeman, M. (1976). The cytogenetic systems of grasshoppers and locusts III. The genus *Tolgadia* (Oxyinae: Aerididae). *Chromosoma*, **55**, 105–19.
John, B. & Henderson, S. A. (1962). Asynapsis and polyploidy in *Schistocerca paranensis*. *Chromosoma*, **13**, 111–47.
John, B. & Hewitt, G. M. (1966a). A polymorphism for heterochromatic supernumerary segments in *Chorthippus parallelus*. *Chromosoma*, **18**, 254–71.
John, B. & Hewitt, G. M. (1966b). Karyotype stability and DNA variability in the Acrididae. *Chromosoma*, **20**, 155–72.
John, B. & King, M. (1977). Heterochromatin variation in *Cryptobothrus chrysophorus* I. Chromosome differentiation in natural populations. *Chromosoma*, **64**, 219–39.
John, B. & King, M. (1980). Heterochromatin variation in *Cryptobothrus chrysophorus* III. Synthetic hybrids. *Chromosoma*, **78**, 165–86.
John, B. & King, M. (1982). Meiotic effects of supernumerary heterochromatin in *Heteropternis obscurella*. *Chromosoma*, **85**, 39–65.
John, B. & King, M. (1985a). Pseudoterminalisation, terminalisation and the origin of terminal associations. *Chromosoma*, **93**, 89–99.
John, B. & King, M. (1985b). The inter-relationship between heterochromatin and chiasma distribution. *Genetica*, **66**, 183–94.
John, B. & Lewis, K. R. (1957). Studies on *Periplaneta americana* I. Experimental analysis of male meiosis. *Heredity*, **11**, 1–9.
John, B. & Lewis, K. R. (1960). Nucleolar controlled segregation of the sex chromosomes in beetles. *Heredity*, **15**, 431–9.
John, B. & Lewis, K. R. (1966). *The Meiotic System*. Protoplasmatologia *VI/F/1, 1–335*. Vienna: Springer Verlag.
Johnston, L. H., Williamson, D. H., Johnson, A. L. & Fennell, D. J. (1982). On the mechanism of premeiotic DNA synthesis in the yeast *Saccharomyces cerevisiae*. *Experimental Cell Research*, **141**, 53–62.
Jones, G. H. (1967). The control of chiasma distribution in rye. *Chromosoma*, **22**, 69–90.
Jones, G. H. (1968). Meiotic errors in rye related to chiasma formation. *Mutation Research*, **5**, 385–95.
Jones, G. H. (1969). Further correlations between chiasmata and U-type exchanges in rye meiosis. *Chromosoma*, **26**, 105–118.
Jones, G. H. (1971). The analysis of exchanges in tritium-labelled meiotic chromosomes II. *Stethophyma grossum*. *Chromosoma*, **34**, 367–82.

Jones, G. H. (1974). Correlated components of chiasma variation and the control of chiasma distribution in rye. *Heredity*, **32**, 375–87.
Jones, G. H. (1978). Giemsa C-banding of rye meiotic chromosomes and the nature of 'terminal chiasmata'. *Chromosoma*, **66**, 45–57.
Jones, G. H. (1984). The control of chiasma distribution. In *Controlling Events in Meiosis*, ed. C. W. Evans & H. G. Dickinson. *Symposia of the Society for Experimental Biology*, **38**, 291–320.
Jones, G. H. (1987). Chiasmata. In *Meiosis*, ed. P. B. Moens, pp. 213–244. Orlando: Academic Press.
Jones, G. H. & Brumpton, R. J. (1971). Sister and non-sister chromatid U-type exchanges in rye meiosis. *Chromosoma*, **33**, 115–28.
Jones, G. H. & Croft, J. A. (1986). Surface spreading of synaptonemal complexes in locusts II. Zygotene pairing behaviour. *Chromosoma*, **93**, 489–95.
Jones, G. H., Stamford, W. K. & Perry, P. E. (1975). Male and female meiosis in grasshoppers II. *Chorthippus brunneus*. *Chromosoma*, **51**, 381–89.
Jones, G. H. & Tease, C. (1984). Analysis of exchanges in differentially stained meiotic chromosomes of *Locusta migratoria* after BrdU-substitution and FPG staining IV. The nature of 'terminal' associations. *Chromosoma*, **89**, 33–6.
Jones, G. H. & Wallace, B. M. N. (1980). Meiotic chromosome pairing in *Stethophyma grossum* spermatocytes studied by a surface-spreading and silver technique. *Chromosoma*, **78**, 187–201.
Jones, K. & Colden, C. (1968). The telocentric complement of *Tradescantia micrantha*. *Chromosoma*, **24**, 135–57.
Jones, R. N. & Rees, H. (1967). Genotypic control of chromosome behaviour in rye XI. The influence of B chromosomes on meiosis. *Heredity*, **22**, 333–47.
Joseph, A. M. & Chandley, A. C. (1984). The morphological sequence of XY pairing in the Norway rat, *Rattus norvegicus*. *Chromosoma*, **89**, 381–86.
Kanda, N. & Kato, H. (1980). Analysis of crossing over in mouse meiotic cells by BrdU labelling technique. *Chromosoma*, **78**, 113–21.
Karp, A. & Jones, R. N. (1982). Cytogenetics of *Lolium perenne* 1. Chiasma frequency variations in inbred lines. *Theoretical and Applied Genetics*, **62**, 177–83.
Karp, A. & Jones, R. N. (1983). Cytogenetics of *Lolium perenne* 2. Chiasma distribution in inbred lines. *Theoretical and Applied Genetics*, **64**, 137–45.
Kassir, Y. & Simchen, G. (1978). Meiotic recombination and DNA synthesis in a new cell cycle mutant of *Saccharomyces cerevisiae*. *Genetics*, **90**, 49–68.
Kayano, H. (1957). Cytogenetic studies in *Lilium callosum* III. Preferential segregation of a supernumerary chromosome in EMCs. *Proceedings of the Japanese Academy*, **B33**, 553–8.
Keats, B. J. B. (1981). *Linkage and Chromosome Mapping in Man*. Honolulu: The University Press of Hawaii.
Kemphues, K. J., Kaufman, T. C., Raff, R. A. & Raff, E. C. (1982). The testis specific β-tubulin subunit in *Drosophila melanogaster* has multiple functions in spermatogenesis. *Cell*, **31**, 655–70.
Kenne, K. & Ljungquist, S. (1984). A DNA-recombinogenic activity in human cells. *Nucleic Acids Research*, **12**, 3057–68.
Kezer, J. & Macgregor, H. C. (1971). A fresh look at meiosis and centromeric heterochromatin in the red-backed salamander *Plethodon cinereus cinereus* (Green). *Chromosoma*, **33**, 146–66.
Kimble, J. (1983). Germ cell development in *Caenorhabditis elegans*: a brief review. In *Current Problems in Germ Cell Differentiation*, ed. A. McClaren & C. C. Wylie, pp. 241–55. Cambridge: Cambridge University Press.

Kimble, J. E. & White, J. G. (1981). On the control of germ cell development in *Caenorhabditis elegans*. *Developmental Biology*, **81**, 208–19.
King, R. C. (1968). The synaptomere-zygosome theory of synaptonemal complex formation. *American Zoologist*, **8**, 822.
Kingwell, B. & Rattner, J. B. (1988). Mammalian kinetochore/centromere composition: a 50 kDa antigen present in the mammalian kinetochore/centromere. *Chromosoma*, **95**, 403–7.
Klapholz, S. & Esposito, R. E. (1980). Recombination and segregation during the single division meiosis in spol2-1 and spol3-1 diploids. *Genetics*, **96**, 589–611.
Klapholz, S., Waddell, C. S. & Esposito, R. E. (1985). The role of the spo-11 gene in meiotic recombination in yeast. *Genetics*, **110**, 187–216.
Klášterská, I. & Natarajan, A. T. (1974a). Cytological studies of the genus *Rosa* with special reference to the section Caninae. *Hereditas*, **76**, 97–108.
Klášterská, I. & Natarajan, A. T. (1974b). The role of the diffuse stage in the cytological problems of meiosis in *Rosa*. *Hereditas*, **76**, 109–16.
Klášterská, I. & Ramel, C. (1980). Sequence and interpretation of meiotic prophase stages in PMCs of *Najas marina* (*Najadaceae*). *Hereditas*, **92**, 171–5.
Klein, H. L. (1984). Lack of association between intrachromosomal gene conversion and reciprocal exchange. *Nature*, **310**, 748–53.
Kmiec, E. B. & Holloman, W. K. (1984). Synapsis promoted by *Ustilago* Rec 1 protein. *Cell*, **36**, 593–8.
Koch, P., Pijnacker, L. P. & Kreke, J. (1972). DNA replication during meiotic prophase in the oocytes of *Carausius morosus* Br. *Chromosoma*, **36**, 313–21.
Konrad, K. D., Engstrom, L., Perrimon, N. & Mahowald, A. P. (1985). Genetic analysis of oogenesis and the role of maternal gene expression in early development. In *Developmental Biology*: *A Comprehensive Synthesis*, *Vol*. 1. *Oogenesis*, ed. L. W. Browder, pp. 577–617. New York: Plenum Press.
Kremer, H., Hennig, W. & Dijkhof, R. (1986). Chromatin organization in the male germ line of *Drosophila hydei*. *Chromosoma*, **94**, 147–61.
Kubai, D. F. (1975). The evolution of the mitotic spindle. *International Review of Cytology*, **43**, 167–227.
Kubai, D. F. (1982). Meiosis in *Sciara coprophila*: structure of the spindle and chromosome behaviour during the first meiotic division. *The Journal of Cell Biology*, **93**, 655–69.
Kubai, D. F. & Wise, D. (1981). Nonrandom segregation in *Neocurtilla* (*Gryllotalpa*) *hexadactyla*: an ultrastructural study. *The Journal of Cell Biology*, **88**, 281–93.
Kunz, B. A. & Haynes, R. H. (1981). Phenomenology and genetic control of mitotic recombination in yeast. *Annual Review of Genetics*, **15**, 57–89.
Kurata, N. & Ito, M. (1978). Electron microscope autoradiography of ^{3}H-thymidine incorporation during the zygotene stage in microsporocytes of lily. *Cell Structure and Function*, **3**, 349–56.
La Fountain, J. (1985). Chromosome movement during meiotic prophase in crane-fly spermatocytes III. Microtubules and the affects of colcemid, nocodazole and vinblastine. *Cell Motility*, **5**, 393–413.
Lajtha, L. G. (1983). Stem cell concepts. In *Stem Cells*, ed. C. S. Potten, pp. 1–11. Edinburgh: Churchill-Livingstone.
Lamb, B. C. (1977). The use of gene conversion to study synaptonemal complex structure and molecular details of chromatid pairing in meiosis. *Molecular and General Genetics*, **157**, 31–7.
Lamb, R. Y. & Willey, R. B. (1979). Are parthenogenetic and related bisexual insects equal in fertility. *Evolution*, **33**, 774–5.

Lamb, R. Y. & Willey, R. B. (1987). Cytological mechanisms of thelytokous parthenogenesis in insects. *Genome*, **29**, 367–9.

Lambert, A.-M. (1980). The role of chromosomes in anaphase trigger and nuclear envelope activity in spindle formation. *Chromosoma*, **76**, 295–308.

Lambie, E. J. & Roeder, G. S. (1986). Repression of meiotic crossing over by a centromere (CEN 3) in *Saccharomyces cerevisiae*. *Genetics*, **114**, 769–89.

Lambie, E. J. & Roeder, G. S. (1988). A yeast centromere acts in *cis* to inhibit meiotic gene conversion of adjacent sequences. *Cell* **52**, 863–73.

Lange, C. S. (1983). Stem cells in planarians. In *Stem Cells*, ed. C. S. Potten, pp. 28–66. Edinburgh: Churchill-Livingstone.

Lawrence, C. W. (1958). Genotypic control of chromosome behaviour in rye VI. Selection for disjunction frequency. *Heredity*, **12**, 127–31.

Lawrence, C. W. (1963). The orientation of multiple associations resulting from interchange heterzygosity. *Genetics*, **48**, 347–50.

Levan, A. (1933). Cytological studies in *Allium IV. Allium fistulosum. Svensk Botanisk Tidskrift*, **27**, 211–32.

Levan, A. (1935). Cytological studies in *Allium VI.* The chromosome morphology of some diploid species of *Allium*. *Hereditas*, **20**, 289–330.

Levan, A. (1940). Meiosis in *Allium porrum*, a tetraploid species with chiasma localization. *Hereditas*, **26**, 454–62.

Levan, A. & Müntzing, A. (1963). Terminology of chromosome numbers. *Portugaliae Acta Biologica*, **A7**, 1–16.

Lewis, K. R. (1961). The genetics of bryophytes. *Transactions of the British Bryological Society*, **4**, 111–30.

Lewis, K. R. (1966). The evolution of chromosome systems. *Indian Journal of Genetics and Plant Breeding*, **26**, 1–11.

Lewis, K. R. & John, B. (1957). The organisation and evolution of the sex multiple in *Blaps mucronata*. *Chromosoma*, **9**, 69–80.

Lewis, K. R. & John, B. (1963). *Chromosome Marker*. London: Churchill.

Lewis, K. R. & John, B. (1966). The meiotic consequences of spontaneous chromosome breakage. *Chromosoma*, **18**, 287–304.

Lewis, K. R. & John, B. (1967). Sex in plants. In *The Biology of Sex*, ed. A. Allison, pp. 59–75. Harmondsworth, Middlesex: Penguin Books.

Liapunova, N. A. & Babadjanian, D. P. (1973). A quantitative study of histone of meiocytes. Investigation of the histone amount in cricket spermatogenesis by interference microscopy. *Chromosoma*, **40**, 387–99.

Lie, T. & Laane, M. M. (1982). Reconstruction analyses of synaptonemal complexes in haploid and diploid pachytene nuclei of *Physarum polycephalum* (Myxomycetes). *Hereditas*, **96**, 119–40.

Lima-de-Faria, A. (1956). The role of the kinetochore in chromosome organisation. *Hereditas*, **42**, 85–160.

Lin, H.-P. P. & Church, K. (1982). Meiosis in *Drosophila melanogaster* III. The effect of orientation disruptor (*ord*) on gonial, mitotic and the meiotic divisions in males. *Genetics*, **102**, 751–70.

Lindsley, D. L. & Sandler, L. (1977). The genetic analysis of meiosis in female *Drosophila melanogaster*. *Philosophical Transactions of the Royal Society of London*, **B277**, 295–312.

Loidl, J. (1986). Synaptonemal complex spreading in *Allium* II. Tetraploid *A. vineale*. *Canadian Journal of Genetics and Cytology*, **28**, 754–61.

Loidl, J. (1987). Heterochromatin and differential chromosome staining in meiosis. *Biologisches Zentralblatt*, **106**, 641–62.

Loidl, J. & Jones, G. H. (1986). Synaptonemal complex spreading in *Allium* I. Triploid *A. sphaerocephala*. *Chromosoma*, **93**, 420–8.

Lokki, H. (1976a). Genetic polymorphism and evolution in parthenogenetic animals VII. The amount of heterozygosity in diploid populations. *Hereditas*, **83**, 57–64.
Lokki, J. (1976b). Genetic polymorphism and evolution in parthenogenetic animals VIII. Heterozygosity in relation to polyploidy. *Hereditas*, **83**, 65–72.
Lokki, J. & Saura, A. (1980). Polyploidy in insect evolution. In *Polyploidy: Biological Relevance*, ed. W. H. Lewis, pp. 277–312. New York: Plenum Press.
Lokki, J., Suomalainen, E., Saura, A. & Lankinen, P. (1975). Genetic polymorphism and evolution in parthenogenetic animals II. Diploid and polyploid *Solenobia triquetrella* (Lepidoptera: Psychidae). *Genetics*, **79**, 513–25.
Longley, A. E. (1945). Abnormal segregation during megasporogenesis in maize. *Genetics*, **30**, 100–13.
Lu, B. C. (1984). The cellular program for the formation and dissolution of the synaptonemal complex in *Coprinus*. *Journal of Cell Science*, **67**, 25–43.
Lu, B. C. & Jeng, D. Y. (1975). Meiosis in *Coprinus* VII. The prekaryogamy S-phase and the post karyogamy DNA replication in *C. lagopus*. *Journal of Cell Science*, **17**, 461–70.
Luykx, P. (1970). Cellular mechanisms of chromosome distribution. *International Review of Cytology*, **Supplement 2**, 1–173.
Luykx, P. & Syren, M. (1979). The cytogenetics of *Incisitermes schwarzi* and other Florida termites. *Sociobiology*, **4**, 191–209.
Luykx, P. & Syren, R. M. (1981a). Experimental hybridization between chromosome races in *Kalotermes approximatus*, a termite with extensive sex-linked translocation heterozygosity. *Chromosoma*, **83**, 563–73.
Luykx, P. & Syren, R. M. (1981b). Multiple sex-linked reciprocal translocations in a termite from Jamaica. *Experientia*, **37**, 819–20.
Lynch, M. (1984). Destabilizing hybridization, general-purpose genotypes and geographic parthenogenesis. *The Quarterly Review of Biology*, **59**, 257–89.
Lynch, M. & Gabriel, W. (1983). Phenotypic evolution and parthenogenesis. *The American Naturalist*, **122**, 745–64.
McClintock, B. (1933). The association of nonhomologous parts of chromosomes in the midprophase of meiosis in *Zea mays*. *Zeitschrift für Zellforschung und mikroskopische Anatomie*, **19**, 191–237.
McGill, M. & Brinkley, B. R. (1975). Human chromosomes and centrioles as nucleating sites for the in vitro assembly of microtubules from bovine tubulin. *The Journal of Cell Biology*, **67**, 189–99.
McIntosh, J. R. (1985). Spindle structure and the mechanisms of chromosome movement. In *Aneuploidy: Etiology and Mechanisms*, ed. V. L. Dellarco, P. E. Voytek & A. Hollaender, pp. 197–227. New York: Plenum Press.
McLaren, A. (1984). Meiosis and differentiation of mouse germ cells. In *Controlling Events of Meiosis*, ed. C. W. Evans & H. G. Dickinson. *Symposia of the Society for Experimental Biology*, **38**, 7–23.
McLeod, M. & Beach, D. (1988). A specific inhibitor of the *ran* 1^+ protein kinase regulates entry into meiosis in *Schizosaccharomyces pombe*. *Nature*, **332**, 509–14.
McQuade, H. A. & Bassett, B. (1977). Synaptonemal complexes in premeiotic interphase of pollen mother cells of *Triticum aestivum*. *Chromosoma*, **63**, 153–9.
Macgregor, H. C. (1980). Recent developments in the study of lampbrush chromosomes. *Heredity*, **44**, 3–35.
Madison, W. P. (1982). XXXY Sex chromosomes in males of the jumping spider genus *Pellenes* (Aranae: Salticidae). Chromosoma, **85**, 23–37.
Maeda, T. (1937). Chiasma studies in *Allium fistulosum*, *Allium cepa* and their F_1, F_2 and backcross hybrids. *Japanese Journal of Genetics*, **13**, 146–59.

Magee, P. T. (1987). Transcription during meiosis. In *Meiosis*, ed. P. B. Moens, pp. 355–82. Orlando: Academic Press.
Maguire, M. P. (1967). Evidence for homologous pairing of chromosomes prior to meiotic prophase in maize. *Chromosoma*, **21**, 221–31.
Maguire, M. P. (1978). Evidence for separate genetic control of crossing over and chiasma maintenance in maize. *Chromosoma*, **65**, 173–183.
Maguire, M. P. (1982). Evidence for a role of the synaptonemal complex in provision of normal chromosome disjunction at meiosis II in maize. *Chromosoma*, **84**, 675–686.
Mahowald, A. P. & Boswell, R. E. (1983). Germ plasm and germ cell development in invertebrates. In *Current Problems in Germ Cell Differentiation*, ed. A. McLaren & C. C. Wylie, pp. 3–17. Cambridge: Cambridge University Press.
Mahowald, A. P. & Hennen, S. (1971). Ultrastructure of the 'germ plasm' in eggs and embryos of *Rana pipiens*. *Developmental Biology*, **24**, 37–53.
Malik, C. P. & Tripathi, R. C. (1970). B chromosomes and meiosis in *Festuca mairei* St. Yav. *Zeitschrift für Biologie*, **116**, 321–6.
Manton, I. (1950). *Problems of Cytology and Evolution in the Pteridophyta.* Cambridge: Cambridge University Press.
Marek, L. F. (1978). Control of spindle form and function in grasshopper spermatocytes. *Chromosoma*, **68**, 367–98.
Marshall, D. R. & Brown, A. H. D. (1981). The evolution of apomixis. *Heredity*, **47**, 1–15.
Masui, Y., Lohka, M. J. & Shibuya, E. K. (1984). Roles of Ca^{2+} ions and ooplasmic factors in the resumption of metaphase-arrested meiosis in *Rana pipiens* oocytes. In *Controlling Events in Meiosis*, ed. C. W. Evans & H. G. Dickinson. *Symposia of the Society for Experimental Biology*, **38**, 45–66.
Mather, K. (1937). The determination of position in crossing-over II. The chromosome length-chiasma frequency relation. *Cytologia*, **Fujii Jubilee Volume**, 514–26.
Mather, K. (1938). Crossing over. *Biological Reviews*, **13**, 252–92.
Mather, K. (1939). Crossing over and heterochromatin in the X chromosome of *Drosophila melanogaster*. *Genetics*, **24**, 413–35.
Matsuda, Y., Imai, H. T., Moriwaki, K., Kondo, K. & Bonhomme, F. (1982). X-Y chromosome dissociation in wild derived *Mus musculus* subspecies, laboratory mice, and their F_1 hybrids. *Cytogenetics and Cell Genetics*, **34**, 241–52.
Matsuda, M., Imai, H. T. & Tobari, Y. N. (1983). Cytogenetic analysis of recombiantion in males of *Drosophila ananassae*. *Chromosoma*, **88**, 286–92.
Matsuura, H. (1950). Chromosome studies on *Trillium kamtschaticum* Pall. and its allies XIX. Chromatid breakage and reunion at chiasmata. *Cytologia*, **16**, 48–57.
Matthews, H. R. (1981). Chromatin patterns and progress through the cell cycle. In *The Cell Cycle*, ed. P. C. L. John, pp. 223–46. Cambridge: Cambridge University Press.
Maudlin, I. & Evans, E. P. (1980). Chiasma distribution in mouse oocytes during diakinesis. *Chromosoma*, **80**, 49–56.
Maynard-Smith, J. (1978). *The Evolution of Sex*. Cambridge: Cambridge University Press.
Mazia, D. (1978). Origin of twoness in cell reproduction. In *Cell Reproduction*, vol. 12, ed. E. R. Dirksen, D. M. Prescott & C. F. Fox, pp. 1–14, ICN-UCLA Symposia on Molecular and Cellular Biology, New York: Academic Press.
Mazia, D. (1984). Centrosomes and mitotic poles. *Experimental Cell Research*, **153**, 1–15.

Meistrich, M. L. & Brock, W. A. (1987). Proteins of the meiotic cells nucleus. In *Meiosis*, ed. P. B. Moens, pp. 333–53. Orlando: Academic Press.
Meselson, M. & Radding, C. M. (1975). A general model for genetic recombination. *Proceedings of the National Academy of Sciences*, **72**, 358–61.
Metz, C. W. (1933). Monocentric mitosis with segregation of chromosomes in *Sciara* and its bearing on the mechanism of mitosis. *Biological Bulletin Woods Hole*, **54**, 333–47.
Meyer, G. F. (1960). The fine structure of the spermatocyte nuclei of *Drosophila melanogaster*. *Proceedings of the European Regional Conference on Electron Microscopy, Delft*, **2**, 951–4.
Meyer, G. F. (1964). A possible correlation between the submicroscopic structure of meiotic chromosomes and crossing over. *Proceedings of the Third European Conference on Electron Microscopy, Prague*, 461–2.
Mitchison, T. J. (1986). The role of microtubule polarity in the movement of kinesin and kinetochores. *Journal of Cell Science*, **Supplement 5**, 121–8.
Mitchison, T. J., Evans, L., Schulze, E. & Kirschner, M. (1986). Sites of microtubule assembly and disassembly in the mitotic spindle. *Cell*, **45**, 515–27.
Mitchison, T. J. & Kirschner, M. W. (1985a). Properties of the kinetochore in vitro I. Microtubule nucleation and tubular binding. *The Journal of Cell Biology*, **101**, 755–65.
Mitchison, T. J. & Kirschner, M. W. (1985b). Properties of the kinetochore in vitro II. Microtubule capture and ATP-dependent translocation. *The Journal of Cell Biology*, **101**, 766–77.
Moens, P. B. (1969). The fine structure of meiotic chromosomes: polarization and pairing in *Locusta migratoria* spermatocytes. *Chromosoma*, **28**, 1–25.
Moens, P. B. (1973). Quantitative electron microscopy of chromosome organization at meiotic prophase. *Cold Spring Harbor Symposia in Quantitative Biology*, **38**, 99–107.
Moens, P. B. (1979). Kinetochore microtubule numbers of different sized chromosomes. *The Journal of Cell Biology*, **83**, 556–61.
Moens, P. B. (1985). Research needs in meiosis, mechanisms of synapsis, and chiasma regulation. In *Aneuploidy: Etiology and Consequences*, ed. V. L. Dellarco, P. E. Voytek & A. Hollaender, pp. 397–407. New York: Plenum Publishing Corporation.
Moens, P. B., Heyting, C., Dietrich, A. J. J., van Raamsdonk, W. & Chen, Q. (1987). Synaptonemal complex antigen location and conservation. *The Journal of Cell Biology*, **105**, 93–103.
Moens, P. B. & Pearlman, R. E. (1988). Chromatin organization at meiosis. *BioEssays*, **9**, 151–3.
Moens, P. B. & Short, S. (1983). Synaptonemal complexes of bivalents with localized chiasmata in *Chloealtis conspersa* (Orthoptera). *Kew Chromosome Conference*, **2**, 99–106.
Moore, P. D. & Holliday, R. (1976). Evidence for the formation of hybrid DNA during mitotic recombination in Chinese hamster cells. *Cell*, **8**, 573–9.
Morgan, G. T. (1978). Absence of chiasmata from the heteromorphic region of chromosome 1 during spermatogenesis in *Triturus cristatus carnifex*. *Chromosoma*, **66**, 269–80.
Moriwaki, D. & Tsujita, M. (1974). Synaptonemal complex and male crossing over in *Drosophila ananassae*. *Cytologia*, **39**, 829–38.
Moses, M. J. (1956). Chromosomal structures in crayfish spermatocytes. *Journal of Biophysical and Biochemical Cytology*, **2**, 215–7.
Moses, M. J. (1958). The relation between the axial complex of meiotic prophase

chromosomes and chromosome pairing in a salamander (*Plethodon cinereus*). *Journal of Biophysical and Biochemical Cytology*, **4**, 633–8.
Moses, M. J. (1977a). Synaptonemal complex karyotyping in spermatocytes of the Chinese hamster (*Cricetulus griseus*) I. Morphology of the autosomal complement in spread preparations. *Chromosoma*, **60**, 99–125.
Moses, M. J. (1977b). The synaptonemal complex and meiosis. In *Molecular Human Genetics*, ed. R. S. Sparkes, D. E. Comings & F. Fox, pp. 101–25. New York: Academic Press.
Moses, M. J., Karatsis, P. A. & Hamilton, A. E. (1979). Synaptonemal complex analysis of heteromorphic trivalents in *Lemur* hybrids. *Chromosoma*, **70**, 141–60.
Moses, M. J. & Poorman, P. A. (1981). Synaptonemal complex analysis of mouse chromosome rearrangements II. Synaptic adjustment in a tandem duplication. *Chromosoma*, **81**, 519–35.
Moses, M J. & Poorman, P. A. (1984). Synapsis, synaptic adjustment and DNA synthesis in mouse oocytes. *Chromosomes Today*, **8**, 90–103.
Moses, M. J., Poorman, P. A. Roderick, T. H. & Davisson, M. T. (1982). Synaptonemal complex analysis of mouse chromosome rearrangements IV. Synapsis and synaptic adjustment in two paracentric inversions. *Chromosoma*, **84**, 457–74.
Moses, M. J., Poorman, P. A., Russell, L. B., Cacheiro, N. L., Roderick, T. H. & Davisson, M. T. (1978). Synaptic adjustment: two pairing phases in meiosis. *The Journal of Cell Biology*, **79**, 123a.
Müller, W. P. (1977). Diplotene chromosomes of *Xenopus* hybrid oocytes. *Chromosoma*, **59**, 273–82.
Murakami, A. & Imai, H. T. (1974). Cytological evidence for holocentric chromosomes of the silkworms *Bombyx mori* and *B. mandarina* (Bombycidae, Lepidoptera). *Chromosoma*, **47**, 167–78.
Murer-Orlando, M. & Richter, C.-L. (1983). Heterochromatin heterogeneity in Chinese hamster sex bivalents. *Cytogenetics and Cell Genetics*, **35**, 195–9.
Murray, A. W. & Szostak, J. W. (1985). Chromosome segregation in mitosis and meiosis. *Annual Review of Cell Biology*, **1**, 289–315.
Narbel, M. (1946). La cytologie de la parthénogénèse chez *Apterona helix* Sieb. (Lep. Psychide). Revue Suisse de Zoologie, **53**, 625–81.
Narbel-Hofstetter, M. (1950). La cytologie de la parthénogénèse chez *Solenobia* sp. (*lichenella* L.?) (Lépidoptères, Psychide). *Chromosoma*, **4**, 56–90.
Neijzing, M. G. (1982). Chiasma formation in duplicated segments of the haploid rye genome. *Chromosoma*, **85**, 287–98.
Nel, P. M. (1973). The modification of crossing over in maize by extraneous chromosomal elements. *Theoretical and Applied Genetics*, **43**, 196–202.
Nelson-Rees, W. A. (1960). A study of sex predetermination in the mealybug *Planococcus citri* (Risso). *Journal of Experimental Zoology*, **144**, 111–37.
Nelson-Rees, W. A. (1963). New observations on lecanoid spermatogenesis in the mealybug *Planococcus citri*. *Chromosoma*, **14**, 1–17.
Nguyen, T. D. & Boyd, J. B. (1977). The meiotic-9 (*mei-9*) mutants of *Drosophila melanogaster* are deficient in repair replication of DNA. *Molecular and General Genetics*, **158**, 141–7.
Nicklas, R. B. (1961). Recurrent pole to pole movements of the sex chromosomes during prometaphase I in *Melanoplus differentialis* spermatocytes. *Chromosoma*, **12**, 97–115.
Nicklas, R. B. (1967). Chromosome micromanipulation II. Induced reorientation and the experimental control of segregation in meiosis. *Chromosoma*, **21**, 17–50.
Nicklas, R. B. (1974). Chromosome segregation mechanisms. *Genetics*, **78**, 205–13.

Nicklas, R. B. (1977). Chromosome movement: facts and hypotheses. In *Mitosis: Facts and Questions*, ed. M. Little, N. Paweletz, C. Petzelt, H. Ponstingl, D. Schroeter, H.-P. Zimmermann, pp. 150–5. Berlin: Springer-Verlag.

Nicklas, R. B. (1985). Mitosis in eukaryotic cells: an overview of chromosome distribution. In *Aneuploidy: Etiology and Consequences*, ed. V. L. Dellarco, P. E. Voytek & A. Hollaender, pp. 183–94. New York: Plenum Publishing Corporation.

Nicklas, R. B. (1988). Chromosomes and kinetochores do more in mitosis than previously thought. In *Chromosome Structure and Function: The Impact of New Concepts*, ed. J. P. Gustafson, R. Appels and R. J. Kaufman, pp 53–74. New York: Plenum Publishing Corporation.

Nicklas, R. B. (1988a). Chance encounters and precision in mitosis. *Journal of Cell Science*, **89**, 283–5.

Nicklas, R. B. (1988b). The forces that move chromosomes in mitosis. *Annual Review of Biophysics and Biophysical Chemistry*, **17**, 431–49.

Nicklas, R. B. & Gordon, G. W. (1985). The total length of spindle microtubules depends on the number of chromosomes present. *The Journal of Cell Biology*, **100**, 1–7.

Nicklas, R. B., Brinkley, B. R., Pepper, D. A., Kubai, D. & Rickards, G. K. (1979). Electron microscopy of spermatocytes previously studied in life: methods and some observations on micromanipulated chromosomes. *Journal of Cell Science*, **35**, 87–104.

Nicklas, R. B. & Koch, C. A. (1969). Chromosome micromanipulation III. Spindle fiber tension and the reorientation of mal-oriented chromosomes. *The Journal of Cell Biology*, **43**, 40–50.

Nicklas, R. B. & Kubai, D. F. (1985). Microtubules, chromosome movement and reorientation after chromosomes are detached from the spindle by micromanipulation. *Chromosoma*, **92**, 313–24.

Noda, S. (1975). Achiasmate meiosis in the *Fritillaria japonica* group 1. Different modes of bivalent formation in the two sex mother cells. *Heredity*, **34**, 373–80.

Nokkala, S. (1983). Segregation mechanism of distance or touch-and-go paired chromosomes. *Kew Chromosome Conference*, **2**, 191–4.

Nokkala, S. & Nokkala, C. (1983). Achiasmatic male meiosis in two species of *Saldula* (Saldidae, Hemiptera). *Hereditas*, **99**, 131–4.

Nokkala, S. & Nokkala, C. (1984). Achiasmate male meiosis in the heteropteran genus *Nabis* (Nabidae, Hemiptera). *Hereditas*, **101**, 31–5.

Nokkala, S. & Puro, J. (1976). Cytological evidence for a chromocenter in *Drosophila melanogaster* oocytes. *Hereditas*, **83**, 265–8.

Nordenskiöld, H. (1962). Studies of meiosis in *Luzula purpurea*. *Hereditas*, **48**, 503–19.

Nordenskiöld, H. (1963). A study of meiosis in the progeny of X-irradiated *Luzula purpurea*. *Hereditas*, **49**, 33–47.

Norrmann, G. A. & Quárin, C. L. (1987). Permanent odd polyploidy in a grass (*Andropogon ternatus*). *Genome*, **29**, 340–44.

Novitski, E. & Puro, J. (1978). A critique of theories of meiosis in the female of *Drosophila melanogaster*. *Hereditas*, **89**, 51–67.

Nur, U. (1971). Parthenogenesis in coccids. *American Zoologist*, **11**, 301–8.

Nur, U. (1978). Asymmetrically heteropycnotic X chromosomes in the grasshopper *Melanoplus femur-rubrum*. *Chromosoma*, **68**, 165–85.

Nur, U. (1979). Gonoid thelytoky in soft scale insects (Coccidae: Homoptera). *Chromosoma*, **72**, 89–104.

Nur, U. (1980). Evolution of unusual chromosome systems in scale insects (Coccoidea: Homoptera). In *Insect Cytogenetics*, ed. R. L. Blackman, G. M. Hewitt & M. Ashburner, pp. 97–117. Oxford: Blackwell Scientific Publishers.

Oakley, H. A. (1982). Meiosis in *Mesostoma ehrenbergii ehrenbergii* II. Synaptonemal complexes, chromosome pairing and disjunction in achiasmate oogenesis. *Chromosoma*, **87**, 133–47.

Oakley, H. A. (1983). Male meiosis in *Mesostoma ehrenbergii ehrenbergii. Kew Chromosome Conference*, **2**, 195–9.

Oakley, H. A. (1985). Meiosis in *Mesostoma ehrenbergii ehrenbergii* (Turbellaria, Rhabdocoela) III. Univalent chromosome segregation during the first meiotic division in spermatocytes. *Chromosoma*, **91**, 95–100.

Oakley, H. A. & Jones, G. H. (1982). Meiosis in *Mesotoma ehrenbergii ehrenbergii* (Turbellaria, Rhabdocoela) I. Chromosome pairing, synaptonemal complexes and chiasma localization in spermatogenesis. *Chromosoma*, **85**, 311–22.

Ogawa, K. (1954). Chromosome studies in the Myriapoda VII. A chain association of the multiple sex chromosomes found in *Otocryptops sexspinosus* (Say). *Cytologia*, **19**, 265–72.

Ogawa, K. (1961). Chromosome studies in the Myriapoda XIII. Three types of sex chromosomes found in *Otocryptops rubiginosis* (L. Koch). *Japanese Journal of Genetics*, **36**, 122–8.

Ohno, S. (1967). *Sex Chromosomes and Sex-Linked Genes*. Heidelberg: Springer-Verlag.

Oliver, J. H. (1967). Cytogenetics of acarines. In *Genetics of Insect Vectors of Disease*, ed. J. W. Wright and R. Pal, pp. 417–39. Amsterdam: Elsevier Co.

Oliver, J. H. (1971). Parthenogenesis in mites and ticks (Arachnida: Acari). *American Zoologist*, **11**, 283–299.

Olson, L. W. & Zimmermann, F. K. (1978). Mitotic recombination in the absence of synaptonemal complexes in *Saccharomyces cerevisiae. Molecular and General Genetics*, **166**, 161–5.

Omodeo, P. (1951). Il fenomeno della restituzione premeiotica in lombricidi partenogenetici. *Bolletino della Societa Italiana di Biologia Sperimentale*, **27**, 1292–3.

Orlando, E. (1974). Sex determination in *Megoura viciae* Buckton (Homoptera, Aphididae). *Monitore Zoologico Italiano (NS)*, **8**, 61–70.

Ortiz, E. (1969). Chromosomes and meiosis in Dermaptera. *Chromosomes Today*, **2**, 33–43.

Östergren, G. (1951). The mechanism of co-orientation in bivalents and multivalents. *Hereditas*, **37**, 85–156.

Palmer, R. G. (1971). Cytological studies of ameiotic and normal maize with reference to premeiotic pairing. *Chromosoma*, **35**, 233–46.

Parker, J. S. (1975). Chromosome-specific control of chiasma formation. *Chromosoma*, **49**, 391–406.

Pastor, J. B. & Callan, H. G. (1952). Chiasma formation in spermatocytes and oocytes of the turbellarian *Dendrocoelum lacteum. Journal of Genetics*, **50**, 449–54.

Pathak, S., Elder, F. F. B. & Maxwell, B. L. (1980). Asynaptic behavior of the X and Y chromosomes in the Virginia opossum and the southern pygmy mouse. *Cytogenetics and Cell Genetics*, **26**, 143–9.

Pathak, S. & Hsu, T. C. (1976). Chromosomes and DNA of *Mus*: the behaviour of constitutive heterochromatin in spermatogenesis of *M. dunni. Chromosoma*, **57**, 227–34.

Payne, F. (1916). A study of the germ cells of *Gryllotalpa borealis* and *Gryllotalpa vulgaris. Journal of Morphology*, **28**, 287–327.

Peacock, W. J. (1971). Cytogenetic aspects of the mechanism of recombination in higher organisms. *Stadler Symposia*, **1 and 2**, 123–52.

Pepper, D. A. (1988). The mammalian kinetochore. In *Heterochromatin: Molecular and Structural Aspects*, ed. R. S. Verma, pp. 187–202. New York: Cambridge University Press.

Perry, P. E. & Jones, G. H. (1974). Male and female meiosis in grasshoppers I. *Stethophyma grossum*. *Chromosoma*, **47**, 227–36.

Petersen, J. B. & Ris, H. (1976). Electron-microscopic study of the spindle and chromosome movement in the yeast *Saccharomyces cerevisiae*. *Journal of Cell Science*, **22**, 219–42.

Pickett-Heaps, J. (1986). Mitotic mechanisms: an alternative view. *Trends in Biochemical Sciences*, **11**, 504–7.

Pickett-Heaps, J. D., Tippit, D. H. & Porter, K. R. (1982). Rethinking mitosis. *Cell*, **29**, 729–44.

Pickett-Heaps, J., Tippit, D. H., Cohn, S. A. & Spurck, T. P. (1986). Microtubule dynamics in the spindle. Theoretical aspects of assembly/disassembly reactions in vivo. *Journal of Theoretical Biology*, **118**, 153–69.

Pijnacker, L. P. & Ferwerda, M. A. (1978). Additional chromosome duplication in female meiotic prophase of *Sipyloidea sipylus* Westwood (Insecta, Phasmida) and its absence in male meiosis. *Experientia*, **34**, 1558–60.

Pijnacker, L. P. & Ferwerda, M. A. (1986). Development of the synaptonemal complex of two types of pachytene in oocytes and spermatocytes *Carausius morosus* Br. (Phasmatodea). *Chromosoma*, **93**, 281–90.

Polani, P. E. (1972). Centromere localization at meiosis and the position of chiasmata in the male and female meiosis. *Chromosoma*, **36**, 343–74.

Polani, P. E. (1982). Pairing of X and Y chromosomes: non-activation of X-linked genes and the maleness factor. *Human Genetics*, **60**, 207–211.

Poorman, P. A., Moses, M. J., Davisson, M. T. & Roderick, T. H. (1981). Synaptonemal complex analysis of mouse chromosomal rearrangements III. Cytogenetic observations on two paracentric inversions. *Chromosoma*, **83**, 419–29.

Porter, D. L. (1971). Oogenesis and chromosomal heterozygosity in the thelytokous midge, *Lundstroemia parthenogenetica* (Diptera, Chironomidae). *Chromosoma*, **32**, 332–42.

Postlethwait, J. H. & Giorgi, F. (1985). Vitellogenesis in Insects. In *Developmental Biology: A Comprehensive Synthesis vol. 1 Oogenesis*, ed. L. W. Browder, pp. 85–119. New York: Plenum Press.

Prescott, D. M. (1976). The cell cycle and the control of cellular reproduction. *Advances in Genetics*, **18**, 99–177.

Procunier, W. S. (1975). A cytological study of two closely related blackfly species: *Cnephia dacotensis* and *Cnephia ornithophilia* (Diptera: Simulidae) *Canadian Journal of Zoology*, **53**, 1622–37.

Puro, J. (1978). Factors affecting disjunction of chromosome 4 in *Drosophila melanogaster* female. *Hereditas*, **88**, 274–6.

Puro, J. & Nokkala, S. (1977). Meiotic segregation of chromosomes in *Drosophila melanogaster* oocytes. A cytological approach. *Chromosoma*, **63**, 273–86.

Rabl, C. (1885). Über Zellteilung. *Morphologisches Jahrbuch*, **10**, 214–330.

Raff, E. C. (1984). Genetics of microtubule systems. *Journal of Cell Biology*, **99**, 1–10.

Rahn, M. I. & Martinez, A. (1983). Chromosome pairing in female and male diploid and polyploid anurans (Amphibia) from South America. *Canadian Journal of Genetics and Cytology*, **25**, 487–501.

Rahn, M. I. & Solari, A. J. (1986). Recombination modules in the oocytes of the chicken *Gallus domesticus*. *Cytogenetics and Cell Genetics*, **43**, 187–93.

Raman, R. & Nanda, I. (1986). Mammalian sex chromosomes I. Cytological changes in the chiasmatic sex chromosomes of the male musk shrew, *Suncus murinus*. *Chromosoma*, **93**, 367–74.

Rao, P. N. & Johnson, R. T. (1970). Mammalian cell fusion: studies on the regulation of DNA synthesis and mitosis. *Nature*, **225**, 159–64.

Rasmussen, S. W. (1973). Ultrastructural studies of spermatogenesis in *Drosophila melanogaster* Meigen. *Zeitschrift für Zellforschung*, **140**, 125–44.

Rasmussen, S. W. (1974). Studies on the development and ultrastructure of the synaptonemal complex in *Drosophila melanogaster*. *Comptes Rendus des Travaux du Laboratoire Carlsberg*, **39**, 443–68.

Rasmussen, S. W. (1975). Ultrastructural studies of meiosis in males and females of the $c(3)G^{17}$ mutant of *Drosophila melanogaster* Meigen. *Comptes Rendus des Travaux du Laboratoire Carlsberg*, **40**, 167–73.

Rasmussen, S. W. (1976). The meiotic prophase in *Bombyx mori* analyzed by three-dimensional reconstructions of synaptonemal complexes. *Chromosoma*, **54**, 245–93.

Rasmussen, S. W. (1977a). The transformation of the synaptonemal complex into the 'elimination' chromatin in *Bombyx mori* oocytes. *Chromosoma*, **60**, 205–221.

Rasmussen, S. W. (1977b). Chromosomal pairing in triploid females of *Bombyx mori* analysed by three-dimensional reconstructions of synaptonemal complexes. *Carlsberg Research Communications*, **42**, 163–97.

Rasmussen, S. W. (1986). Initiation of synapsis and interlocking of chromosome during zygotene in *Bombyx* spermatocytes. *Carlsberg Research Communications*, **51**, 401–32.

Rasmussen, S. W. (1987). Chromosome pairing in autotetraploid *Bombyx* males. Inhibition of multivalent correction by crossing over. *Carlsberg Research Communications*, **52**, 211–42.

Rasmussen, S. W. & Holm, P. B. (1979). Chromosome pairing in autotetraploid *Bombyx* females. Mechanism for exclusive bivalent formation. *Carlsberg Research Communications*, **44**, 101–25.

Rasmussen, S. W. & Holm, P. B. (1980). Mechanics of meiosis. *Hereditas*, **93**, 187–216.

Rasmussen, S. W. & Holm, P. B. (1982). The meiotic prophase in *Bombyx mori*. In *Insect Ultrastructure, vol. 1*, ed. R. C. King & H. Akar, pp. 61–85. New York: Plenum Publishing Company.

Rasmussen, S. W. & Holm, P. B. (1984). The synaptonemal complex, recombination nodules and chiasmata in human spermatocytes. In *Controlling Events in Meiosis*, ed. C. W. Evans & H. G. Dickinson. *Symposia of the Society for Experimental Biology*, **38**, 273–92.

Rasmussen, S. W., Holm, P. B., Lu, B. C., Zickler, D. & Sage, J. (1981). Synaptonemal complex formation and distribution of recombination nodules in pachytene trivalents of triploid *Coprinus cinereus*. *Carlsberg Research Communications*, **46**, 347–60.

Rattner, J. B., Goldsmith, M. & Hamkalo, B. A. (1980). Chromatin organization during meiotic prophase of *Bombyx mori*. *Chromosoma*, **79**, 215–24.

Reddy, O. S., Reddy, G. M. & Rao, M. S. (1965). Induction of crossing-over in *Drosophila* males by means of ovarian extract. *Nature*, **208**, 203.

Rees, H. (1961). Genotypic control of chromosome behaviour. *Botanical Reviews*, **27**, 288–318.

Rees, H. (1962). Developmental variation in the expressivity of genes causing chromosome breakage in rye. *Heredity*, **17**, 427–37.

Rees, H. (1984). Nuclear DNA variation and the homology of chromosomes. In *Plant Biosystematics*, ed. W. F. Grant, pp. 87–96. Toronto: Academic Press.
Rees, H. & Durrant, A. (1986). Recombination and genome size. *Theoretical and Applied Genetics*, **73**, 72–6.
Rees, H. & Narayan, R. K. J. (1989). Chromosome constraints: chiasma frequency and genome size. *3rd Kew Chromosome Conference*, 231–9.
Rees, H. & Sun, S. (1965). Chiasma, frequency and the disjunction of interchange associations in rye. *Chromosoma*, **16**, 500–10.
Rees, H. & Thompson, J. B. (1955). Localisation of chromosome breakage at meiosis. *Heredity*, **9**, 399–407.
Reitalu, J. (1970). Observations on the behavioural pattern of the sex chromosome complex during spermatogenesis in man. *Hereditas*, **64**, 283–90.
Resnick, M. A. (1987). Investigating the genetic control of biochemical events in meiotic recombination. In *Meiosis*, ed. P. B. Moens, pp. 157–209. Orlando: Academic Press.
Revell, S. H. (1947). Controlled X-segregation at meiosis in *Tegenaria*. *Heredity*, **1**, 337–47.
Rhoades, M. M. (1952). Preferential segregation in maize. In *Heterosis*, ed. J. W. Gowen, pp. 66–80. Ames: Iowa State College Press.
Ribbert, D. & Kunz, W. (1969). Lampenbürstenchromosomen in der Oocyten Kernen von *Sepia officinalis*. *Chromosoma*, **28**, 93–106.
Rick, C. M. (1974). The Tomato. In *Handbook of Genetics*, vol. 2, ed. R. C. King, pp. 247–80. New York: Plenum Press.
Rickards, G. K. (1975). Prophase chromosome movements in living house cricket spermatocytes and their relationship to prometaphase, anaphase and granule movements. *Chromosoma*, **49**, 407–55.
Rickards, G. K. (1981). Chromosome movements within prophase nuclei. In *Cell Biology: Mitosis/Cytokinesis*, ed. A. M. Zimmerman & A. Forer, pp. 103–31. New York: Academic Press.
Rickards, G. K. (1983). Orientation behaviour of chromosome multiples of interchange (reciprocal translocation) heterozygotes. *Annual Review of Genetics*, **17**, 443–98.
Rieder, C. L. (1982). The formation, structure and composition of the mammalian kinetochore and kinetochore fibre. *International Review of Cytology*, **79**, 1–58.
Rieder, C. L. & Nowogrodzki, R. (1983). Intranuclear membranes and the formation of the first meiotic spindle in *Xenos peckii* (*Aroschismus wheeleri*) oocytes. *The Journal of Cell Biology*, **97**, 1144–55.
Riley, R. & Flavell, R. B. (1977). A first view of the meiotic process. *Philosophical Transactions of the Royal Society of London*, **B277**, 191–9.
Ripoll, P., Pimpinelli, S., Valdivia, M. M. & Avila, J. (1985). A cell division mutant of *Drosophila* with a functionally abnormal spindle. *Cell*, **41**, 907–12.
Robbins, L. G. (1971). Non exchange alignment: a meiotic process revealed by a synthetic meiotic mutant of *Drosophila melanogaster*. *Molecular and General Genetics*, **110**, 144–66.
Rohloff, H. (1970). Die spermatocyten-teilungen der Tipuliden. IX Mitteilung. Analyse der Orientierung röntgenstrahleninduzierter Quadrivalente bei *Pales ferruginea*. Ph.D. Thesis, Eberhard-Karls-Universität zu Tübingen.
Roman, H. & Fabre, F. (1983). Gene conversion and associated reciprocal recombination are separable events in vegetative cells of *Saccharomyces cerevisiae*. *Proceedings of the National Academy of Sciences*, **80**, 6912–6.
Rosen, J. M. & Westergaard, M. (1966). Studies on the mechanisms of crossing over II. Meiosis and the time of meiotic chromosome replication in the

Ascomycete *Neottiella rutilans* (Fr.) Dennis. *Comptes Rendus des Travaux du Laboratoire Carlsberg*, **35**, 233–60.

Roth, T. F. & Ito, M. (1967). DNA-dependent formation of the synaptinemal complex at meiotic prophase. *The Journal of Cell Biology*, **35**, 247–55.

Roth, T. F. & Parchman, L. G. (1971). Alteration of meiotic chromosomal pairing and synaptonemal complexes by cyclohexamide. *Chromosoma*, **35**, 9–27.

Rothfels, K. H. & Mason, G. F. (1975). Achiasmate meiosis and centromere shift in *Eusimulium aureum* (Diptera-Simulidae). *Chromosoma*, **51**, 111–24.

Rouyer, F., Simmler, M.-C., Johnson, C., Vergnaud, G., Cooke, H. J. & Weissenbach, J. (1986). A gradient of sex linkage in the pseudoautosomal region of the human sex chromosomes. *Nature*, **319**, 291–5.

Rowell, D. M. (1985). Complex sex linked fusion heterozygosity in the Australian huntsman spider *Delena cancerides* (Araneae: Sparassidae). *Chromosoma*, **93**, 169–76.

Rowell, D. M. (1987). The population genetics of the huntsman spider *Delena cancerides*. Ph.D. Thesis, Australian National University.

Rückert, J. (1892). Zur Entwicklungsgeschichte des Ovarioleies bei Selachien. *Anatomischer Anzeiger*, **7**, 107–58.

Ruthmann, A. & Dahlberg, R. (1976). Pairing and segregation of the sex chromosomes in X_1X_2 males of *Dysdercus intermedius* with a note on the kinetic organization of heteropteran chromosomes. *Chromosoma*, **54**, 89–97.

Ruthmann, A. & Permantier, Y. (1973). Spindel und Kinetochoren in der Mitose und Meiose der Baumwollwanze *Dysdercus intermedius* (Heteroptera). *Chromosoma*, **41**, 271–88.

Sachs, L. (1953). The giant sex chromosomes in the mammal *Microtus agrestis*. *Heredity*, **7**, 227–38.

Sandler, L. & Szauter, P. (1978). The effect of recombination-defective meiotic mutants on fourth-chromosome crossing over in *Drosophila melanogaster*. *Genetics*, **90**, 699–712.

Saura, A., Lokki, J., Lankinen, P. & Suomalainen, E. (1976a). Genetic polymorphism and evolution in parthenogenetic animals III. Tetraploid *Otiorrhyncus scaber* (Coleoptera: Curculionidae). *Hereditas*, **82**, 79–100.

Saura, A., Lokki, J., Lankinen, P. & Suomalainen, E. (1976b). Genetic polymorphism and evolution in parthenogenetic animals IV. Triploid *Otiorhynchus salicis* (Coleoptera: Curculionidae). *Entomologica Scandinavica*, **7**, 1–6.

Scali, V. (1982). Evolutionary biology and speciation of the stick insect *Bacillus rossius* (Insecta, Phasmatodea). In *Mechanisms of Speciation*, ed. C. Barigozzi, pp. 393–410. New York: Alan R. Liss Inc.

Schibler, M. J. & Pickett-Heaps, J. D. (1987). The kinetochore fiber structure in the acentric spindles of the green alga *Oedogonium*. *Protoplasma*, **137**, 29–44.

Schmid, M., Solleder, E. & Haaf, T. (1984). The chromosomes of *Micromys minutus* (Rodentia, Murinae). *Cytogenetics and Cell Genetics*, **38**, 221–6.

Scholl, H. (1956). Die Chromosomen parthenogenetischer Mücken. *Naturwissenschaften*, **43**, 91–2.

Scholl, H. (1960) Die Oogenese eineiger Parthenogenetischer Orthocladiinen (Diptera). *Chromosoma*, **11**, 380–401.

Schrader, F. (1940). Touch-and-go pairing in chromosomes. *Proceedings of the National Academy of Sciences*, **26**, 634–6.

Schuetz, A. W. (1985). Local control mechanisms during oogenesis and folliculogenesis. In *Developmental Biology: A Comprehensive Synthesis, vol. 1. Oogenesis*, ed. L. W. Browder, pp. 3–83. New York: Plenum Press.

Schultz, J. & Redfield, H. (1951). Interchromosomal effect on crossing over in *Drosophila*. *Cold Spring Harbor Symposia in Quantitative Biology*, **16**, 175–97.
Schultz, R. J. (1980). Role of polyploidy in the evolution of fishes. In *Polyploidy: Biological Relevance*, ed. W. H. Lewis, pp. 313–40. New York: Plenum Press.
Schultz, R. J. (1982). Competition and adaptation among diploid and polyploid clones of unisexual fishes. In *Evolution and Genetics of Life Histories*, ed. H. Dingle & J. P. Hegmann, pp. 103–19. New York: Springer Verlag.
Schwarzacher, T., Mayr, B. & Schweizer, D. (1984). Heterochromatin and nucleolus-organizer region behaviour at male pachytene of *Sus scrofa domestica*. *Chromosoma*, **91**, 12–9.
Sears, E. R. (1976). Genetic control of chromosome pairing in wheat. *Annual Review of Genetics*, **10**, 31–51.
Seiler, J. & Schäffer, K. (1960). Untersuchungen über die Entstehung der Parthenogenese bei *Solenobia triquetrella* F.R. (Lepidoptera, Psychidae) II. Analyse der diploidparthenogenetischen *S. triquetrella*, Verhalten, Aufzuchtresultate und Zytologie. *Chromosoma*, **11**, 29–102.
Sen, S. K., Hazra, S. K., Iyengar, A. S. & Banerjee, R. S. (1981). A comparison of intra- and inter-gene recombination in the meiotic mutant *c(3)G* of *Drosophila melanogaster*. *Genetica*, **55**, 47–9.
Serrano, J. (1981). Male achiasmatic meiosis in Caraboidea (Coleoptera: Adephaga). *Genetica*, **57**, 131–7.
Shanahan, C. M. (1986). Cytogenetic studies on Australian scorpions. Ph.D. Thesis, University of Adelaide.
Sharp, P. (1982). Sex chromosome pairing during male meiosis in marsupials. *Chromosoma*, **86**, 27–47.
Shaw, D. D. (1970). Pseudomultiple production in *Agenotettix deorum deorum*. *Chromosoma*, **31**, 421–33.
Shaw, D. D. (1971a). The supernumerary segment system of *Stethophyma* II. Heterochromatin polymorphism and chiasma variation. *Chromosoma*, **34**, 19–39.
Shaw, D. D. (1971b). Genetic and environmental components of chiasma control I. Spatial and temporal variation in *Schistocerca* and *Stethophyma*. *Chromosoma*, **34**, 281–301.
Shaw, D. D. (1972). Genetic and environmental components of chiasma control II. The response to selection in *Schistocerca*. *Chromosoma*, **37**, 297–308.
Shaw, D. D. & Knowles, G. R. (1976). Chiasma analysis using a computerised optical digitiser. *Chromosoma*, **59**, 103–27.
Shaw, D. D. & Wilkinson, P. (1978). 'Homologies' between non-homologous chromosomes in the grasshopper *Caledia captiva*. *Chromosoma*, **68**, 241–59.
Shchapova, A. I. (1971). On the karyotype pattern and the chromosome arrangements in the interphase nucleus. *Tsitologiya*, **13**, 1157–64.
Shepherd, J., Boothroyd, E. R. & Stern, H. (1974). The effect of colchicine on synapsis and chiasma formation in microsporocytes of *Lilium*. *Chromosoma*, **44**, 423–37.
Sheridan, W. F. & Stern, H. (1967). Histones of meiosis. *Experimental Cell Research*, **45**, 323–35.
Sillers, P. J. & Forer, A. (1981). Autosomal spindle fibres influence subsequent sex-chromosome movements in crane-fly spermatocytes. *Journal of Cell Science*, **49**, 51–67.
Simchen, G. (1974). Are mitotic functions required in meiosis? *Genetics*, **76**, 745–53.
Simchen, G. (1978). Cell cycle mutants. *Annual Review of Genetics*, **12**, 161–91.
Simchen, G., Kassir, Y., Horesh-Cabilly, O. & Friedmann, A. (1981). Elevated

recombination and pairing structures during meiotic arrest in yeast of the nuclear division mutant *cdc 5*. *Molecular and General Genetics*, **184**, 46–51.
Simchen, G., Piñon, R. & Salts, Y. (1972). Sporulation in *Saccharomyces cerevisiae*: premeiotic DNA synthesis, readiness and commitment. *Experimental Cell Research*, **75**, 207–18.
Simmler, M. C., Rouyer, F., Vergnaud, G., Nystrom-Lahti, M., Ngo, K. Y., Chapelle, A. de la & Weissenbach, J. (1985). Pseudoautosomal DNA sequences in the pairing region of the human sex chromosomes. *Nature*, **317**, 692–7.
Sluder, G., Rieder, C. L. & Miller, F. (1985). Experimental separation of pronuclei in fertilized sea urchin eggs: chromosomes do not organize a spindle in the absence of centrosomes. *The Journal of Cell Biology*, **100**, 897–903.
Smith, L. D., Michael, P. & Williams, M. A. (1983). Does a predetermined germ cell line exist in amphibians? In *Current Problems in Germ Cell Differentiation*, ed. A. McLaren & C. C. Wylie, pp. 19–39. Cambridge: Cambridge University Press.
Smith, S. G. & Virrki, N. (1978). Coleoptera. In *Animal Cytogenetics*, vol. 3 (5), ed. B. John, pp. 1–366. Stuttgart: Gebrüder Borntraeger.
Smith-White, S. (1948). Polarised segregation in the pollen mother cells of a stable triploid. *Heredity*, **2**, 119–29.
Smith-White, S. (1955). The life history and genetic system of *Leucopogon juniperinum*. *Heredity*, **9**, 79–91.
Solari, A. J. (1970). The behaviour of chromosomal axes during diplotene in mouse spermatocytes. *Chromosoma*, **31**, 217–30.
Solari, A. J. (1974). The behavior of the XY pair in mammals. *International Review of Cytology*, **38**, 273–317.
Solari, A. J. (1977). Ultrastructure of the synaptic autosomes and the ZW bivalent in chicken oocytes. *Chromosoma*, **64**, 155–65.
Solari, A. J. (1979). Autosomal synaptonemal complexes and sex chromosomes without a SC in *Triatoma infestans* (Reduviidae: Hemiptera). *Chromosoma*, **72**, 225–240.
Solari, A. J. (1980). Synaptonemal complexes and associated structures in microspread human spermatocytes. *Chromosoma*, **81**, 315–37.
Solari, A. J. & Ashley, T. (1977). Ultrastructure and behaviour of the achiasmatic telosynaptic XY pair of the sand rat (*Psammomys obesus*). *Chromosoma*, **62**, 319–36.
Solari, A. J. & Rahn, M. I. (1985). Asymmetry and resolution of the synaptonemal complex in the XY pair of *Chinchilla laniger*. *Genetica*, **67**, 63–71.
Sorsa, M. & Suomalainen, E. (1975). Electron microscopy of chromatin elimination in *Cidaria* (Lepidoptera). *Hereditas*, **80**, 35–40.
Southern, D. I. (1967a). Pseudo-multiple formation as a consequence of prolonged non-homologous chromosome association in *Metrioptera brachyptera*. *Chromosoma*, **21**, 272–84.
Southern, D. I. (1967b). Chiasma distribution in truxaline grasshoppers. *Chromosoma*, **22**, 164–91.
Southern, D. I. (1968). Persistent heterochromatic association in *Metrioptera brachyptera* I. Variation in the frequency of multiple formation, chiasma production and chromosome morphology. *Chromosoma*, **25**, 303–18.
Southern, D. I. (1980). Chromosome diversity in tsetse flies. In *Insect Cytogenetics*, ed. R. L. Blackman, G. M. Hewitt & M. Ashburner, pp. 225–43. Oxford: Blackwell Scientific Publishers.
Southern, D. I. & Pell, P. E. (1973). Chromosome relationships and meiotic mechanisms of certain *Morsitans* group tsetse flies and their hybrids. *Chromosoma*, **44**, 319–34.

Speed, R. M. (1986). Oocyte development in XO foetuses of man and mouse: the possible role of heterologous X-chromosome pairing in germ cell survival. *Chromosoma*, **94**, 115–24.
Spring, H., Scheer, U., Franke, W. W. & Trendelenburg, M. F. (1975). Lampbrush type chromosomes in the primary nucleus of the green alga *Acetabularia mediterranea*. *Chromosoma*, **50**, 25–43.
Spyropoulos, B. & Moens, P. B. (1984). The synaptonemal complex: does it have contractile proteins? *Canadian Journal of Genetics and Cytology*, **26**, 776–81.
Stack, S. M. (1984). Heterochromatin, the synaptonemal complex and crossing over. *Journal of Cell Science*, **71**, 159–76.
Stack, S. M. & Anderson, L. K. (1986a). Two-dimensional spreads of synaptonemal complexes from solanaceous plants II. Synapsis in *Lycopersicon esculentum* (tomato). *American Journal of Botany*, **73**, 264–81.
Stack, S. M. & Anderson, L. K. (1986b). Two-dimensional spreads of synaptonemal complexes from solanaceous plants III. Recombination nodules and crossing over in *Lycopersicon esculentum* (tomato). *Chromosoma*, **94**, 253–8.
Stack, S. M. & Brown, S. V. (1969). Somatic and premeiotic pairing of homologues in *Plantago ovata*. *Bulletin of the Torrey Botanical Club*, **96**, 143–9.
Stack, S. M. & Soulliere, D. L. (1984). The relation between synapsis and chiasma formation in *Rhoeo spathacea*. *Chromosoma*, **90**, 72–83.
Staiger, H. (1954). Der Chromosomendimorphismus beim Prosobranchier *Purpura lapillus* in Beziehung zur Ökologie der Art. *Chromosoma*, **6**, 419–78.
Stalker, H. D. (1956). A case of polyploidy in the Diptera. *Proceedings of the National Academy of Sciences*, **42**, 194–8.
Stearns, S. C. (1982). On fitness. In *Environmental Adaptation and Evolution*, eds. D. Mossakowski & G. Roth, pp. 3–17. New York: Gustav Fischer.
Stebbins, G. L. (1960). The comparative evolution of genetic systems. In *Evolution After Darwin Vol. I., The Evolution of Life*, ed. S. Tax, pp. 197–226. Chicago, University of Chicago Press.
Stebbins, G. L. (1980). Polyploidy in plants: unsolved problems and prospects. In *Polyploidy: Biological Relevance*. ed. W. H. Lewis, pp. 495–520. New York: Plenum Press.
Steffen, W., Fuge, H., Dietz, R., Bastmeyer, M. & Müller, G. (1986). Aster-free spindle poles in insect spermatocytes: evidence for chromosome-induced spindle formation. *The Journal of Cell Biology*, **102**, 1679–87.
Stern, C. (1936). Somatic crossing over and segregation in *Drosophila melanogaster*. *Genetics*, **21**, 625–730.
Stern, H. (1986). Meiosis: some considerations. *Journal of Cell Science*, **Supplement 4**, 29–43.
Stern, H. & Hotta, Y. (1969). Biochemistry of meiosis. In *Handbook of Molecular Cytology*, ed. A. Lima-de-Faria, pp. 520–39. Amsterdam: North Holland.
Stern, H. & Hotta, Y. (1978). Regulatory mechanisms in meiotic crossing-over. *Annual Review of Plant Physiology*, **29**, 415–36.
Stern, H. & Hotta, Y. (1980). The organization of DNA metabolism during the recombinational phase of meiosis with special reference to humans. *Molecular and Cellular Biochemistry*, **29**, 145–58.
Stern, H. & Hotta, Y. (1983). Meiotic aspects of chromosome organization. *Stadler Genetics Symposia*, **15**, 25–41.
Stern, H. & Hotta, Y. (1984). Chromosome organization in the regulation of meiotic prophase. In *Controlling Events in Meiosis*, ed. C. W. Evans & H. G. Dickinson. *Symposia of the Society for Experimental Biology*, **38**, 161–75.
Stern, H. & Hotta, Y. (1985). Molecular biology of meiosis: synapsis-associated

phenomena. In *Aneuploidy: Etiology and Consequences*, ed. V. L. Dellarco, P. E. Voytek, & A. Hollaender, pp. 305–316. New York: Plenum Press.

Stern, H. & Hotta, Y. (1987). The biochemistry of meiosis. In *Meiosis*, ed. P. B. Moens, pp. 303–31. Orlando: Academic Press.

Stern, H., Westergaard, M. & von Wettstein, D. (1975). Presynaptic events in meiocytes of *Lilium longiflorum* and their relation to crossing over: a preselection hypothesis. *Proceedings of the National Academy of Sciences*, **72**, 961–5.

Stille, B. & Dävring, L. (1980). Meiosis and reproductive strategy in the parthenogenetic gall wasp *Diplolepis rosae* (L.) (Hymenoptera, Cynipidae). *Hereditas*, **92**, 353–62.

Strokov, A. A., Bogdanov, Yu. & Reznickova, S. A. (1973). A quantitative study of histones of meiocytes II. Polyacrylamide gel electrophoresis of isolated histones from *Lilium* microsporocytes. *Chromosoma*, **43**, 247–60.

Sun, S. & Rees, H. (1967). Genotypic control of chromosome behaviour in rye IX. The effect of selection on the disjunction frequency of interchange associations. *Heredity*, **22**, 249–54.

Suomalainen, E. (1966). Achiasmatische Oogenese bei Trichopteran. *Chromosoma*, **18**, 201–7.

Suomalainen, E., Cook, L. M. & Turner, J. R. G. (1973). Achiasmatic oogenesis in the Heliconiine butterflies. *Hereditas*, **74**, 302–3.

Suomalainen, E. & Saura, A. (1973). Genetic polymorphism and evolution in parthenogenetic animals I. Polyploid Curculoinidae. *Genetics*, **74**, 489–508.

Suomalainen, E., Saura, A. & Lokki, J. (1976). Evolution of parthenogenetic insects. *Evolutionary Biology*, **9**, 209–57.

Sybenga, J. (1981). Specialisation in the behaviour of chromosomes on the meiotic spindle. *Genetica*, **57**, 143–51.

Sybenga, J. & Rickards, G. K. (1988). The orientation of multivalents at meiotic metaphase I: a workshop report. *Genome*, **29**, 612–20.

Syren, R. M. & Luykx, P. (1981). Geographic variation of sex-linked translocation heterozygosity in the termite *Kalotermes*. *Chromosoma*, **82**, 65–88.

Szauter, P. (1984). An analysis of regional constraints on exchange in *Drosophila melanogaster* using recombination defective meiotic mutants. *Genetics*, **106**, 45–71.

Szöllösi, A., Ris, H., Szöllösi, D. & Debec, A. (1986). A centriole-free *Drosophila* cell line. A high voltage EM study. *European Journal of Cell Biology*, **40**, 100–4.

Szöllösi, D., Calarco, P. G. & Donahue, R. P. (1972). Absence of centrioles in the first and second meiotic spindles of the mouse oocyte. *Journal of Cell Science*, **11**, 521–41.

Szostak, J. W., Orr-Weaver, T. L. & Rothstein, R. J. (1983). The double strand break repair model for recombination. *Cell*, **33**, 25–35.

Taylor, J. H. (1965). Distribution of tritium-labelled DNA among chromosomes during meiosis I. Spermatogenesis in the grasshopper. *The Journal of Cell Biology*, **25**, 57–67.

Tease, C. (1978). Cytological detection of crossing-over in BudR substituted meiotic chromosomes using the fluorescent plus Giemsa technique. *Nature*, **272**, 823–4.

Tease, C. & Jones, G. H. (1976). Chromosome-specific control of chiasma formation in *Crepis capillaris*. *Chromosoma*, **57**, 33–49.

Tease, C. & Jones, G. H. (1978). Analysis of exchanges in differentially stained meiotic chromosomes of *Locusta migratoria* after BrdU substitution and FPG staining I. Crossover exchanges in monochiasmate bivalents. *Chromosoma*, **69**, 163–78.

Tease, C. & Jones, G. H. (1979). Analysis of exchanges in differentially stained meiotic chromosomes of *Locusta migratoria* after BrdU-substitution and FPG staining II. Sister chromatid exchanges. *Chromosoma*, **73**, 75–84.

Telzer, B. R., Moses, M. J. & Rosenbaum, J. L. (1975). Assembly of microtubules onto kinetochores of isolated mitotic chromosomes of HeLa cells. *Proceedings of the National Academy of Sciences*, **72**, 4023–7.

Templeton, A. R. (1979). The parthenogenetic capacities and genetic structures of sympatric populations of *Drosophila mercatorum* and *Drosophila hydei. Genetics*, **92**, 1283–93.

Templeton, A. R. (1982). The prophecies of parthenogenesis. In *Evolution and Genetics of Life Histories*, ed. H. Dingle & J. P. Hegmann, pp. 75–101. New York: Springer Verlag.

Templeton, A. R. (1983). Natural and experimental parthenogenesis. In *The Genetics and Biology of Drosophila*, vol. 3c, ed. M. Ashburner, H. L. Carson & J. N. Thompson, pp. 343–97. London: Academic Press.

Therman, E., Denniston, C. & Sarto, G. E. (1978). Mitotic chiasmata in human chromosomes. *Human Genetics*, **45**, 131–5.

Therman, E. & Kuhn, E. M. (1976). Cytological demonstration of mitotic crossing over in man. *Cytogenetics and Cell Genetics*, **17**, 254–67.

Therman, E. & Kuhn, E. M. (1981). Mitotic crossing over and segregation in man. *Human Genetics*, **59**, 93–100.

Therman, E. & Sarto, G. E. (1977). Premeiotic and early meiotic stages in the pollen mother cells of *Eremurus* and in human embryonic oocytes. *Human Genetics*, **35**, 137–51.

Thomas, J. B. & Kaltsikes, P. J. (1976). A bouquet-like attachment plate for telomeres in leptotene of rye revealed by heterochromatin staining. *Heredity*, **36**, 155–62.

Thuriaux, P. (1977). Is recombination confined to structural genes on the eukaryotic genome? *Nature*, **268**, 460–2.

Timmis, J. N. & Rees, H. (1971). A pairing restriction at pachytene upon multivalent formation in autotetraploids. *Heredity*, **26**, 269–75.

Ting, Y. C. (1966). Duplications and meiotic behaviour of the chromosomes in haploid maize. (*Zea mays* L.). *Cytologia*, **31**, 324–9.

Ting, Y. C. (1973). Synaptonemal complex of haploid maize. *Cytologia*, **38**, 497–500.

Tippit, D. H., Pickett-Heaps, J. D. & Leslie, R. (1980). Cell division in two large pennate diatoms *Hantzschia* and *Nitzschia* III. A new proposal for kinetochore function during prometaphase. *The Journal of Cell Biology*, **86**, 402–16.

Toledo, L. A., Bennett, M. D. & Stern, H. (1979). Cytological investigations of the effect of colchicine on meiosis in *Lilium* hybrid cv. 'Black Beauty' microsporocytes. *Chromosoma*, **72**, 157–73.

Torvik-Greb, M. (1935). The chromosomes of *Habrobacon. Biological Bulletin of Woods Hole*, **68**, 25–34.

Tres, L. L. (1977). Extensive pairing of the XY bivalent in mouse spermatocytes as visualised by whole mount electron microscopy. *Journal of Cell Science*, **25**, 1–15.

Tsafriri, A. & Pomerantz, S. H. (1984). Regulation of the development of meiotic competence and the resumption of oocyte maturation in the rat. In *Controlling Events in Meiosis*, ed. C. W. Evans & H. G. Dickinson. *Symposia of the Society for Experimental Biology*, **38**, 25–43.

Tschermak-Woess (1971). Endomitose. *Handbuch der Allgemeinen Pathologie*, **2**, 569–625.

Turner, B. J. (1982). The evolutionary genetics of a unisexual fish, *Poecilia formosa*. In *Mechanisms of Speciation*, ed. C. Barigozzi, pp. 265–305. New York: Alan R. Liss Inc.
Uemura, T., Ohkura, H., Adachi, Y., Morino, K., Shiozaki, K. & Yanagida, M. (1987). DNA topoisomerase II is required for condensation and separation of mitotic chromosomes in *S. pombe*. *Cell*, **50**, 917–25.
Ueshima, N. (1963). Chromosome behaviour of the *Cimex pilosellus* complex (Cimicidae: Hemiptera). *Chromosoma*, **14**, 511–21.
Ueshima, N. (1979). Hemiptera II: Heteroptera. In *Animal Cytogenetics*, vol. 3, Insecta 6, ed. B. John, pp. 1–117. Stuttgart: Gebrüder Borntraeger.
Ullerich, F.-H. (1961). Achiasmatische Spermatogenese bei der Skorpionsfliege *Panorpa* (Mecoptera). *Chromosoma*, **12**, 215–32.
Upcott, M. (1936). The origin and behaviour of chiasmata XII. *Eremurus spectabilis*. *Cytologia*, **7**, 118–30.
Vaarama, A. (1954). Cytological observations on *Pleurozium schreberi*, with special reference to centromere evolution. *Annales Botanici Societatis Zoologicae Botanicae Fennicae 'Vanamo'*, **28**, 1–59.
Valentin, J. (1984). Genetic control of meiosis: the *eq* gene in *Drosophila melanogaster*. *Hereditas*, **101**, 115–7.
Ved Brat, S. & Rai, K. S. (1973). An analysis of chiasma frequencies in *Aedes aegypti*. *The Nucleus*, **16**, 184–93.
Virrki, N. (1967). Orientation and segregation of asynaptic multiple sex chromosomes in the male *Omophoita clerica* Erichson (Coleoptera: Alticidae). *Hereditas*, **57**, 275–88.
Virrki, N. (1971). Formation and maintenance of the distance sex bivalent in *Oedionychina* (Coleoptera, Alticidae). *Hereditas*, **68**, 305–12.
Virrki, N. (1972). Contraction stage and formation of the distance sex bivalent in *Oedionychina* (Coleoptera, Alticidae). *Hereditas*, **71**, 259–88.
Virrki, N. (1989a). Proximal vs. distal collochores in Coleopteran chromosomes. *Hereditas*, **110**, 101–7.
Virrki, N. (1989b). What happens in the 'clump' stage of meiosis? *Proceedings of the International Symposium on Recent Advances in Cytogenetic Research* (in press).
Vistorin, G., Camberl, R. & Rosenkranz, W. (1977). Studies on sex chromosomes of four hamster species. *Cytogenetics and Cell Genetics*, **18**, 24–32.
Von Wettstein, D. (1977). The assembly of the synaptonemal complex. *Philosophical Transactions of the Royal Society of London*, **B277**, 235–43.
Von Wettstein, D. (1984). The synaptonemal complex and genetic segregation. In *Controlling Events in Meiosis* ed. C. W. Evans & H. G. Dickinson. *Symposia of the Society for Experimental Biology*, **38**, 195–231.
Von Wettstein, D., Rasmussen, S. W. & Holm, P. B. (1984). The synaptonemal complex in genetic segregation. *Annual Review of Genetics*, **18**, 331–413.
Vosa, C. G. & Barlow, P. W. (1972). Meiosis and B chromosomes in *Listera ovata* (Orchidaceae). *Caryologia*, **25**, 1–8.
Wagner, W. H. & Wagner, F. S. (1980). Polyploidy in Pteridophytes. In *Polyploidy: Biological Relevance*, ed. W. H. Lewis, pp. 199–218. New York: Plenum Press.
Wahrman, J. (1981). Synaptonemal complexes – origin and fate. *Chromosomes Today*, **7**, 105–13.
Wahrman, J. & Nezer, R. (1976). Disjunction and ultrastructure of a sex-chromosome complex. *Chromosomes Today*, **5**, 464 (abst.).
Wahrman, J., Nezer, R. & Freund, O. (1973). Multiple sex chromosome mechanisms with 'segregation bodies'. *Chromosomes Today*, **4**, 434 (abst.).

Wahrman, J., Richter, C., Neufeld, E. & Friedmann, A. (1983). The origin of multiple sex chromosomes in the gerbil *Gerbillus gerbillus* (Rodentia: Gerbillinae). *Cytogenetics and Cell Genetics*, **35**, 161–80.

Wallace, B. M. N. & Hulten, M. A. (1983). Triple chromosome synapsis in oocytes from a human foetus with trisomy 21. *Annals of Human Genetics*, **47**, 271–6.

Wallace, B. M. N. & Jones, G. H. (1978). Incomplete chromosome pairing and its relation to chiasma localization in *Stethophyma grossum* spermatocytes. *Heredity*, **40**, 385–96.

Walters, J. (1956). Spontaneous breakage in natural populations of *Paeonia californica*. *American Journal of Botany*, **43**, 342–54.

Walters, M. S. (1970). Evidence on the time of chromosome pairing from the preleptotene spiral stage in *Lilium longiflorum* 'Croft'. *Chromosoma*, **29**, 375–418.

Walters, M. S. (1976). Variation in preleptotene chromosome contraction among three cultivars of *Lilium longiflorum*. *Chromosoma*, **57**, 51–80.

Walters, M. S. (1978). Meiosis readiness in *Lilium longiflorum* 'Croft'. *Chromosoma*, **67**, 365–91.

Walters, M. S. (1985). Meiosis readiness in *Lilium*. *Canadian Journal of Genetics and Cytology*, **27**, 33–8.

Wandall, A. & Svendsen, A. (1985). Transition from somatic to meiotic pairing and progressional changes in the synaptonemal complex in spermatocytes of *Aedes aegypti*. *Chromosoma*, **92**, 254–64.

Wang, X. (1988). Chromosome pairing analysis in haploid wheat by spreading of meiotic nuclei. *Carlsberg Research Communications*, **53**, 135–66.

Watson, J. D. & Callan, H. G. (1963). The form of bivalent chromosomes in newt oocytes at first metaphase of meiosis. *Quarterly Review of Microscopical Science*, **104**, 281–95.

Weatherbee, J. A., May, G. S., Gambino, J. & Morris, N. R. (1985). Involvement of a particular species of beta-tubulin (beta 3) in conidial development in *Aspergillus nidulans*. *The Journal of Cell Biology*, **101**, 706–11.

Weismann, A. (1892). *Das Keimplasma*. Jena: Fischer.

Weith, A. & Traut, W. (1986). Synaptic adjustment, non-homologous pairing and non-pairing of homologous segments in sex chromosome mutants of *Ephestia kuehniella* (Insecta, Lepidoptera). *Chromosoma*, **94**, 125–31.

Welsch, B. (1973). Synaptonemal complex und Chromosomenstruktur in der achiasmatischen Spermatogenese von *Panorpa communis* (Mecoptera). *Chromosoma*, **43**, 19–74.

Westergaard, M. & von Wettstein, D. (1970). Studies on the mechanism of crossing over IV. The molecular organisation of the synaptonemal complex in *Neottiella* (Cooke) Saccardo (Ascomycetes). *Comptes Rendus des Travaux du Laboratoire Carlsberg*, **37**, 239–68.

Westergaard, M. & von Wettstein, D. (1972). The synaptonemal complex. *Annual Review of Genetics*, **6**, 71–116.

White, M. J. D. (1954). An extreme form of chiasma localisation in a species of *Bryodema* (Orthoptera, Acrididae). *Evolution*, **8**, 350–8.

White, M. J. D. (1961). Cytogenetics of the grasshopper *Moraba scurra* VI. A spontaneous pericentric inversion. *Australian Journal of Zoology*, **9**, 784–90.

White, M. J. D. (1965). Chiasmatic and achiasmatic meiosis in African eumastacid grasshoppers. *Chromosoma*, **16**, 271–307.

White, M. J. D. (1970). Heterozygosity and genetic polymorphism in parthenogenetic animals. *Evolutionary Biology*, **Supplementary Volume,** 237–62.

White, M. J. D. (1973). *Animal Cytology and Evolution*, 3rd edition. Cambridge: Cambridge University Press.

White, M. J. D. (1977). Karyotypes and meiosis of the morabine grasshoppers I. Introduction and genera *Moraba*, *Spectriforma* and *Filoraba*. *Australian Journal of Zoology*, **25**, 567–80.
White, M. J. D. (1979). Karyotypes and meiosis of the morabine grasshoppers II. The genera *Culmacris* and *Stiletta*. *Australian Journal of Zoology*, **27**, 109–33.
White, M. J. D. (1980). Meiotic mechanisms in a parthenogenetic grasshopper species and its hybrids with related bisexual species. *Genetica*, **52/53**, 379–83.
White, M. J. D. (1981). Karyotypes and meiosis of the morabine grasshoppers III. The genus *Hastella*. *Australian Journal of Zoology*, **29**, 461–70.
White, M. J. D., Cheney, J. & Key, K. H. L. (1963). A parthenogenetic species of grasshopper with complex structural heterozygosity (Orthoptera: Acridoidea). *Australian Journal of Zoology*, **11**, 1–19.
White, M. J. D., Contreras, N., Cheney, J. & Webb, G. C. (1977). Cytogenetics of the parthenogenetic grasshopper *Warramaba* (formerly *Moraba virgo*) and its bisexual relatives II. Hybridization studies. *Chromosoma*, **61**, 127–48.
Whitehouse, H. L. K. (1963). A theory of crossing over by means of hybrid deoxyribonucleic acid. *Nature*, **199**, 1034–40.
Whitehouse, H. L. K. (1982). *Genetic Recombination*. Chichester: John Wiley and Sons.
Wiens, D. & Barlow, B. A. (1979). Translocation heterozygosity and the origin of dioecy in *Viscum*. *Heredity*, **42**, 201–22.
Williams, G. C. (1975). *Sex and Evolution*. New Jersey: Princeton University Press.
Williamson, D. H., Johnson, L. H., Fennell, D. J. & Simchen, G. (1983). The timing of the S phase and other nuclear events in yeast meiosis. *Experimental Cell Research*, **145**, 209–17.
Wilson, E. B. (1925). *The Cell in Development and Heredity*. New York: Macmillan.
Wischmann, B. (1986). Chromosome pairing and chiasma formation in wheat plants trisomic for the long arm of chromosome 5B. *Carlsberg Research Communications*, **51**, 1–25.
Wise, D. (1983). An electron microscope study of the karyotypes of two wolf spiders. *Canadian Journal of Genetics and Cytology*, **25**, 161–8.
Wise, D. (1984). The ultrastructure of an intraspindle membrane system in meiosis of spider spermatocytes. *Chromosoma*, **90**, 50–6.
Wise, D. & Rickards, G. K. (1977). A quadrivalent studied in living and fixed grasshopper spermatocytes. *Chromosoma*, **63**, 305–15.
Wise, D. A. & Shaw, R. G. (1984). The mechanism of non-random chromosome segregation in lycosid spiders. *The Journal of Cell Biology*, **99**, 246a.
Wise, D., Sillers, P. J. & Forer, A. (1984). Non-random chromosome segregation in *Neocurtilla hexadactyla* is controlled by chromosomal spindle fibres: an ultraviolet microbeam analysis. *Journal of Cell Science*, **69**, 1–17.
Wise, D. & Wolniack, S. M. (1984). A calcium-rich intraspindle membrane system in the spermatocytes of wolf spiders. *Chromosoma*, **90**, 156–61.
Wolf, B. E. (1950). Die Chromosomen in der Spermatogenese der dipteren *Phryne* und *Mycetobia*. *Chromosoma*, **4**, 148–204.
Wolfe, S. L. & John, B. (1965). The organisation and ultrastructure of male meiotic chromosomes in *Oncopeltus fasciatus*. *Chromosoma*, **17**, 85–103.
Wolniak, S. M., Hepler, P. K. & Jackson, W. T. (1980). Detection of the membrane-calcium distribution during mitosis in *Haemanthus* endosperm with chlorotetracycline. *The Journal of Cell Biology*, **87**, 23–32.
Wolniak, S. M., Hepler, P. K. & Jackson, W. T. (1981). The coincident

distribution of calcium-rich membranes and kinetochore fibers at metaphase in living cells of *Haemanthus*. *European Journal of Cell Biology*, **25**, 171–4.

Woodland, H. R. (1980). Histone synthesis during the development of *Xenopus*. *FEBS Letters*, **121**, 1–7.

Yacobi, Y. Z., Mello-Sampayo, T. & Feldman, M. (1982). Genetic induction of bivalent interlocking in common wheat. *Chromosoma*, **87**, 165–75.

Yamamoto, M. & Miklos, G. L. G. (1978). Genetic studies on heterochromatin in *Drosophila melanogaster* and their implications for the functions of satellite DNA. *Chromosoma*, **66**, 71–98.

Zarchi, Y., Hillel, J. & Simchen, G. (1974). Supernumerary chromosomes and chiasma distribution in *Triticum speltoides*. *Heredity*, **33**, 173–80.

Zarchi, Y., Simchen, G., Hillel, J. & Schaap, T. (1972). Chiasmata and the breeding system in wild populations of diploid wheats. *Chromosoma*, **38**, 77–94.

Zečević, L. J. & Paunović, D. (1969). The effect of B chromosomes on chiasma frequency in wild populations of rye. *Chromosoma*, **27**, 198–200.

Zenzes, M. T. & Wolf, U. (1971). Paarungsverhalten der Geschlechtsechromosomen in der männlichen Meiose von *Microtus agrestis*. *Chromosoma*, **33**, 41–7.

Zickler, D. (1973). Fine structure of chromosome pairing in ten ascomycetes: meiotic and premeiotic (mitotic) synaptonemal complexes. *Chromosoma*, **40**, 401–16.

Zickler, D. (1977). Development of the synaptonemal complex and the 'recombination nodules' during meiotic prophase in the seven bivalents of the fungus *Sordaria macrospora* Auersw. *Chromosoma*, **61**, 289–316.

Index